Contents

List of Examples

CHAPTER 11: VBA PROGRAMMING

Preface

The goal of this book is to provide a well-organized, systematic introduction to the use of spreadsheets and their associated programming in the problem-solving activities that engineers and scientists, both professionals and students, encounter on a day-to-day basis. There is a need for such a text based on an up-to-date treatment of the Microsoft Excel spreadsheet software along with its companion, the Visual Basic for Applications (VBA) programming language. Most instructional materials on spreadsheets are based on business/financial applications, and there is a scarcity of these focused on the needs of engineers and scientists.

Although spreadsheet problem-solving is taught in many academic undergraduate engineering and science programs, it is not consistent across all institutions. Most students pick up skills in spreadsheets during their studies, as they find these valuable in their various academic courses. Many professional users pick up the use of spreadsheets in a haphazard, situational manner. Over enough time, they may develop broad knowledge and skills, but it is common for there to be gaps. Frequently absent from their skillset is anything about programming in VBA to streamline and improve spreadsheet applications. The purpose of this book is to fill those gaps.

The authors have been teaching and using spreadsheets for over 35 years. As with many faculty and engineering/science practitioners during the early days of the late 1970s and 1980s, we were at first skeptical of spreadsheets. Having a background in scientific programming with the Fortran language, we did not initially see the potential that spreadsheets provide for technical problem-solving. As happens with some frequency, practicing professionals took the lead in discovering the utility of spreadsheets, with programs like VisiCalc and Lotus 1-2-3. They brought us along to see the utility of spreadsheet problem-solving.

We began teaching the use of Lotus 1-2-3 in the mid-1980s, carving out three weeks in our introduction to engineering computing course. As they say, from there, the rest is history. Eventually, spreadsheet problem-solving and programming occupied two-thirds of the introductory course. We used various text materials during this evolution but never found the right combination of content and organization. That has brought us to today and our efforts to create this book.

Many academics are still skeptical of spreadsheets and whether they should be taught formally in an undergraduate curriculum. They tend to be more focused on coding, programming languages, and less on the results that programs produce. Spreadsheets turn the tables because they focus on the presentation of results with the formulas and programs hidden in the background. Alternately, practicing professionals have gravitated to spreadsheets precisely for this last reason. They are focused on results and enjoy the live presentation that spreadsheets make, allowing "what if" explorations in real-time.

One of us (Clough) has a lengthy history of teaching spreadsheets and associated programs to engineers, primarily chemical and mechanical. Since 1989, he has taught over 170 in-person short courses in the U.S. and beyond through the continuing education programs of the American Institute of Chemical Engineers (AIChE) and the American Society of Mechanical Engineers (ASME). Also, he provides online versions of his courses via the AIChE.

The other of us (Chapra) has used spreadsheets extensively in environmental modeling and has conducted over 100 associated workshops around the world. In addition, he has developed several Excel/VBA-based environmental software packages for the U.S. Environmental Protection Agency (EPA). Because of ubiquity of Microsoft Office, his free and open-source software is now extensively used to model the fate and transport of water pollutants in the world's rivers and lakes.

Together, we originated the introduction to engineering computing first-year course at the University of Colorado, and Chapra later carried this course to Tufts University. Clough has developed numerous Excel/VBA applications in various industrial/governmental settings. Evidence suggests we have come a long way since our skepticism of the 1980s.

This text is organized into 12 chapters with a logical progression from spreadsheet fundamentals to specific application topics to VBA programming. Each chapter is generally 20-to-40 pages long accompanied by scores of figures, numerous examples, and a dozen or so problems at the end of each chapter. Problems tend to focus on practical scenarios.

The authors' observations of spreadsheet users over the years have identified that many users do not use efficient techniques. The purpose of Chapter 1 is to introduce fundamentals and skills that allow the learner to complete their spreadsheet tasks in less time and with more reliability. Although Excel's charts are designed primarily for business and financial needs, Chapter 2 shows how these charting alternatives can be used to produce the graphs that engineers and scientists employ most frequently. Also, with the introduction of chart templates, users are shown how to create charts in a less tedious way.

Chapter 3 introduces implementation of the mathematical formulas often employed by engineers and scientists. This includes the use of mathematical functions such as trigonometric and exponential. The chapter first presents the details of cell addressing and the use of names for cells and ranges of cells. Learners often do not know about the latter and are missing out on a great opportunity to improve the readability and reliability of their spreadsheets. Later in the text, they find that naming cells is essential in VBA applications. The chapter also introduces array formulas with which most learners are unfamiliar.

Practicing professionals, more so than beginning students, typically manage sets of information, often data, which are arranged in tables. For this reason, Chapter 4 focuses on management of table data. The chapter introduces the features of Excel, such as lookup functions and pivot tables that professionals find convenient when working with table-based information.

In Chapter 5, we introduce two problem-solving techniques used frequently by engineers and scientists. First is targeting, sometimes called back-solving, where one or more input values to a calculation are adjusted automatically to achieve a desired value or optimum in a result. Users are often familiar with the Goal Seek feature of Excel, but the details on the Solver add-in are new to many. Their use is illustrated with relevant examples. We have observed that most learners are unfamiliar with the use of the Data Table feature of Excel to implement case studies and view this as a revelation and significant addition to their spreadsheet skill set.

Of course, the origin of spreadsheets is in financial calculations, and that is still prevalent today. Engineers and scientists also encounter these, often in proposals and profitability studies. In the past, it was commonplace for some universities to offer entire courses in Engineering Economics. Chapter 6 reviews typical financial concepts, such as the time value of money, and applies these to the financial applications that engineers encounter. This includes project evaluation and cash flow studies. As with Chapter 4, this material may be considered less important for first-year students, but it is important for practitioners and of value to engineering students later when they complete design projects.

Chapter 7 introduces the implementation of numerical methods to solve the mathematical equations that describe phenomena. These include algebraic equations, both nonlinear and linear, and differential equations. In the solution of the latter, techniques include computing "area under the curve," also called quadrature, and simple, yet effective, numerical methods for the general solution of differential equations. Learners will be surprised how effective Excel can be in implementing numerical methods.

Chapter 8 provides an overview of the implementation of statistical methods on the spreadsheet. This chapter is not intended to present a comprehensive coverage of statistics but rather focuses on common applications in engineering and science. These include sample statistics, histograms, and hypothesis tests. There is also a section on curve-fitting, AKA regression analysis.

Chapter 9 opens the world of VBA programming. Many experienced Excel users have no knowledge of VBA, and learning VBA elevates their spreadsheet capabilities significantly. This chapter introduces the VBA environment and the first-line, bread-and-butter of VBA, recording of short-cut macros. Basic VBA syntax is covered along with useful debugging techniques.

Chapter 10 introduces learners to developing their own functions. Opportunities are many for user-defined functions in engineering and scientific spreadsheet applications. They provide increased efficiency, reliability, and portability in custom add-ins.

Chapters 11 and 12 complete the picture with VBA. Chapter 11 provides a general basis for developing VBA programs beyond macros and user-defined functions. It includes the elements of structure that are common to programming languages: decisions, loops, and modules. Communication to and from the spreadsheet environment is also covered. This chapter opens the door to larger VBA applications, possibly containing hundreds or thousands of program statements.

Finally, Chapter 12 introduces the various *graphical-user-interface* (*GUI*) elements of VBA. These include on-sheet buttons, message and input boxes, and userforms. The GUI elements are essential in the development of spreadsheet applications that are intended for many users including those with little or no programming skills. The chapter leads the learner through the creation of an example application illustrating these various features, thus equipping the learner to develop their own applications.

There are two appendices. The first provides an approachable review of matrix algebra for those who have not seen these concepts or, more frequently, need to brush up on their knowledge in this area. A second brief appendix provides a useful reference to the various shortcut keys and key sequences introduced in the text.

A common question that instructors have is how to use a text like this in the context of an academic course or term. In many cases, the teaching of spreadsheets and spreadsheet programming will only take up part of an introductory course. We have designed this book to accommodate various course durations. The following table describes several of these. Although the most common academic term is the 15-week semester, many institutions are organized into 12-week quarters. This difference is taken into account.

Chapter Allocation to Academic Term Weeks

Week	6-week introduction to spreadsheets	10-week spreadsheets and programming	15-week semester course
1	1	1	1
2	2	2	2
3	3	3	3
4	4	4	4
5	5	5	5
6	7	7	6
7		8	7
8		9	7
9		10	8
10		11	8
11			9
12			10
13			11
14			12
15			Review

The 6-week introduction focuses on spreadsheet problem-solving, leaving out VBA programming and the topics of financial calculations and applied statistics. The pace is one chapter per week, although complete coverage of Chapter 7 may not be feasible. In the latter case, solving differential equations might be left out.

The 10-week format, corresponding to two-thirds of a semester-long introductory course or to an academic quarter, expands the coverage to include applied statistics and three chapters on VBA programming. Chapter 6 on financial calculations is again left out along with Chapter 12 on user interfaces. The full 15-week plan provides complete coverage of the text and allows two weeks for each of Chapters 7 (numerical methods) and 8 (applied statistics). It also allows for some flexibility with a review week at the end. In fact, the 10-week plan allows flexibility for the 12-week quarter term.

Academics may consider that several chapters, or topics within chapters, are best left to courses later in the curriculum. Four prime examples are:

- Chapter 6 (financial calculations) delayed to an engineering economics course or senior design course
- Chapter 7 (numerical methods) included in a junior/senior course in numerical methods
- Chapter 8 (applied statistics) delayed to a course in applied statistics
- Chapter 12 (user interfaces) included in a senior capstone course

An important aspect of this text is that it is inspired by our experience of teaching spreadsheets and programming in a hands-on manner. Our approach is based on the adage "crawl-walk-run." First, it is important to read and understand the text material. Second, it is important to "do" the text, that is, work through all the material with Excel and VBA. Third, solve the problems at the end of the chapters. The latter step will confirm to the student/reader that they have mastered the text material.

There are several acknowledgments we provide regarding our development and writing of this text. Clough taught over 100 in-person short courses with F. Miles Julian, formerly of the Engineering Department of the DuPont Company. Our partnership provided the basis for a considerable portion of the content of this text. Of course, the sponsorship of these courses by the AIChE and ASME is worthy of mention.

Chapra was introduced to VBA by his wife, Cynthia, when they lived in London in 1997. She was developing Excel/VBA software to expedite common gas-exploration calculations for Amoco and British Petroleum. After seeing the power and accessibility of her applications, Steve immediately began developing user-friendly environmental software with the Excel/VBA environment. Over the following decades, he was helped (and educated) by some very clever students including Mike Glazner, Dr. James Martin, Dr. Jean Marie Boyer, Dr. Rob Runkel, Dr. Hua Tao, Prof. Luis Camacho, Dr. Kyle Flynn, Greg Coyle, Dr. Jingshui Huang, and Dr. Anika Kuscinski. Last but not least, Lee Minardi was a great collaborator with Steve when they taught first-year computing at Tufts.

We both acknowledge the participation and contribution of thousands of our students over the past decades. The experience has been much more than our delivering knowledge and skills to the students. Their feedback and interactions have added much to our overall background and effectiveness of communication in spreadsheet problem solving and programming. Also, the professionals we have taught have provided many inputs along the lines of "Didn't you know you could do this?" that we have integrated into our teaching and writing here.

This book would not be possible without the sponsorship and collaboration of CRC Press/Taylor & Francis Group, and, in particular Nicola Sharpe, Editor, Mechanical Engineering. Also, Shatakshi Singh, Editorial Assistant, Robert Sims, Production Editor, and Dueata Menon of KGL, Ltd. provided invaluable help in bringing the final version of the manuscript to press.

Despite the best of our efforts, the authors are certain that this book is not perfect and not completely free of occasional errors, although we hope these are not in the major category. We encourage you to contact us with questions for clarification, to report any errors you have found. In addition, suggestions for material to include in future editions are welcomed. We can be reached at the emails below.

Finally, if you find this book to be valuable to you, we would like to hear from you too. Such will gratify us and help us confirm that our efforts have been worthwhile.

David Clough
Boulder, Colorado
david.clough@colorado.edu

Steve Chapra
Medford, MA
steven.chapra@tufts.edu

About the Authors

David E. Clough joined the faculty of the Department of Chemical and Biological Engineering at the University of Colorado in 1975 after a brief career with DuPont in Wilmington, Delaware. He retired from Colorado in 2017 and holds the position of Professor Emeritus. He remains active at the university by assisting faculty and students in teaching and research. Notably, he teaches a series of workshops on process modeling and computer simulation as part of the senior design course sequence. In addition to this book, he has collaborated with Steve Chapra on two other texts: *Applied Numerical Methods with Python for Engineers and Scientists* (WCB/McGraw-Hill, 2022) and *Introduction to Engineering and Scientific Computing with Python* (CRC Press, 2023).

Dr. Clough received degrees in chemical engineering from Case Western Reserve University and the University of Colorado. He has extensive experience in applied computing, process automation, and the modeling of various processes with emphasis on dynamic behavior, including polymerization, high-temperature catalytic reactors, fluidized beds, open-channel flow, biomedical instrumentation, and solar-thermal reactors.

Clough first learned to program in the original Fortran language while he was in high school in the early 1960s. Since then, he has gained experience in a wide array of programming languages and computing tools and has applied his expertise through his teaching, research, and industrial applications. In the 1980s, Clough and Chapra originated the Introduction to Engineering Computing course for first-year students at the University of Colorado. The course is still taught, and it has included content on spreadsheets (originally Lotus 1-2-3, then Excel) and spreadsheet programming (VBA) for three decades.

Over his career, Clough has taught hundreds of in-person short courses to practicing professionals on applied computing and problem-solving. His courses on spreadsheet problem-solving, offered through the American Institute of Chemical Engineers, have been among the most popular offered by AIChE, spanning three decades. He also offers online versions of his spreadsheet courses via AIChE Academy.

Steven C. Chapra is Emeritus Professor and Louis Berger Chair in Civil and Environmental Engineering at Tufts University. His best-selling books include *Surface Water-Quality Modeling, Numerical Methods for Engineers*, 8th Ed. (WCB/McGraw-Hill, 2022), and *Applied Numerical Methods with MATLAB for Engineers and Scientists*, 5th Ed (WCB/McGraw-Hill, 2022). He has also co-authored two texts with Dr. Clough on the Python language.

Dr. Chapra received engineering degrees from Manhattan College and the University of Michigan. Before joining Tufts, he worked for the U.S. Environmental Protection Agency and the National Oceanic and Atmospheric Administration, and taught at Texas A&M University, the University of Colorado, and Imperial

College London. His general research interests focus on surface water-quality modeling and advanced computer applications in environmental engineering.

He is a Fellow and Distinguished Member of the American Society of Civil Engineering (ASCE) and has received several awards for his scholarly and academic contributions, including the Rudolph Hering Medal (ASCE), and the Meriam-Wiley Distinguished Author Award (American Society for Engineering Education). He has also been recognized as an outstanding teacher and advisor among the engineering faculties at Texas A&M University, the University of Colorado, and Tufts University. As a strong proponent of continuing education, he has taught more than 90 workshops around the world introducing professionals to numerical methods, computer programming, and environmental modeling.

1 Spreadsheet Basics

CHAPTER OBJECTIVES

- Learn about the origins and development of spreadsheet software
- Be able to navigate the Excel spreadsheet environment and appreciate how the cursor changes personality at different locations
- Understand cell addressing and the extents of the worksheet
- Be able to carry out worksheet navigations, like jumps, selections, copies, and moves
- Learn how to enter numeric and other quantities, such as text and dates
- Learn how to enter simple formulas, including the use of some built-in functions
- See how the formatting of spreadsheets can be improved, including the use of color

Many, perhaps most, spreadsheet users have picked up their skills in a haphazard manner, lacking any formal instruction. Further, much available instruction is oriented to business and financial applications. The purpose of this introductory chapter is to establish a baseline of skills and good practices that will empower engineering and science students and practitioners to achieve efficiency in their day-to-day spreadsheet activities. Of course, habits are often established and difficult to change. We can only encourage the reader to be open to learning and accepting better basic spreadsheet skills.

To start, we consider the origin of the electronic spreadsheet and why this business/financial software became a favorite for technical calculations. Part of the story is how spreadsheet software evolved over the decades to become the tool of choice for many engineers and scientists. Next, the layout and personality of the current Excel worksheet environment is described along with how cells are referenced by column-row addresses. How numeric quantities are entered, stored, and formatted is important, so that is discussed. The entry and formatting of text and dates are also presented.

Basic everyday skills related to manipulation of the information on and navigation around the spreadsheet are keys to efficiency. Various techniques are presented via relevant examples. Although Chapter 3 deals with engineering and scientific formulas in detail, here we get a start with simple formulas and a few common built-in functions. Finally, spreadsheet "cosmetics" are considered in terms of the use of color and other formatting techniques.

Even if you have significant background in spreadsheets, we suggest you take to heart the information and techniques introduced in this chapter. We are

DOI: 10.1201/9781003361053-1

forming a common basis for what is to come in the following chapters. Now, let's start with the answer to the spreadsheet question, "Where did I come from?"

1.1   A LITTLE HISTORY

Financial spreadsheets preceded the electronic era by ages. They could also be called ledgers and were sets of figures arranged in rows and columns. Originally, they would be scribed by hand. If done so in ink on paper, any changes would usually require that the entire ledger be rewritten. Other options would be to paste over individual entries with tags so that new numbers could be entered without starting over. But then, if these spreadsheets included calculated values, such as sums, any changes would propagate through the ledger and many modifications would be required. Expansive ledgers could be displayed on chalkboards mounted on walls. Then, modifications could be made via erasure, and results could eventually be transcribed.

With the advent of digital computers, programs could be written to generate spreadsheet-formatted output. The two most common programming languages in the 1960s were Fortran for scientific calculations and Cobol for business calculations.[1] Cobol programs were used to provide tabular output on wide-format line printers that was akin to a spreadsheet. Any changes would require the program to be rerun, and new printed output to be generated. These applications were financial in nature.

It is important to place the evolution of spreadsheet software in the context of the development of computer technology. From the 1950s through the mid-1960s, large mainframe computers dominated the scene, and smaller minicomputers appeared toward the end of the 1960s. For mainframes, common input media were punched cards and magnetic tape. Typical interfaces for minicomputers included typewriter-like terminals, like the Teletype ASR-33 shown in Figure 1.1. This device was primitive by today's standards, noisy, and used punched paper tape as a storage medium. It was also used to access online computing resources via a 110 bits/second serial, telephone connection.

In the mid-1960s, the BASIC[2] programming language was developed by Profs. John Kemeny and Thomas Kurtz at Dartmouth College. This simpler, utilitarian language grew in acceptance and was the cornerstone of one of the first successful online computing services by General Electric. By the late 1960s, GE BASIC was widely used by engineers and scientists. There are remnants of the original BASIC language present in today's Visual Basic for Applications (VBA) programming language that accompanies the Excel spreadsheet.

The early 1970s saw the development of microcomputers pioneered by Intel. Human interface devices based on cathode-ray-tube displays (CRTs) became common. These were monochrome, text-based displays often with integrated keyboards. Although the Apple I computer was introduced in 1975, it was in kit form and required the purchase of additional elements including a keyboard, display, power supply, and even an enclosure. The Apple II computer arrived in 1977. It had an integrated keyboard and required a separate display.

FIGURE 1.1 Teletype ASR-33 terminal. (From https://commons.wikimedia.org/wiki/
File:Teletype-IMG_7287.jpg, Rama & Musée Bolo, CC BY-SA 2.0 FR <https://creative-
commons.org/licenses/by-sa/2.0/fr/deed.en>, via Wikimedia Commons.).

The *Apple II* computer provided the architecture for the first electronic spreadsheet, *VisiCalc*. This computer is shown in Figure 1.2 with its separate display
and two 5-1/4" floppy disk drives. It was based on the 6502 microprocessor from
MOS Technology.

The origin of the first electronic spreadsheet is credited to Daniel Bricklin
in 1978 while he was a student at Harvard Business School. Bricklin then collaborated with Bob Frankston to create the *VisiCalc* program, short for "Visible
Calculator." The original program was written by Bricklin in the BASIC language
and was very limited in scope, with 5 columns and 20 rows. Folklore has it that
he was preparing a financial analysis for a Harvard Business School case study
and saw the need for an easy-to-use program with which users could see their
results as they entered the data. After he developed the prototype, he recruited
Bob Frankston from MIT, who created an assembly-language-based code with
expanded capabilities. Developed in the era when computer memory was greatly
limited, VisiCalc was extremely compact, using less than 30k of memory.

FIGURE 1.2 Apple II computer. (From https://commons.wikimedia.org/wiki/File:
Apple_II-IMG_7064.jpg, Rama & Musée Bolo, CC BY-SA 2.0 FR <https://creativecommons.org/licenses/by-sa/2.0/fr/deed.en>, via Wikimedia Commons.).

An example VisiCalc display is shown in Figure 1.3. This is a monochrome, character-based display with an arrangement of lettered columns and numbered rows. Apart from entering numbers and text in the "cells" of the display, one could enter simple formulas to calculate and display results involving other cells, such as

+D3 + D4 + D5 + D6

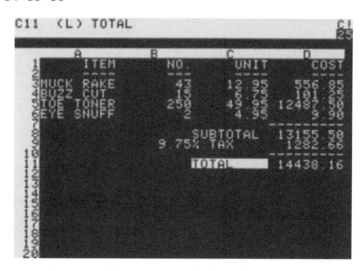

FIGURE 1.3 VisiCalc display. (From https://commons.wikimedia.org/wiki/File:Visicalc.
png, User:Gortu, Public domain, via Wikimedia Commons.).

as a formula in cell D8. VisiCalc also included a family of built-in functions, like @SUM(list), @PI, and @LN(value). To select a cell for entry, one generally used the arrow keys on the keyboard. Additional commands, like clearing the screen with /C, were preceded by the /character.

Although it may be obvious, it is worth mentioning that this early spreadsheet and others in the 1980s were based on character displays accompanied by keyboards. There was neither a pointing device (mouse) nor a *graphical user interface* (or GUI).[3]

It is estimated that more than one million copies of VisiCalc were sold during its predominance. In fact, the accompanying purchases of Apple II computers were said to have saved Apple from the failure of the Apple III. When the IBM PC was introduced in 1991, a version of VisiCalc was developed for that architecture. Concurrently, a new spreadsheet program, Lotus 1-2-3, was introduced, specifically designed for the PC. Lotus 1-2-3 had expanded capabilities, including chart creation and keyboard macros, and soon dominated the spreadsheet market.

It was with Lotus 1-2-3 that spreadsheet use caught on with engineers and scientists during the 1980s. Given the spreadsheet's origins in financial ledgers, this was somewhat of a surprise. Lotus Development recognized this and developed tools for technical users. One of interest was an add-in called Measure that allowed for data acquisition and signal output directly to and from the spreadsheet. This allowed spreadsheet use to move into the manufacturing and laboratory environments.

There were many competitive spreadsheet products introduced in the 1980s. Microsoft introduced Multiplan, Lotus Development released Symphony, and Borland produced Quattro Pro. These and others captured segments of the spreadsheet market but never posed significant threats to Lotus 1-2-3 domination. This changed later in the 1980s with the introduction of the GUI and the mouse pointing device.

Earlier in the 1980s, Apple introduced the Lisa computer that used a GUI and mouse. There was legal action against Apple by Xerox claiming infringement of the latter's invention. Later, Apple introduced the Macintosh computer that became a great success.

An original version of Excel was developed by Microsoft for the Apple Macintosh computer. When Microsoft introduced the first version of the Windows operating system in 1987, this was followed shortly by the release of Excel for Windows. Excel quickly became the main competitor to Lotus 1-2-3 and reduced the latter's market share significantly.

As the spreadsheet products developed, the original macro[4] capabilities that were based on literal keystrokes were gradually expanded to add programming-capable features. Then, in 1993, Microsoft introduced Excel 5.0 with the VBA programming language companion. The VBA package was a separate, object-oriented programming environment that markedly expanded Excel's capabilities. This spelled the end for Lotus 1-2-3 and other competitors. Since then, through its evolving versions, Excel has come to dominate the spreadsheet market.

Another key aspect of its success was its inclusion in the Microsoft Office software package. This meant that users would generally have access to Excel without having to purchase additional software, such as was the case with Lotus 1-2-3. Further, it allowed users to link Excel to other Windows software like Word and PowerPoint with VBA as the common macro language.

From the late 1970s through the 1990s, there were a series of lawsuits in the computer technology and software areas, and these included spreadsheet software. Also, the careers and movements of the key persons are fascinating to follow. Some will say that, since the mid-1990s, the development of the spreadsheet has been stymied by the lack of competition. There were "buggy features" in Excel 5.0 that persist today. But, over a 20-year dynamic period, we now are used to spreadsheets and Excel as synonymous descriptors. The history of spreadsheets is an interesting one, and there are many sources available on the Internet if you wish to read further.

1.2 NAVIGATING THE SPREADSHEET[5]

The initial display screen of the Excel spreadsheet program is shown in Figure 1.4. This view has not changed significantly with recent versions of Excel. The version shown is 2205 from the Office 365 software package running on Windows 10. It is difficult to predict the future, but, for now, we don't believe this will change drastically in upcoming versions.[6]

Key elements of the spreadsheet display are noted in the figure. The major regions of the display from the top on down are as follows:

- **Title bar:** Includes quick launch commands, workbook file name, search box, and window controls.
- **Menu:** These selections open different segments of the Ribbon below. All the menu items here, for example those having to do with Aspen, may not appear on your display.
- **Ribbon:** Contains groups of commands for the menu tab selected. Depending on screen size and resolution, this display may be abbreviated.
- **Formula bar:** Includes the Name Box and the Insert Function command icon.
- **Column indicators:** Columns A through O displayed here. Extends out to XFD.
- **Row indicators:** Along the left side. Rows 1 through 17 displayed here. Extends down to 1048576 (2^{20}).
- **Main spreadsheet:** The number of cells visible will depend on screen size, resolution, and the zoom setting.
- **Sheet tabs and sliders:** Initially, there is one sheet. Sliders allow manual movement on larger spreadsheets.
- **Status bar and zoom control:** Status bar shows entry mode: Ready, Enter, Edit, Point, and whether a macro is recording or not.

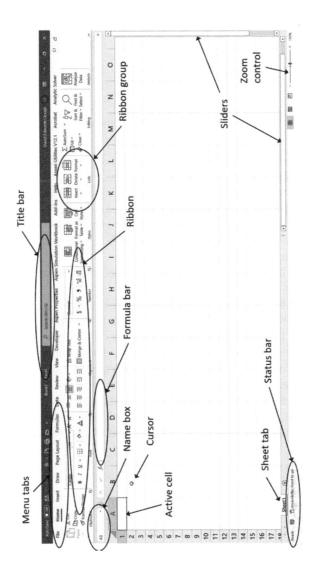

FIGURE 1.4 Excel initial window.

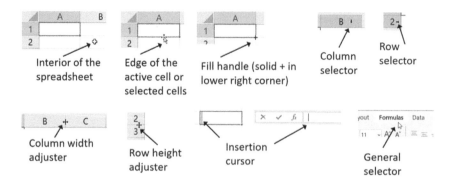

FIGURE 1.5   Different cursor personalities.

The early version, Excel 5.0, included 16 worksheets by default. All of these were rarely needed, so many users just left them unused. It was possible to make an adjustment in the number of initial worksheets via the File ⇨ Options ⇨ General menu selection. Later versions of Excel reduced the default number to three and finally to one. It is easy to add worksheets as needed using the plus icon to the right of the sheet tabs.

As you move the cursor around the Excel window, it takes on multiple personalities. Figure 1.5 illustrates these. You should exercise the cursor to identify all of them.

Cells are identified by their column letter and row number, as in B2. This is called the *cell address*. As we will understand in Chapter 3, it is also referred to as a relative address. There is an alternate, legacy method for identifying cells called R1C1. This mode relates more closely to mathematical array subscripting. Cell B2 is also referenced as R2C2. Although this method is rarely used today, it is available via File ⇨ Options ⇨ Formulas ⇨ Working with formulas ⇨ R1C1 reference style. This legacy method impacts the naming of cells, as we will see in Chapter 3.

To explore navigation on the spreadsheet, we will enter a few numbers into adjacent cells. These are shown in Figure 1.6. Any similar set of numbers will do.

	A	B	C	D	E
1					
2		6.49	8.37	0.46	2.67
3		0.38	9.83	1.95	4.61
4		3.33	0.72	1.95	0.9
5		2.66	6.99	3.85	2.79
6		5.4	0.25	9.65	8.53
7		8.85	7.32	7.79	1.32
8		3.14	3.77	8.63	3.87
9					

FIGURE 1.6   Spreadsheet with cells filled with numbers.

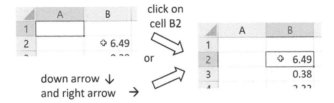

FIGURE 1.7   Moving the selection/ActiveCell to a cell within view.

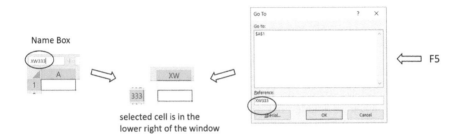

FIGURE 1.8   Two methods to select cells that are out of view.

The easiest way to select any cell within view is to click on the cell. Alternately, one can use the arrow keys to move to the cell. These options are shown in Figure 1.7.

When you need to select a cell that is not in view, there are several options. If it is close by, you can use the PageDown key (and PageUp, if you go too far) to move vertically. You can also use the vertical and horizontal sliders. If you know the cell location, and if it is perhaps far from your current location, you can type the cell address into the Name Box or use the GoTo command via the F5 key. Figure 1.8 illustrates these options. To return quickly to the A1 cell, you can press the Ctrl-Home key combination.

1.2.1   CELL JUMPS AND SELECTIONS

The next navigational skills have to do with jumps and selections. By jumps, we mean moving quickly through a range of cells, either within a block of filled cells or from a blank cell. There are two common techniques:

1. **Ctrl-Arrow:** Hold down the Ctrl key and press an arrow key for the direction desired
2. **Cell boundary:** Double-click on the boundary of a cell in the direction desired

These are not quite equivalent, and we will illustrate with various examples.

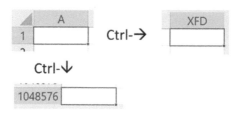

FIGURE 1.9   Ctrl-Arrow jumps to the spreadsheet extremes from the Home cell (A1).

From the Home cell (A1) with empty row 1 and column A, we can use Ctrl-→ and Ctrl-↓ to jump to the extremes of the spreadsheet. This is illustrated in Figure 1.9. Column XFD corresponds to the 16,384 or 2^{14}, and Row 1,048,576 corresponds to 2^{20}. This allows us to calculate the total number of cells on a worksheet, 2^{34} or 17,179,869,184 cells. If numbers were stored in every cell that would require eight bytes per cell, and the total memory required would be 137,438,953,472 bytes, or 128 Gb, exceeding most computers' capacity. Alternately, when we double-click a boundary of the Home cell, nothing happens.

If we select cell B2 and press Ctrl-↓, the selection jumps to cell B8, the last filled cell in the block. Also, if we double-click the lower boundary of cell B2, the same thing happens. So, the two methods are equivalent here. Similarly, with B2 selection, pressing Ctrl-→ will cause the selection to jump to E2 and likewise by double-clicking the right boundary of the cell. These jumps can be easily reversed. Perhaps an obvious question is why use jumps for cells that are within view on the spreadsheet. Of course, these are not needed as we can simply click on the desired cell. The value of these techniques comes to the fore when we are dealing with large blocks of filled cells, perhaps 100s to 1,000s of rows or columns.

A question might arise as to which of the jump techniques is preferable. The answer to this may be to consider where your hands are when you want to initiate a jump. If they are on the keyboard, the Ctrl-Arrow method may be better. If you have a hand on a mouse or a finger on a touchpad, the double-click technique may be preferable.[7]

When a jump starts on a blank cell and encounters a filled cell, the personalities of the two methods vary. Figure 1.10 illustrates this.

In contrast to a jump, a selection expands a single selected ActiveCell into a range of cells, perhaps a column, row, or block range(s). The first and simplest method is to adjoin the use of the Shift key with the jump methods above, either

FIGURE 1.10   Jumps from a blank cell toward a filled cell.

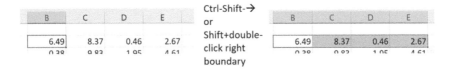

FIGURE 1.11   Row selection by using the Shift key.

the Ctrl-Arrow or double-click on boundary method. Figure 1.11 illustrates this with a row selection.

Two techniques are available for selecting all cells in single range blocks of cells: Ctrl-A and Ctrl-*. The latter is often achieved via Ctrl-Shift-8 unless you have a full-size keyboard with a numeric keypad to the right having a separate * key. You can select the block starting with a selection of any cell within the block. Figure 1.12 shows these methods and how they differ. The Ctrl-A method doesn't change the location of the ActiveCell, whereas the Ctrl-* method moves it to the upper left of the block selection. For most situations, this difference will not matter, but occasionally it will, so it is important to take note of it.

Notice in Figure 1.12 that the ActiveCell, highlighted with a white background, is in different positions. A curiosity is that it is possible to move the ActiveCell within the selection. This is done with the Tab and BackTab (Shift-Tab) keys. The ActiveCell is rotated across rows forward or backward. When it reaches the upper-left or lower-right corner, it rotates off to the opposite corner. You should experiment with this to get a feel for it.

Selections for long ranges of data present another small challenge. This is illustrated in Figure 1.13. After selecting the long column of numbers with Ctrl-Shift-↓, only the lower end of the column is in view, and the ActiveCell is at the top, out of view. Commonly, we want to continue with the upper part of the selection in view. One can move the vertical slider to get there, but it is more efficient to press the Tab key. This moves the ActiveCell down one and brings it into view. Then, the BackTab returns the ActiveCell to the top and also brings it into view.

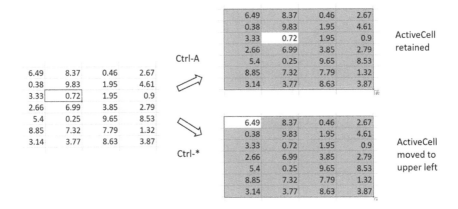

FIGURE 1.12   Selection of a block of cells using two techniques.

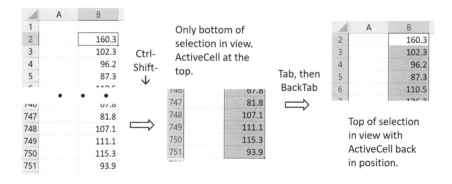

FIGURE 1.13 Returning the view to the top of a long selection.

We will now illustrate how to select several cells or blocks of cells that are not contiguous. This arises in two scenarios:

1. Applying a common format to sets of cells
2. Selecting non-adjacent columns (or rows) of cells in tables for plotting purposes

The essence of the method is to make an initial selection and then hold down the Ctrl key for the remaining selections. Figure 1.14 shows the steps for selecting two non-adjacent columns.

6.49	8.37	0.46	2.67		6.49
0.38	9.83	1.95	4.61	Ctrl-	0.38
3.33	0.72	1.95	0.9	Shift-	3.33
2.66	6.99	3.85	2.79	↓	2.66
5.4	0.25	9.65	8.53		5.4
8.85	7.32	7.79	1.32	⇨	8.85
3.14	3.77	8.63	3.87		3.14

Ctrl-Click

6.49	8.37	0.46		6.49	8.37	0.46
0.38	9.83	1.95	Ctrl-	0.38	9.83	1.95
3.33	0.72	1.95	Shift-	3.33	0.72	1.95
2.66	6.99	3.85	↓	2.66	6.99	3.85
5.4	0.25	9.65		5.4	0.25	9.65
8.85	7.32	7.79	⇦	8.85	7.32	7.79
3.14	3.77	8.63		3.14	3.77	8.63

FIGURE 1.14 Selection of two non-adjacent columns.

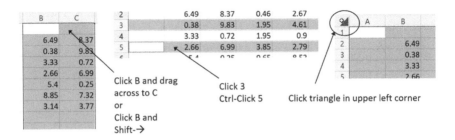

FIGURE 1.15   Selecting entire columns, rows, or the entire worksheet.

The last selections mentioned here are more global in nature. They are to select entire columns, rows, or to select the entire worksheet. Three examples are shown in Figure 1.15. These include selecting adjacent columns by dragging the cursor across the column labels, selecting two non-adjacent rows using the Ctrl key, and selecting the entire worksheet with the triangle in the upper left corner. These large-scale selections are often used to delete or insert rows or columns. Also, they are convenient for applying a common format.

1.2.2   COPYING AND MOVING CELLS

In Excel, there are several ways to copy the contents or move cells from one location to another. The variety is, in part, because of the evolving versions of the software over the decades. Earlier techniques were carried over as new ones were added. Figure 1.16 illustrates the various ways to initiate the copy of a cell. These include using the command icon on the Ribbon, right-clicking the cell to show the context-sensitive menu, the traditional keystroke combination, Ctrl-C, and preparing for a drag copy with the Ctrl key held down. The copy can also be initiated on a selection of cells.

Once the copy has been initiated, the cell or cells to be copied are surrounded by a moving, dashed border. This is shown in the lower left corner of Figure 1.16.

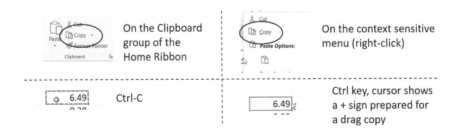

FIGURE 1.16   Four ways to initiate the copy of a cell.

The next step is to select the destination for the copy and execute it. There are three common methods:

1. Select the destination cell and press Enter
2. Select the destination cell and press Ctrl-V (the Paste command)
3. Hold the Ctrl key down and drag the cell(s) to the destination

Generally, we recommend the first method over the second. The Paste command carries out the copy but retains the source cell(s) for subsequent copies. The Enter key terminates the process with the copy. So, if you are going to copy one or more cells to multiple locations, repeated Paste commands (Ctrl-V) are the way to go. For just a single copy, use the Enter key. Apart from Ctrl-V, you can see the Paste command on the upper items in Figure 1.16. The third method is useful when the destination cell(s) are in view and, in particular, when the destination might be varied during the drag to achieve the best visual arrangement on the spreadsheet.

One can also copy a cell by dragging the fill handle (the + in lower right corner) in any direction and any number of cells. It is also possible to double-click the fill handle for an "autofill" operation that continues downward to the end of an adjacent filled column.

Moving cells is similar to copying except the source cells are removed. The move can be initiated in the following ways:

- Select the source cell(s) and execute the Cut command from the Ribbon or context-sensitive menu
- Select the source cell(s) and press Ctrl-X, the shortcut for the Cut command
- Select the source cell(s) and implement the drag operation

The move is completed by selecting the destination cell and pressing the Enter (or Paste) keys. When the destination is within view, it is often convenient to drag the source selection to the destination.

1.3　HOW INFORMATION IS ENTERED, STORED, AND EDITED IN CELLS

1.3.1　How Information is Entered in Cells

There are several types of information that are contained in cells on the spreadsheet, including numbers, Boolean (T/F) values, text, and dates. Numerical quantities are entered in cells in two formats: number and scientific notation and they are right justified by default. Figure 1.17 shows several examples. What is apparent here is that there can be a difference between the numeric entry and what is displayed in the cell. The Formula Bar often reveals the actual quantity. The cell display may change if the column width is altered. In the second example,

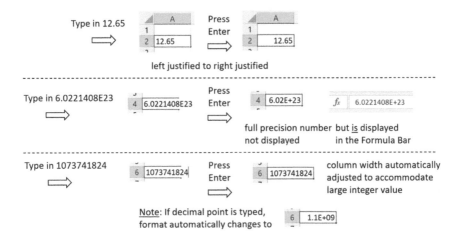

FIGURE 1.17 Examples of numerical entry.

Avogadro's number is entered directly using scientific notation, 6.0221408×10^{23} as 6.0221408E23, where the *"E" notation* signifies "10 to the power."

1.3.2 How Information is Stored in Memory

Now that we have distinguished that the number displayed may be abbreviated from the actual number stored, it is important to understand how Excel stores numerical quantities in computer memory.

Quantities internal to the computer are stored as binary digits and organized in groups. The fundamental grouping is called a *word*. In the early days of digital computers, words were groups of 8, 12, or 16 binary digits, or *bits*. A group of 8 bits is commonly called a *byte*.[8] As computer architectures became more sophisticated, word lengths of 32 and then 64 bits became common. Words in the computer typically represent either instructions or information, and they are of finite length.

Most computer programming languages and software packages have different numerical types, such as integer vs. floating point, and single precision vs. double precision. That is not the case with Excel. All numbers on the spreadsheet, whether they appear as integers, floating point (numbers with a decimal fraction), or in scientific notation, are stored in the same way with eight bytes (64 binary digits or bits). A standard method, called *IEEE 754*, is used to store the parts of the number (sign, exponent of 2, mantissa). This is also called *double precision* floating point. We do not go into detail on this here but refer you to Clough and Chapra (2023) for more information.

The characteristics of numerical storage that are of interest are *range* and *precision*. These are illustrated in Figure 1.17 and are approximately:

- **Range:** $-9.0 \times 10^{307} \leftrightarrow -2.2 \times 10^{-308}, 0, 2.2 \times 10^{-308} \leftrightarrow 9.0 \times 10^{308}$
- **Precision:** 15-to-16 significant figures

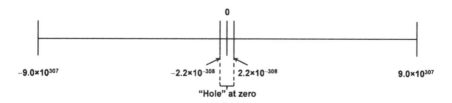

FIGURE 1.18 Range of numerical storage.

Notice that there are tiny gaps around zero. As indicated in Figure 1.18, in computer parlance, these are together dubbed the *hole at zero*. It is evident that this representation is more than adequate for most engineering and scientific applications. In fact, if the range is exceeded, that usually indicates problems in calculations. If you attempt to enter a number outside the range, it will be entered as text. If you attempt to enter a number in the "hole," it will appear as zero.

Boolean values are entered and displayed as TRUE and FALSE in a cell. These constants are stored in memory in two bytes (16 bits). A stored value of zero is considered FALSE, and one is TRUE (In general, all nonzero values are considered TRUE.) A formula can provide a logical result.

That brings us to *text entries*.[9] If the entry is obviously text, it will be displayed as such in the cell, left-justified. If a string needs to be entered as text but would be interpreted as a number, the entry needs to be preceded with an apostrophe. Figure 1.19 shows several examples. You note that you cannot start text entry with a + or − sign but must precede that with an apostrophe, '.

Text is stored by Excel in a similar manner to most programming languages and software packages. One byte is used to store each character, and a standard code scheme is used called the "American Standard Code for Information Interchange" or ASCII. There are extensive tables of the ASCII codes available. As an example, see https://www.asciitable.com. One byte allows for 2^8 or 256 codes. For example, the code for "A" is 65_{10} or 01000001_2. Note that the subscript following the number represents the number's base. Thus, the 10 indicates that 65 is in the familiar base-10 or decimal system whereas the 2 indicates the base-2 or binary system.

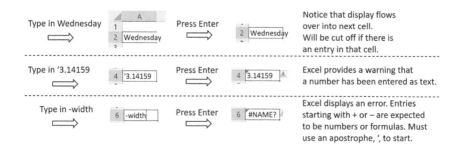

FIGURE 1.19 Examples of text entry.

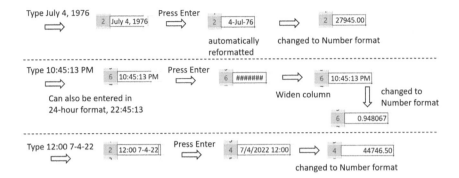

FIGURE 1.20 Example of date and time entries showing actual numbers stored.

Dates and times are in a special category in Excel. They can be entered in several formats and will be automatically interpreted as such. Figure 1.20 illustrates several and also shows the numeric value stored. As you can see, dates are stored as integers and times as fractional parts. The convention used by Excel is as follows:

- The year January 1, 1900, is considered the base year, month, and day: one (1). The days are counted forward from the base, and that is the integer stored.
- The time is stored as a decimal fraction. It is the number of seconds since midnight divided by 86,400 seconds in a day.
- A combined date and time is stored as the integer date + the fractional time.

If a date or time is out of range, e.g., July 4, 1776, it is stored automatically as text. There are many options for displaying date and time available via the options in the Number group on the Home Ribbon and via the Format Cells selection on the context-sensitive (right-click) menu.

1.3.3 How Information is Edited in Cells

As quantities are entered in cells, the Status Bar displays Enter in the lower left of the display. Once in place, there are two common options for editing (modifying) the entry in the cell.

The first is to double-click the cell or press the Edit key, F2 (this is most common), and the second is to click in the Formula Bar. In either case, the mode indicator on the Status Bar changes to Edit. When in Edit mode, you can move the insertion cursor back and forth within the cell and make changes. You can also use the Home and End keys to move to the beginning or end. You cannot do this while you first enter the quantity, that is, in Enter mode. If you move the arrow keys while in Enter mode, you will go into Point mode and cell addresses will be added to or inserted in the entry. Point mode will be described and illustrated when we discuss formula entry in the next section. Finally, you can switch back and forth between Edit and Enter mode with the F2 key. If you are unsure about the current mode, look at the Status Bar.

1.4   ENTERING SIMPLE FORMULAS AND USING BUILT-IN FUNCTIONS

Much of the "heavy lifting" performed by spreadsheets is via formulas entered in cells that reference values in other cells. The latter may be information (numbers, text, dates, etc.) or the results of other formulas. Formulas frequently include functions that are built into Excel, such as sums, averages, and mathematical calculations. Excel has hundreds of such built-in functions.

To enter a formula into a cell, one begins with the equal sign, =. In the prior competitor and market leader, Lotus 1-2-3, the start symbols were the plus, +, or minus, −, signs. To make it easy for Lotus users switching to Excel, some 30 years ago, Microsoft also allowed formulas to begin with the latter signs. That persists to the present. After you enter the formula with a + or − sign, Excel converts it to begin with an = sign.

The simplest formula refers to one or more adjacent cells by their addresses. An example placed in cell F2 might be

$$= B2 + C2 + D2 + E2$$

If done on our example spreadsheet, after pressing the Enter key, the displayed result is 17.99. This is what we call a "live" formula. If we change the value in B2 from 6.49 to 6.2, the formula result in cell F2 updates automatically to 17.7. This live presentation[10] is the main characteristic that sold users on adopting the electronic spreadsheet.

One would assume we typed in the formula above, but there is a preferred way to create it called *pointing*. Figure 1.21 illustrates this.

Pointing is preferred to typing in cell addresses because its visual aspects provide a more reliable result, avoiding any typographical errors. If you edit the

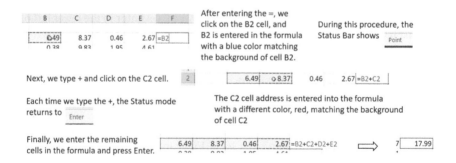

FIGURE 1.21   Entry of a simple formula using pointing.

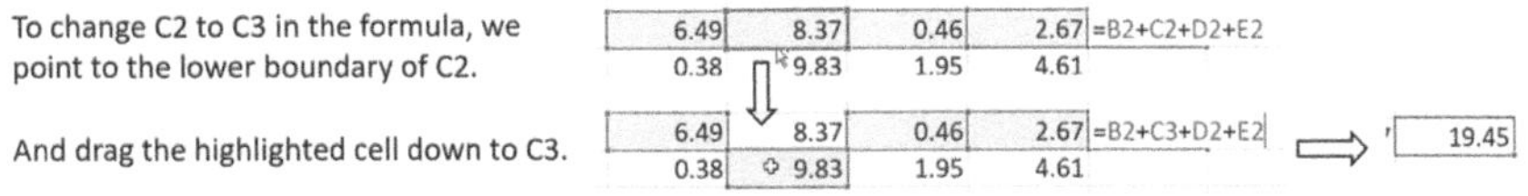

To change C2 to C3 in the formula, we
point to the lower boundary of C2.

6.49	8.37	0.46	2.67 =B2+C2+D2+E2
0.38	9.83	1.95	4.61

And drag the highlighted cell down to C3.

6.49	8.37	0.46	2.67 =B2+C3+D2+E2
0.38	9.83	1.95	4.61

⟹ | 19.45 |

FIGURE 1.22 Graphical editing of a formula.

formula in the cell by double-clicking on it or pressing the F2 key, the same color
coding appears as illustrated in Figure 1.21. Since you are in Edit mode (Status
Bar), you can move the insertion cursor back and forth to make any changes. An
alternate way to modify a cell address is shown in Figure 1.22. This is called
"graphical editing."

By invoking a built-in function, SUM, we can replace our formula with

= SUM(B2,C2,D2,E2)

Alternately, we can take advantage of using a range cell address, B2:E2, to have

= SUM(B2:E2)

Range cell addresses have the format

[upper left corner address]:[lower right corner address][11]

This includes defining a rectangular range, such as a formula for the sum of all the
numbers in our example spreadsheet (Figure 1.6),

= SUM(B2:E8)

Table 1.1 lists a few common built-in functions that we use in formulas. Others
more relevant to engineering and scientific calculations will be introduced
in Chapter 4. We will also introduce the Insert Function command, f_x, on the
Formula Bar.

TABLE 1.1
Common Built-in Functions

Function Name[12]	Description
SUM(•)	sum of a range of cells
AVERAGE(•)	average of a range of cells
ABS(•)	absolute value of a cell value
MAX(•)	maximum value of a range of cells
MIN(•)	minimum value of a range of cells
SQRT(•)	square root of a cell value
PI()	value of π

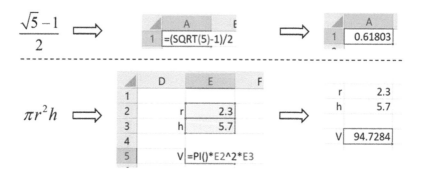

FIGURE 1.23   Example formulas involving built-in functions.

Many formulas involve simple arithmetic calculations. Figure 1.23 shows two examples. The first is a numerical calculation of the *Golden Ratio*. It makes use of the SQRT function, encloses the numerator in parentheses, and employs the subtraction (–) and division (/) arithmetic operators. The second formula is for the volume of a right circular cylinder of radius r and height h. The PI function is used along with the multiplication (*) and exponentiation (^) operators. This formula references values for r and h that are stored in nearby cells.

The different arithmetic operators are listed in Table 1.2 in order from highest to lowest precedence. Notice that there is a distinction between negation and subtraction. The former applied to a single numerical quantity, while the latter requires two. Also, for most other programming languages and software packages, including Excel's VBA, exponentiation has a higher priority than negation.

The general procedure that Excel uses to evaluate a formula is to proceed from left to right. However, functions are evaluated first, and parentheses are used to force a particular order of evaluation. If we want to enter an Excel formula to evaluate

$$\frac{14 - \sqrt{8}}{1 + \dfrac{1}{6.4}}$$

TABLE 1.2

Arithmetic Operators

Operator	Description
–	unary minus, negation
^	exponentiation
*, /	multiply, divide
+, –	add, subtract

the following would be incorrect:

14-SQRT(8)/1+1/6.4

because the order of evaluation would be

SQRT(8) ⇨ result1
result1/1 ⇨ result2
1/6.4 ⇨ result3
14-result2 ⇨ result4
result3 + result4 ⇨ final result

where the final result would be 11.3278, which is clearly wrong. A corrected version is

(14-SQRT(8))/(1+1/6.4)

which yields the correct result of 9.6619.

The next consideration is what happens when cells containing formulas are moved/copied or cells referenced by formulas are moved. Figure 1.24 illustrates what happens when the *r* and *h* value cells in the second formula in Figure 1.23 are shifted one column to the right and what happens when the formula is either

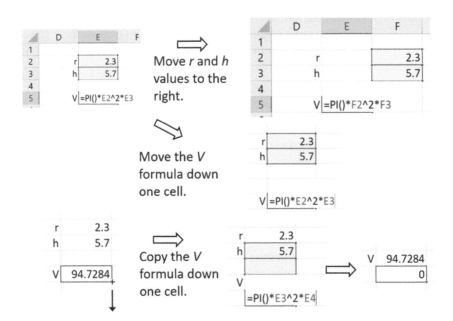

FIGURE 1.24   Examples of moving and copying cells.

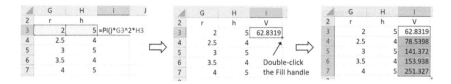

FIGURE 1.25   Copying a formula down to carry out a case study.

moved or copied down one cell. In the first example in Figure 1.24, the two cells containing the *r* and *h* values are moved one column to the right. The cell references in the formula automatically follow this move with the formula now referencing column F. The second example leaves the *r* and *h* value cells in place but moves the *V* formula down one cell. Nothing changes with the correct cell references in the formula.

The third example in Figure 1.24 illustrates what happens when the formula in cell E5 is copied down to cell E6. The cell references in the formula also move down one cell. This causes a zero result because the formula now contains cell E4, which is zero. The important takeaway here is that, when a formula is copied, its cell addresses shift, tracking the copy. This is called a *relative copy* because the referenced cells travel with the cell being copied.

A way that relative copies are typically used is to compute a case study of a formula for a set of different input values. This is shown in Figure 1.25. First, the volume formula is entered in cell I3, referring to *r* and *h* values in cells G3 and H3, respectively. Then, an "autofill" is created by double-clicking the fill handle (solid +). This copies the formula down adjacent to the values for the four other cases, and the corresponding volumes, *V*, are computed and displayed.

We'll introduce further features of cell addressing and naming cells in Chapter 3.

1.5   FORMATTING THE SPREADSHEET

It is common for spreadsheets that have extensive formatting to make them appear attractive. Creating these takes significant time and effort. Our emphasis in this text is on the day-to-day problem solving carried out by engineers and scientists. Generally, there is not enough time to invest in the cosmetic appearance of the spreadsheets we create. However, a few elementary formatting features can be added quickly to a spreadsheet to make it more readable for you and other users. We endeavor to introduce these here.

First, let's concentrate on the contents of the cell. A problem with Excel's default display of numerical results is that too many digits often appear. These clutter the spreadsheet, hinder readability, and misrepresent the number of significant figures. There are two common ways to adjust the number of digits displayed. These are illustrated in Figure 1.26. The quickest technique is clicking the Decrease Decimal command in the Number group on the Home Ribbon.

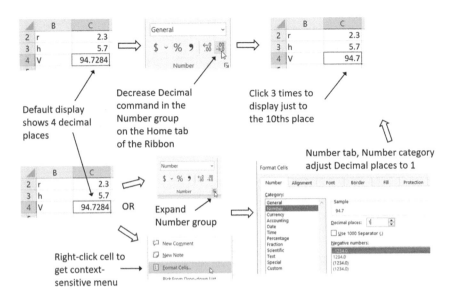

FIGURE 1.26 Techniques to decrease the number of decimal digits displayed.

As the Format Cells dialog in the lower right of Figure 1.26 shows, the Category of the display can be changed. Often, it is preferred to use the Scientific category and display an exponent. Given the E format, the number of decimal digits displayed can still be adjusted using the Decrease Decimal (or Increase Decimal) commands.

Changing to or adjusting other formats, such as Currency, Date, Time, Fraction, and Percentage, is available via the Format Cells Number tab. There are also quick commands for currency, percentage, and comma styles in the Number group on the Home Ribbon.

We often want to change the alignment of numeric or other displays in cells. Numbers are right-justified, and text is left-justified by default. Options for changing the alignment are available in the Alignment group on the Home Ribbon. One cosmetic change is to move the numeric display away from the right boundary of the cell by one or more indentations. This is shown in Figure 1.27. Notice that the right indent has been applied to three cells with one command.

Other alignment changes might be to move text entries away from the left border of the cell. Also, of course, you can quickly center or move cell entries from one border to another with the buttons in the Alignment group.

Another formatting need that arises frequently has to do with column headings. First, they are often too wide for the column width appropriate to the table entries below. Second, we may want a heading across more than one column. The changes required for both these are illustrated in Figure 1.28. There is a Wrap text command in the Alignment group on the Home Ribbon, but it doesn't provide the

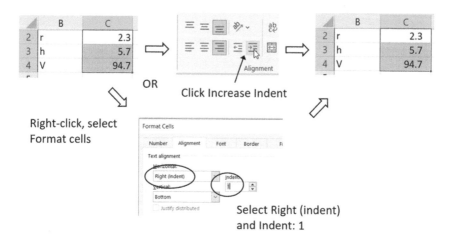

FIGURE 1.27 Two methods to set right indent equal to one.

options available on the Format Cells dialog window. By centering horizontally and vertically and checking Wrap text, a decent heading is produced and can be formatted further, as illustrated, if needed. You note that the height of the row is adjusted automatically to accommodate the heading. A table heading can be distributed across the columns of the table with the Merge & Center command. This is shown in the lower part of the figure.

Adding borders to cells is common. This can be accomplished with the Apply Borders drop-down command in the Font group of the Home Ribbon. Figure 1.29 shows how All Borders are added to the table of Figure 1.28, and then a Bottom Double Border is added to separate the headings for the table content. This is typical formatting for a table with headings.

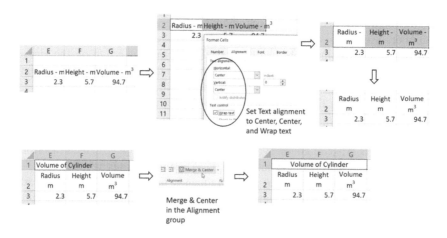

FIGURE 1.28 Formatting column and table headings.

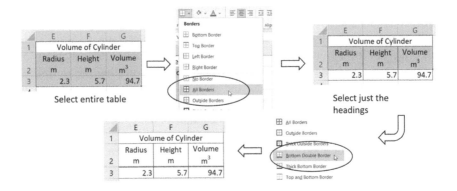

FIGURE 1.29 Adding borders to a table.

We often see the use of color on spreadsheets that are going to be used by others. It is easy to go overboard with color and spend too much time adding it. If spreadsheets are to be included in other documentation that will be printed in black and white (such as this text), color will only appear as shades of gray. When engineers and scientists prepare spreadsheets for others to use as tools, they often use a background (or fill) color, such as pale green or blue, for cells employed for user entry. Figure 1.30 illustrates the application of color to the table in Figure 1.29. First, the cell or cells are selected. Then, the Fill Color dropdown in the Font group on the Home Ribbon is displayed. From there, one of the Theme Colors is chosen, or, if those are not acceptable, many more colors are available via the More Colors… choice.

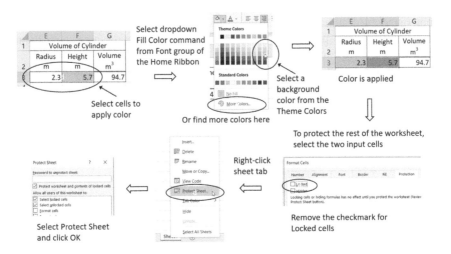

FIGURE 1.30 Setting the background color of input cells and protecting the rest of the worksheet.

For colored input cells, protection is often applied to the remaining cells to prevent modification by the user. Figure 1.30 shows how to accomplish that. The procedure is a bit backward. First, you remove the Locked status for those cells (they are not locked until you protect the worksheet) via the Protection tab of the Format Cells dialog window. This is followed by protecting the worksheet using the context-sensitive menu for the sheet tab. When complete, the user is only allowed to modify the colored input cells. There is an option to provide a password for the sheet protection. We offer caution for doing this since passwords are often forgotten.

It is necessary or appropriate to widen columns or heighten rows on spreadsheets. We saw the latter with the table headings in Figure 1.28. There are two techniques used to adjust these. One is to double-click the border to the right of a column letter (or group of column letters) or the border to the bottom of a row number (or group of row numbers). Excel will then adjust the width or height automatically. The second way is to set the width or height manually. This is illustrated in Figure 1.31 for the width of three adjacent columns, changing the width from 8.11 to 10 points.

A consideration when changing column widths or row heights is that these extend throughout the worksheet. For larger spreadsheets, these changes may interfere with the appearance in other regions. There are two techniques used to counter this. The first is to segment the workbook into separate worksheets. The second, which is more traditional, is to create a diagonal arrangement of the worksheet in rectangular blocks of cells. Then, any change within a block doesn't affect the other blocks. This arrangement provides the challenge of how to move quickly from one diagonal block to another. This is usually accomplished using named cells. We introduce cell names in Chapter 3.

It is easy to change the character of the font displays in cells. This is using the Font group on the Home tab of the Ribbon. The changes include the font type, point size, bold, italic and underline styles, and font color. Although we do not take the time and space to illustrate all these, you can experiment with them easily.

We must mention a final formatting technique, removing the gridlines on the worksheet, as it is seen in many spreadsheet applications. For day-to-day problem solving, we have observed that engineers and scientists prefer to "drive between

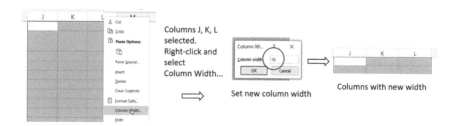

FIGURE 1.31 Example of manual setting of the width of three columns.

the lines" and not remove the visible gridlines. Removing gridlines is more common for spreadsheet "products" to be used by others or even commercially. Gridlines are removed (and reinstalled) easily in the Show group of the View tab of the Ribbon.

Throughout this text, you will find that we do not spend much time and effort in formatting spreadsheets. Our emphasis is on day-to-day problem solving for the benefit of the primary user, and engineer or scientist or student of those disciplines. In larger organizations, there may be staff members who are skilled in taking basic spreadsheets and formatting them according to the standards and needs of the organization. Our final advice is to spend more time on the substance of your spreadsheets than the cosmetic appearance. Some formatting will always be appropriate, but don't overdo it.

PROBLEMS

1.1 In the 1960s, 4k words of 12-bit word core[13] memory cost $4,000, or $1/word. Compare this cost, without accounting for inflation, with that of a modern USB 1 TB memory drive. What is the ratio in cost per bit? You may need scientific notation.

1.2 The final column on the right of the Excel worksheet is labeled XFD, and it is claimed in this chapter that this accounts for $2^{14} = 16384$ columns. Show the calculations that demonstrate this result.

1.3 Experiment on the spreadsheet with the two jump techniques, double-clicking the boundary of a cell and using the Ctrl and arrow keys. Start with the spreadsheet in Figure 1.6. Carry out the following operations with both techniques and comment on differences and similarities.
   a. Select the A1 cell and try the techniques to the right and downward.
   b. Select the B2 cell. Repeat the techniques to the right and downward as far as they will take you. When you reach that limit, repeat the techniques in an reverse fashion as far as they go.

1.4 Select an empty cell on a spreadsheet with all empty cells to its left. Compare what happens when you press the Home key and when you hold down the Ctrl key and press the Home key.

1.5 Using the spreadsheet of Figure 1.6,
   a. Select the cell in the upper left corner. Hold the Shift key down and click the cell in the lower right corner. What do you observe? Can you reverse this process?
   b. With the entire block of filled cells selected and the ActiveCell in the upper left corner, press Shift-Tab (BackTab). Describe what happens. Continue the BackTabs. Describe this.

1.6 On a blank worksheet, enter the numbers shown in Figure P1.6. Carry out the following and document your steps in detail.
   a. Select the three blocks of numbers (no other cells).
   b. With the selection in (a), add All Borders and indent all the entries one space from the right of the respective cells.

	A	B	C	D
1				
2		14	2	
3		9	-7	
4	1			
5	3		2	6

FIGURE P1.6   Spreadsheet with numbers in separate blocks.

1.7 Using the spreadsheet of Figure 1.6,
   a. Copy cells C4:D5 to G2:H3 using the Ctrl-C and Enter method. Reverse this with the Undo shortcut, Ctrl-Z.
   b. Copy cells C4:D5 to G2:H3 using the drag-copy (Ctrl) method. Reverse this with the Undo shortcut, Ctrl-Z.
   c. Move cells C4:D5 to G2:H3 using the drag-move method. Reverse this with the Undo shortcut, Ctrl-Z.
   d. Move cells C4:D5 to G2:H3 using the Cut method, Ctrl-X and Enter. Reverse this with the Undo shortcut, Ctrl-Z.

Be prepared to demonstrate these to others.

1.8 Enter the formula =PI() in a cell. Note the value of π that is displayed. Use the Increase Decimal command in the Number group on the Home tab of the Ribbon to show more and more digits. What do you notice happens at some point? How does this relate to the way numbers with decimal fractions, also known as *floating point numbers*, are stored in memory by Excel?

1.9 Figure P1.9 shows a triangle, its largest inscribed circle, and its smallest circumscribed circle. The formulas for the radii of these circles, r and R, in terms of the sides of the triangle, a, b, and c, are

$$r = \sqrt{\frac{(s-a)(s-b)(s-c)}{s}} \qquad R = \frac{abc}{4\sqrt{s(s-a)(s-b)(s-c)}} \qquad \text{where}$$

$$s = \frac{1}{2}(a+b+c)$$

   a. Create a spreadsheet calculation of the radii where there are entry cells for a, b, and c, a separate formula to calculate s, and then formulas in two cells for r and R.
   b. Format the spreadsheet so the entry cells have a light green background and protect the worksheet so that only those cells can be modified by the user.
   c. Experiment with different values for the sides. Discover what happens when you enter values that do not form a triangle. What requirements need to be met for the calculation to give a valid result?

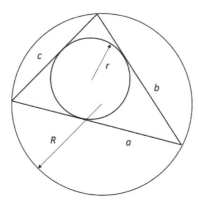

FIGURE P1.9 Triangle with inscribed and circumscribed circles.

1.10 Enter the number 6890.4583333 in a cell on a spreadsheet.
    a. Format this cell for a date and time display. What is displayed? What does this date represent in history?
    b. Use the fill handle to drag-copy this formula down one cell. What is displayed? What happened here?

1.11 In two stacked cells, one below the other, enter the numbers 1 and 2. Select both cells. Use the fill handle to drag copy down three cells or so. Comment on what happens. Try the same operation with the text entries a and b. What happens there? Enter January in a cell. Drag copy this cell down one cell. Describe what happens.

1.12 On a blank spreadsheet, starting in B3, create a column of numbers from −50 to 50 in steps of 10. Place a heading "degC" in cell B2 and a heading "degF" in cell C2.
    a. In cell C3, enter a formula referencing cell B3 that converts the value from degrees Celsius to Fahrenheit. Double-click the fill handle to copy the formula down adjacent to the values in column B.
    b. In cell A1 enter a title "Temperature Table." Format the resulting table including borders, and merge the title across columns A and B. Adjust column widths as needed.

1.13 In this chapter, we noted that for spreadsheet formulas, negation has higher priority than exponentiation, whereas for most other programming languages and software packages, including Excel's VBA, the reverse is true. Here is an equation where this distinction would matter:

$$y = -3^2$$

Evaluate the formula
    a. as you would with pen and paper,
    b. using the spreadsheet priority rules, and
    c. using the VBA priority rules.

Based on your analysis, which of the priority rule schemes is superior?

NOTES

1. The name Fortran was derived from "**For**mula **Tran**slator" and Cobol from "**Com**mon **B**usiness-**O**riented **L**anguage."
2. **B**eginners **A**ll-purpose **S**ymbolic **I**nstruction **C**ode
3. You can obtain a version of VisiCalc from www.bricklin.com that will run in a command window (MS-DOS like) in Windows, but it will only run conveniently in the 32-bit version of Windows.
4. A *macro*, short for macro instruction, is a saved sequence of commands (or keystrokes) that can be stored and then recalled with a single command or keyboard stroke. In the context of spreadsheets, Microsoft's choice of a programming language (VBA) for composing macros effectively transformed Excel into a programming environment. This allowed engineers and scientists to efficiently implement the complex mathematics required for their work.
5. It is important, as you read this text, that you "work" the material. That is, carry out all the illustrations and examples and practice all the skills that are described. In the long run, this practice will absolutely accelerate your mastery of the spreadsheet environment.
6. The entire Excel file is called a *Workbook* and the individual spreadsheets are called *Worksheets*. Later we will describe another type of sheet, called a *Chart*, which contains only a chart.
7. In teaching hundreds of classroom and short courses in a hands-on manner, we have found that individuals who use a touch panel have a hard time keeping up with those who use a mouse, especially since the instructor commonly uses the latter.
8. Although it might seem like a "bit" of nerd humor, 4 bits have been dubbed a *nibble*. The term originated from the fact that 4 bits are "half a byte," as byte is a homophone for bite. To add insult to injury, some have adopted an alternative spelling, nybble, to be consistent with the spelling of byte. If, after reading this, you are laughing uncontrollably (as we still do), you could have a future in computing!
9. In computer science, these are called *alphanumeric*, *string*, or *character* values.
10. This automatic updating is commonly called a "What if" calculation. Using our very simple example, the spreadsheet allows us to immediately learn "**What** happens to our bottom line (F2), **if** we change the value in cell B2."
11. A curiosity: You can type in the range address in reverse order, but Excel will change it back once you press Enter.
12. Most built-in functions require input(s) enclosed in parentheses as indicated by (•). These are referred to as the function's *argument*(s). Obviously, some functions, like PI(), do not require an argument but still need the parentheses as indicated.
13. Each bit was implemented with a tiny iron torus through which four wires were threaded.

2 Charts and Graphs

CHAPTER OBJECTIVES

- Be able to create graphs of engineering and scientific data
- Know how to import larger data sets from external files, such as text and comma-delimited files
- Learn how to enhance graphs with marker and line styles, multiple series, legends, multiple axes, and annotations
- Understand how to create and interpret contour and surface plots of two-dimensional data and functions
- Be able to create charts of categorical data including bar and pie charts
- Learn how to streamline the creation of charts with templates

---

We have all heard the adage, "a picture tells a thousand words." Engineers and scientists gain a deeper understanding of numbers and phenomena through charts and graphs. It has been observed that many engineers and scientists have trouble carrying on a conversation without having a pad of graph paper at hand. Excel's charting features have been designed primarily for business and financial applications. We often have to adapt these to our needs, and that is the purpose of this chapter.

This chapter first introduces the creation of typical, elementary graphs of data. These are the plots that engineers and scientists create on a day-to-day basis. As needed, features are added to enhance the appearance and communication of the charts. We introduce the most common of these here. We also show how larger sets of data can be imported into the spreadsheet for the purpose of creating plots and carrying out calculations.

When response data depend on more than one factor or independent variable, contour and surface plots can be used to depict their behavior. Excel has decent capabilities to produce contour plots, but surface plots are more limited. Both will be illustrated in this chapter. Finally, you will find that creating good graphs with Excel is often time-consuming and tedious. When a particular graph style is to be repeated, chart templates can increase efficiency markedly. Further, families of chart templates can be assembled and shared to standardize their appearance within a workgroup or organization. Templates to do this will be introduced here.

2.1 ENGINEERING AND SCIENTIFIC GRAPHS

Graphs frequently involve one or more quantities plotted against a single factor or independent variable. The quantities will be numerical and may represent continuous or discrete variables. The independent variables may also be continuous

DOI: 10.1201/9781003361053-2

or discrete. Experimental data may be the basis for the graphs, but we will also want to plot mathematical functions and will illustrate that in later chapters. Also, experimental data may contain a significant random or "noise" component. The nature of what we are plotting will influence the type of graph we create in Excel.

Among the most important factors governing chart style is the quantity of data being plotted. Simply put, small and large datasets demand different styles to achieve the best clarity. The following two sections deal with each of these situations separately to illustrate these styles.

2.1.1 PLOTTING SMALLER DATA SETS

For cases where each measurement is intrinsically difficult or costly, there might be a small number of data ($n < 20$). The following example provides a simple illustration of a small dataset generated from a laboratory experiment.

Example 2.1   Plotting Small Data Sets

Data sets may represent careful measurements made in the laboratory, such as those for chemical or physical properties. Table 2.1 presents the density of solutions of common salt in water for different concentrations of salt at 25°C. With these data entered on a worksheet, Figure 2.1 shows how to create the first version of a scatter plot.

We have work to do to make the plot more presentable, but first we consider the options in the Insert Scatter drop-down group. These are illustrated in Figure 2.2.

TABLE 2.1
Density of Aqueous Solutions of
Salt (NaCl) at 25°C

Concentration (wt% NaCl)	Density (g/cm³)
1	1.00409
2	1.01112
4	1.02530
8	1.05412
12	1.08365
16	1.11401
20	1.14533
24	1.17776
26	1.19443

Source: http://butane.chem.uiuc.edu/pshapley/
                    genchem1/l21/1.html

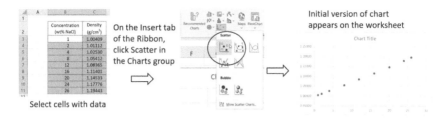

FIGURE 2.1 Creating the first version of a scatter plot.

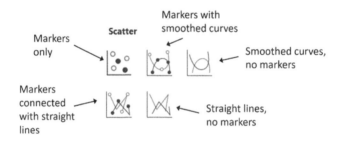

FIGURE 2.2 Insert Scatter drop-down choices.

We make the following recommendations in terms of the plot choices in Figure 2.2:

- **Markers only:** Use for plots of experimental data with tens of points.
- **Markers connected with straight lines:** Add straight lines to illustrate patterns or trends in the data.
- **Straight lines, no markers:** Use for experimental data when there are 100s of points where markers will clutter the plot. Also, lines without markers are used for plots of mathematical functions.
- **Both smoothed curve options:** Do not use. May add artificial curvature to the plot that leads to erroneous interpretations.

Also, we note that other plot options, like Line Chart, are not appropriate for most engineering and scientific plots because they are designed for business and financial plots with categorical scales on the x-axis, e.g., months.

Starting with the chart created in Figure 2.1, a few common modifications are in order. These include:

- Move the chart to its own sheet
- Change the marker color to black
- Add axes labels and a chart title
- Change the font sizes for the axes labels and the title, and change color to black
- Change the font sizes for the axes tick labels and change color to black

These are illustrated in Figures 2.3–2.5.

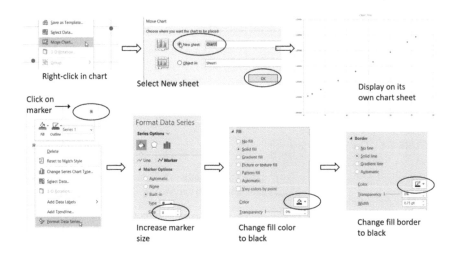

FIGURE 2.3　Moving chart to its own sheet and changing markers.

There are alternate choices for many of the steps in Figures 2.3–2.5. Font styles, sizes, bold, or italic can all be changed to suit your needs. The size and style of the markers along with interior and border can be adjusted. Users generally adopt an overall appearance that they consider to be clear and appealing. For engineering and scientific plots, it is usually preferable to keep things simple and not divert attention from the substance of the plot.

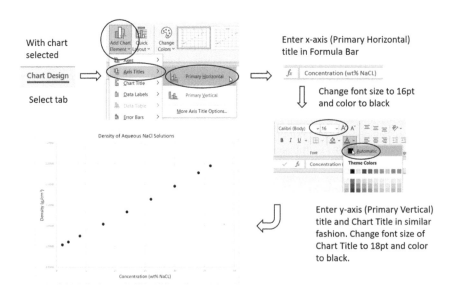

FIGURE 2.4　Adding and formatting axis labels and chart titles.

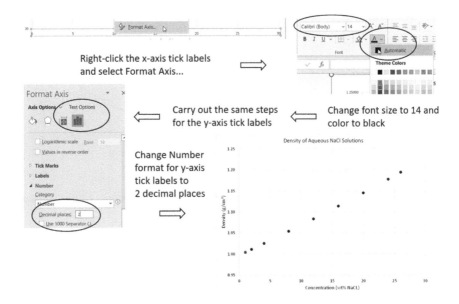

FIGURE 2.5 Formatting the axis tick labels.

The chart in Figure 2.5 could be changed by adding straight lines inter-connecting the data marks. This procedure is described in Figure 2.6 and the resulting plot in Figure 2.7.

If the gridlines appear too faint in Figure 2.7, it is possible to change their intensity from light to darker gray. It is also possible to remove the vertical or horizontal gridlines, or both. These options are available via the Gridlines option on the Add Chart Element drop-down menu.

You will note that, to produce an acceptable plot, many steps were taken. If we are going to repeat that procedure many times, it will become tedious and

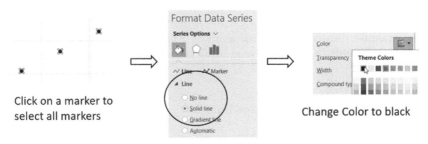

FIGURE 2.6 Adding interconnecting straight lines between the data markers.

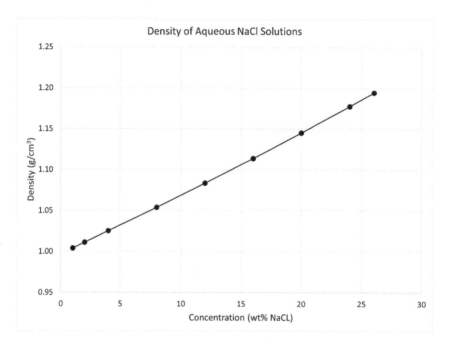

FIGURE 2.7   Plot with straight lines interconnecting the data markers.

time-consuming. As we will see in Section 2.5, it is possible to streamline the process with templates. Next, we consider creating a plot of a much larger data set.

2.1.2   Plotting Large Data Sets

Because of several factors, today's engineers and scientists frequently need to make sense of very large datasets. For example, advances in measurement technology such as probes and remote sensing allow the collection of high-frequency time series. Problems ranging from climate change to demographics often require many decades of observations to understand and substantiate long-term patterns that are critical for anticipating the future. The following example illustrates one such analysis where an Excel-generated graph provides insights that cannot be gleaned from merely examining a table of numbers.

Example 2.2   Plotting a Large Data Set

Data are available on the history of global average temperature deviation from the average of the 20th century from NOAA's *National Centers for Environmental Information (NCEI) – Climate at a Glance* website. These can be

TABLE 2.2
Global Average Temperature Deviation from 1880 to 2019 (°C) from the National Centers for Environmental Information (NCEI) – Climate at a Glance

Year	Value	Year	Value	Year	Value	Year	Value	Year	Value	Year	Value	Year	Value
1880	−0.12	1900	−0.08	1920	−0.23	1940	0.16	1960	0.05	1980	0.28	2000	0.43
1881	−0.09	1901	−0.16	1921	−0.16	1941	0.27	1961	0.10	1981	0.33	2001	0.57
1882	−0.10	1902	−0.26	1922	−0.25	1942	0.11	1962	0.11	1982	0.19	2002	0.62
1883	−0.18	1903	−0.38	1923	−0.25	1943	0.11	1963	0.12	1983	0.36	2003	0.64
1884	−0.27	1904	−0.46	1924	−0.25	1944	0.28	1964	−0.14	1984	0.17	2004	0.58
1885	−0.26	1905	−0.28	1925	−0.18	1945	0.18	1965	−0.07	1985	0.16	2005	0.67
1886	−0.25	1906	−0.21	1926	−0.08	1946	−0.01	1966	−0.01	1986	0.24	2006	0.64
1887	−0.29	1907	−0.39	1927	−0.17	1947	−0.04	1967	0.00	1987	0.39	2007	0.62
1888	−0.14	1908	−0.44	1928	−0.18	1948	−0.05	1968	−0.03	1988	0.40	2008	0.55
1889	−0.10	1909	−0.45	1929	−0.33	1949	−0.07	1969	0.11	1989	0.30	2009	0.65
1890	−0.36	1910	−0.41	1930	−0.11	1950	−0.15	1970	0.06	1990	0.45	2010	0.73
1891	−0.27	1911	−0.45	1931	−0.06	1951	0.00	1971	−0.06	1991	0.39	2011	0.58
1892	−0.32	1912	−0.34	1932	−0.13	1952	0.05	1972	0.04	1992	0.24	2012	0.64
1893	−0.34	1913	−0.32	1933	−0.26	1953	0.13	1973	0.20	1993	0.29	2013	0.68
1894	−0.32	1914	−0.14	1934	−0.11	1954	−0.10	1974	−0.06	1994	0.35	2014	0.74
1895	−0.25	1915	−0.10	1935	−0.16	1955	−0.13	1975	0.02	1995	0.47	2015	0.93
1896	−0.10	1916	−0.33	1936	−0.12	1956	−0.18	1976	−0.07	1996	0.33	2016	1.00
1897	−0.11	1917	−0.40	1937	−0.01	1957	0.07	1977	0.21	1997	0.52	2017	0.91
1898	−0.28	1918	−0.31	1938	−0.02	1958	0.13	1978	0.12	1998	0.65	2018	0.83
1899	−0.16	1919	−0.25	1939	0.01	1959	0.08	1979	0.23	1999	0.44	2019	0.95

downloaded as a comma-delimited text file (.csv) that can be opened in Excel. A table of these 140 data points is shown in Table 2.2. If we create a plot of these data using markers (or markers with lines), the plot looks too cluttered. Consequently, we use the scatter plot choice with straight lines, no markers, as shown in Figure 2.9.

The data in Table 2.2 are entered on a worksheet in a single pair of columns. Using steps like those in Example 2.1, the plot is generated. One extra change required for this plot is to modify the crossing point on the vertical axis of the horizontal axis to place it at the bottom of the plot. Figure 2.8 shows how this change is implemented.

As in Figure 2.9, the plot shows a relatively flat pattern up to the 1940s (WWII), where there is a peak. Then, starting in the 1970s, there is a linear, upward trend. There is considerable variability of about ±0.1°C that underlies these trends. Our "picture" has revealed some interesting observations that are not obvious in Table 2.2.

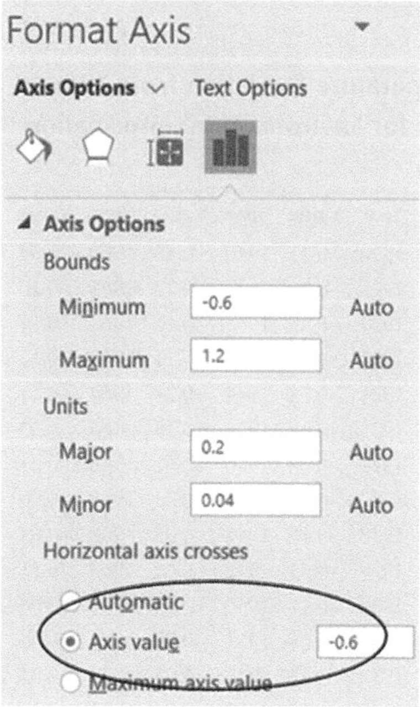

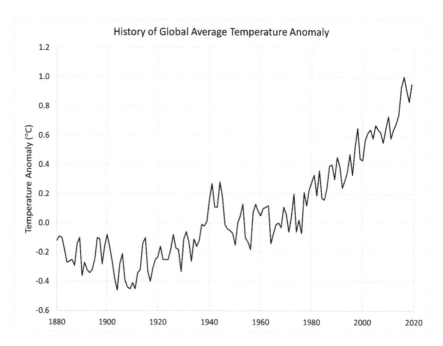

FIGURE 2.8   Changing the crossing point of the horizontal axis.

History of Global Average Temperature Anomaly

FIGURE 2.9   History of global average temperature anomaly from 1880 to 2019.

2.2   PLOTTING MORE THAN ONE SERIES

In adding a second series to a plot, we consider two common scenarios: the additional series is similar to the one already plotted, or it is quite different in magnitude and range. For the first case, we consider adding to the data in Table 2.1 to include different temperatures. For the second case, we include a series of world population data, which is very different from the temperature anomaly data.

Example 2.3   Plotting the Density of Salt Solutions at Different Temperatures

In Figure 2.7, we created a plot of salt solution density versus concentration at 25°C. If we are interested in the effect of temperature on density, we might consider Table 2.3 that presents additional data at 0°C and 80°C. This table provides us the basis to create a plot with three series using the markers-with-lines style of Figure 2.7.

We start by selecting all four columns of data, including the column headings, and inserting a markers-with-straight lines scatter plot as illustrated in Figure 2.10. Although the initial plot needs work, you will notice right away that a legend has been created at the bottom with the labels from the column headings. Also, default colors have been used to differentiate the three series, although they may only be visible here as shades of gray.

To improve the appearance of the plot, we carry out the following steps:

- Move the chart to its own sheet
- Change the y-axis scale to Bounds of 0.95 to 1.25 with Major Units 0.05
- Change the y-axis tick labels to Number with 2 decimal places

---

TABLE 2.3
Density of Aqueous Solutions of Salt (NaCl) at 0°C, 25°C, and 80°C

Concentration (wt% NaCl)	Density (g/cm³)		
	0°C	25°C	80°C
1	1.00747	1.00409	0.97850
2	1.01509	1.01112	0.98520
4	1.03038	1.02530	0.99880
8	1.06121	1.05412	1.02640
12	1.09224	1.08365	1.05490
16	1.12419	1.11401	1.08420
20	1.15663	1.14533	1.11460
24	1.18999	1.17776	1.14630
26	1.20709	1.19443	1.16260

Source: http://butane.chem.uiuc.edu/pshapley/genchem1/l21/1.html

---

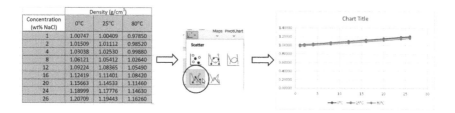

FIGURE 2.10   Creating the initial scatter plot for three series of density data.

- Change the x-axis and y-axis tick labels font to 14pt, black
- Add axis labels and a title identical to Figure 1.2

The resulting plot is illustrated in Figure 2.11.

The next issue we face is the use of color by default to distinguish the three series. To keep a black-and-white (B&W) format, we convert the three series to black and change the marker and line style for two of the series to differentiate their appearance. The steps are illustrated in Figure 2.12 for the 80°C series.

When similar changes are made to the other two series, the plot shown in Figure 2.13 results.

The final modification we suggest is optional, but we think it improves the appearance of the legend. The legend is first moved to the right of the plot. Then, its format is changed, and finally, it is moved to the interior of the plot. Figure 2.14 shows how to do this. Once these changes have been made, the legend box can be dragged into the plot area and placed appropriately. The final plot is displayed in Figure 2.15.

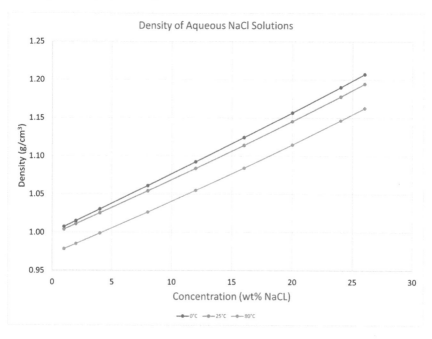

FIGURE 2.11   Improved version of plot as it would appear on its own chart sheet.

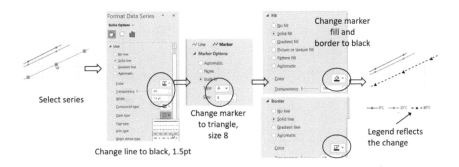

FIGURE 2.12 Changing the 80°C series line and marker style and color.

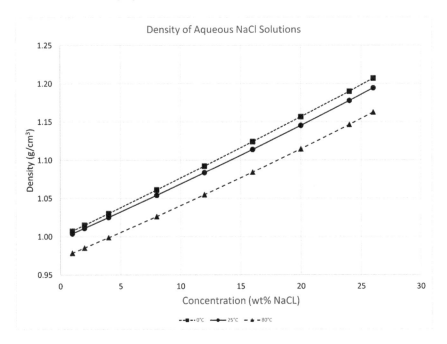

FIGURE 2.13 Plot with B&W series lines and markers.

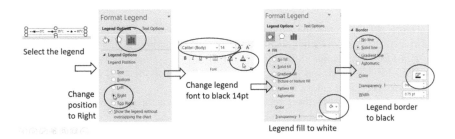

FIGURE 2.14 Modifying the legend for placement within the plot area.

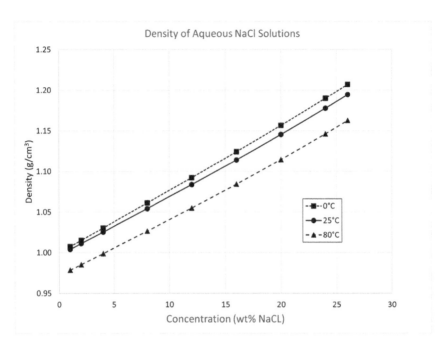

FIGURE 2.15 Final plot of salt solution density versus concentration for three different temperatures.

If the purpose or publication vehicle of your spreadsheet requires or supports color, you can certainly incorporate that into the plot. We have found the default colors provided by Excel to be lacking, and, when color is permissible, we have generally changed those, most often to brighter, more distinctive, and more readable colors.

The next example deals with the situation where multiple series plotted have very different numerical ranges.

Example 2.4   Adding Global Atmospheric CO_2 Concentration to the Temperature Anomaly Plot

Figure 2.9 illustrated the history of the global average temperature anomaly from 1880 to 2019. Often, with such plots, we are interested in adding another series for comparative study. Here, we are going to restrict the year range from 1980 to 2019 and add global annual mean atmospheric CO_2 concentration (ppm).[1] See Table 2.4 for the new data.

First, we recreate the temperature anomaly plot for the new range from 1980 to 2019. This is shown in Figure 2.16. Since there are fewer data, we have included markers.

Now, we want to add the CO_2 series. Since the years match, we select just the CO_2 values from a single column on the spreadsheet and initiate a

TABLE 2.4
Global Annual Mean Atmospheric CO_2 Concentration (ppm)

Year	CO_2 (ppm)	Year	CO_2 (ppm)	Year	CO_2 (ppm)	Year	CO_2 (ppm)
1980	338.91	1990	354.05	2000	368.96	2010	388.76
1981	340.11	1991	355.39	2001	370.57	2011	390.63
1982	340.86	1992	356.09	2002	372.59	2012	392.65
1983	342.53	1993	356.83	2003	375.15	2013	395.40
1984	344.07	1994	358.33	2004	376.95	2014	397.34
1985	345.54	1995	360.17	2005	378.98	2015	399.65
1986	346.97	1996	361.93	2006	381.15	2016	403.06
1987	348.68	1997	363.05	2007	382.90	2017	405.22
1988	351.16	1998	365.70	2008	385.02	2018	407.61
1989	352.78	1999	367.80	2009	386.50	2019	410.07

copy (Ctrl-C). Then, switch to the chart sheet and click to activate it. And, finally, press Enter. The result is shown in Figure 2.17.

Adding the CO_2 series to the chart was simple and efficient, but there are problems with the result. First, the scales of the two series are markedly different, so the temperature anomaly line is suppressed with an awkward amount of space in between it and the CO_2 series line. Second, the left axis would now have to reflect titles for both series.

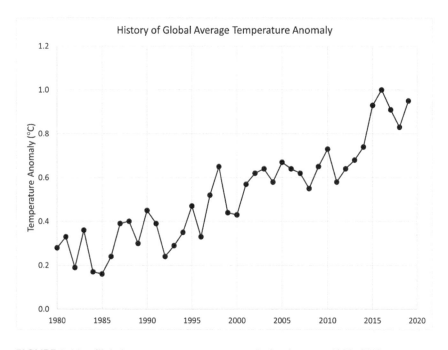

FIGURE 2.16 Global average temperature anomaly for the years 1980–2019.

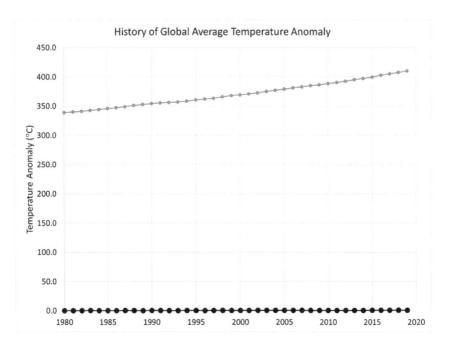

FIGURE 2.17   Temperature anomaly plot with CO_2 data added.

To resolve these problems, we introduce a second axis scale on the right of the plot for the CO_2 series. The steps to do this are described in Figure 2.18. Once completed, the series marker and line styles are changed, the secondary axis tick labels are reformatted to 14pt and black, and a secondary vertical axis label is added.

The remaining element needed for the chart is a legend. This was not automatically generated when the CO_2 series was added. Figure 2.19 illustrates the addition and formatting of the legend. Changing the legend labels is a bit tricky.

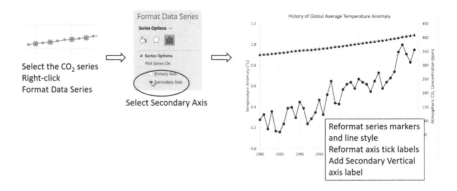

FIGURE 2.18   Moving the CO_2 series to the secondary axis on the right.

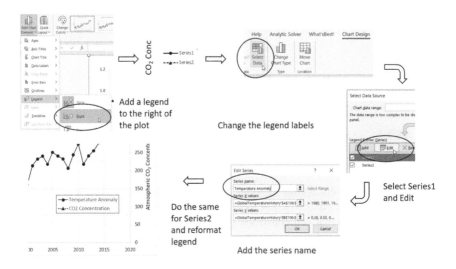

FIGURE 2.19 Adding and formatting the legend to the plot.

The final version of the plot is shown in Figure 2.20. You will note that the main title has been modified. One additional possible modification would be to reset the bounds of the secondary axis. Doing this might cause the two series to overlap on the plot, so this should be handled with care to achieve a result pleasing to the user and others.

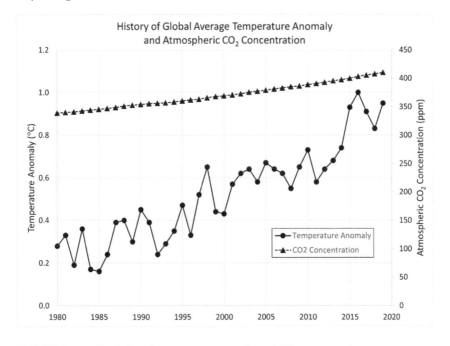

FIGURE 2.20 Final plot of temperature anomaly and CO_2 concentration.

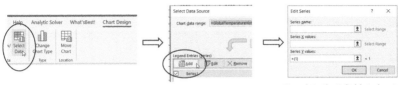

Complete the 3 fields to locate
the new series and provide a name

FIGURE 2.21   Adding a series to the plot where X and Y values are specified.

Of course, there are scenarios where we want to plot numerous series, and they may be of different character, not as in Example 2.3. Excel is limited to the left and right axes, and that is a constraint on how these scenarios might be handled. If this is a critical need, it may be necessary to obtain a supplementary graphical program or Excel add-in. Systat's *SigmaPlot* is one example.

In Example 2.4, the second series was simply copied from the worksheet onto the chart sheet. This was possible because the years of the two series were the same. If the new series is different in both x-axis and y-axis values, a different procedure must be used to add the series to the plot. This general procedure is illustrated in Figure 2.21.

2.3   CREATING PLOTS OF TWO-DIMENSIONAL DATA

In Example 2.3, we plotted the density of salt solutions versus salt concentration for different temperatures. This used a separate curve for each temperature. From that example, we recognize that density is a function of two different factors or independent variables, concentration and temperature. We can get some idea of the effect of temperature from Figure 2.15 by the spacing of the curves, but a better way would be to provide a three-dimensional depiction. This is done in two typical manners. First, a contour plot is used, and second, a surface plot can be created.

Contour plots are similar to topographic maps, such as those shown in Figure 2.22. The curved lines on the plot are called *isobars* because they correspond to constant levels of atmospheric pressure and thus elevation. *Surface plots* attempt to depict the response or dependent variable as an elevation above a plane of the two independent variables. Figure 2.23 provides an example of a surface plot. Another style of surface plot presents a mesh instead of a continuous surface. The contour plot may not provide as much "feel" for the shape of the data as does the surface plot; however, the contour plot is better suited to extract numerical values.

In this section, we will display the salt density data as both contour and surface plots in Excel.

FIGURE 2.22   Example of a topographic map with isobars.

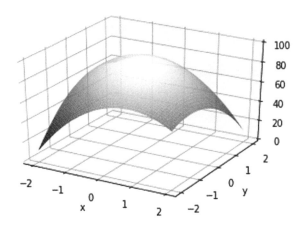

FIGURE 2.23   Example of a surface plot.

Example 2.5   Contour and Surface Plots for Salt Density Data

To provide a richer set of salt density data, we include more temperatures than in Table 2.3. Table 2.5 provides the complete data set for our plots.

Contour and surface plots in Excel are intended for use with *categorical factors* (independent variables). Engineers and scientists often have *continuous numerical factors*, such as with the salt data. It takes some work to adapt Excel's plots to our needs.

TABLE 2.5

Density of Aqueous Solutions of Salt (NaCl) versus Concentration and Temperature

Concentration (wt% NaCl)	Density (g/cm³)						
	0°C	10°C	25°C	40°C	60°C	80°C	100°C
1	1.00747	1.00707	1.00409	0.99908	0.9900	0.9785	0.9651
2	1.01509	1.01442	1.01112	1.00593	0.9967	0.9852	0.9719
4	1.03038	1.02920	1.02530	1.01977	1.0103	0.9988	0.9855
8	1.06121	1.05907	1.05412	1.04798	1.0381	1.0264	1.0134
12	1.09224	1.08946	1.08365	1.07699	1.0667	1.0549	1.0420
16	1.12419	1.12056	1.11401	1.10688	1.0962	1.0842	1.0713
20	1.15663	1.15254	1.14533	1.13774	1.1268	1.1146	1.1017
24	1.18999	1.18557	1.17776	1.16971	1.1584	1.1463	1.1331
26	1.20709	1.20254	1.19443	1.18614	1.1747	1.1626	1.1492

Source: http://butane.chem.uiuc.edu/pshapley/genchem1/l21/1.html

We start by selecting the interior of the data shown in Table 2.5 and choosing a Wireframe Contour plot from the Surface category of the Insert Waterfall, Funnel, Stock, Surface, or Radar chart group. The steps and resulting plot are shown in Figure 2.24. Obviously, we have work to do to improve the plot. First, we move the plot to its own chart sheet.

There are two main steps to improve the plot and make it more expressive. These are to assign our table ranges to the axis categories and change the contour interval range and specifications. It is not that clear how to do this, so it is detailed in Figure 2.25.

Several final changes are made to the plot displayed in Figure 2.25. These include moving and reformatting the legend. The legend here describes the contour intervals. Also, the tick labels are changed, and axis labels plus a chart title are added. The final plot is shown in Figure 2.26.

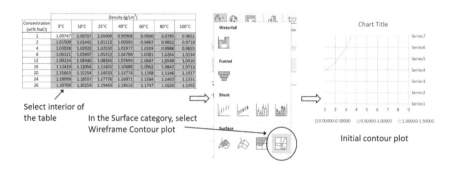

FIGURE 2.24   Creating the initial contour plot.

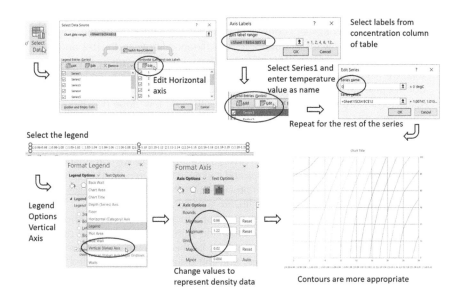

FIGURE 2.25   Changes to improve the contour plot.

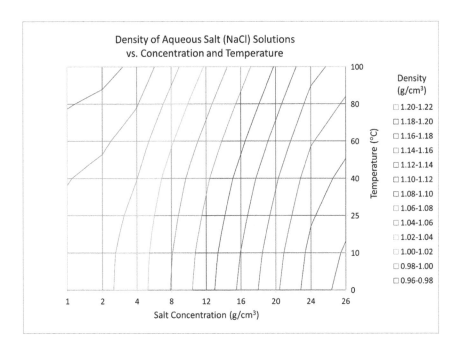

FIGURE 2.26   Final version of contour plot.

There are limitations evident in the final contour plot, and they are related to the capabilities of Excel's contour plotting. First, the intervals on the axes are categorical. That means that there is no difference between the 1-2 and 2-4 intervals of concentration. The plot considers all intervals equal. To adapt to that, we would have to present the data only with even intervals, and our table doesn't provide for that. Second, the contours are plotted only with the data values. They are not interpolated and smooth, so a few slope discontinuities are observable.

There is no convenient way to label the contours; rather, we must refer to the intervals in the legend. Also, the contours are colored, so the plot would not render in a useful way in B&W. Given these limitations, we can still use the plot to get a feel for the relationships and can even use it to estimate intermediate values. However, contour plots provided by other software packages, such as MATLAB® and Python's Matplotlib, are superior.

Creating a surface plot follows a similar procedure. Instead of choosing the Wireframe Contour plot, we choose the Wireframe 3-D Surface option. Note: For either the previous contour plot or this one, we could have chosen the options without Wireframe, and the results would be shaded plots. We have chosen the Wireframe options here, but you will explore the shaded options in problems at the end of this chapter. In Figure 2.27, the first version of the plot is illustrated on the left, and, after similar modifications, the final version is shown on the right.

One lingering problem in the last plot is that the view of the surface isn't the best. The surface view folds on itself. By selecting the interior of the plot, we can get the Format Walls panel to the right and change the X and Y Rotation to 10° and 20°, respectively. The resulting plot is shown in Figure 2.28.

The same issues of axis spacing are present with the surface plot as we encountered with the contour plot. Surface plots do a better job of communicating shape, but they are more difficult to read quantitatively. Also, the use of color is essential as it codes the intervals described by the legend. In the end, despite the limitations, Excel's surface and contour plots can aid in understanding the relationships between factors and responses. When there are more than

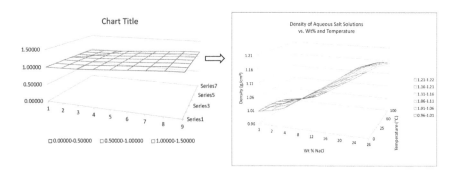

FIGURE 2.27 Initial and formatted surface plots.

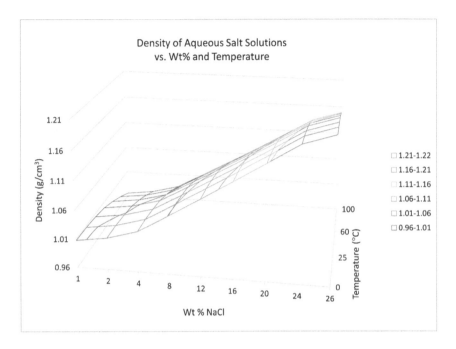

FIGURE 2.28   Final version of surface plot.

two factors, they can be chosen in pairs for plots to help paint a picture of the relationships. We will see the use of contour and surface plots with two-way case studies in Chapter 5.

2.4   BAR AND PIE CHARTS

There are scenarios where engineers and scientists find it convenient to display data in the format of bar and pie charts. Most often, these charts are categorical in nature. Also, there are situations with bar charts where the factor (independent variable) is quantitative. We do not treat that in this section, but rather in Chapter 8 with histogram charts. Bar and pie charts often take advantage of color, but we will emphasize using B&W patterns here.

Example 2.6   Depicting Sources of Energy Worldwide with a Pareto Bar Chart

There is a form of a bar chart, called a *Pareto chart*,[2] that sorts the categories (and corresponding bars) in either ascending or descending order. Table 2.6 summarizes the different sources of energy. You will note that the entries have been sorted from highest to lowest.

TABLE 2.6

Energy Sources by Consumption in 2021

Source	Energy Consumption (TWh)
Oil	51170
Coal	44473
Gas	40375
Hydro	11183
Nuclear	7031
Wind	4872
Solar	2702
Biofuels	1140
Geothermal	763

Source: ourworldindata.org

Figure 2.29 illustrates the steps to create a traditional bar chart. In contrast to the X-Y scatter chart, we do not select the category column and will deal with that later.

Additional formatting steps are:

• moving the chart to its own sheet
• formatting the vertical axis tick labels
• incorporating and formatting the category labels instead of the numbers 1 through 9
• changing the format of the bars (many options here)
• adding axis titles and a chart title

Key steps are illustrated in Figures 2.30 and 2.31.

The final version of the Pareto bar chart is shown in Figure 2.32. The chart clearly communicates the predominance of fossil fuel sources on the current world stage. It is also interesting that coal is more predominant than natural gas. This is problematic from a climate change perspective, as coal emits about 1.7 times more CO_2 per unit of energy output (https://www.nrel.gov/analysis/life-cycle-assessment.html).

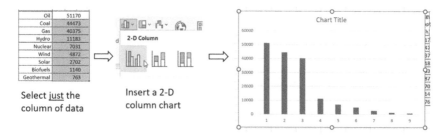

FIGURE 2.29 Creating the initial version of a Pareto bar chart for the energy data.

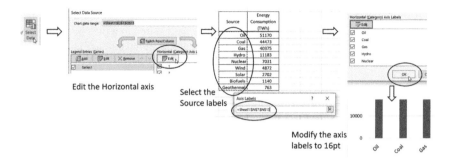

FIGURE 2.30 Adding the Source labels to the x-axis of the bar chart.

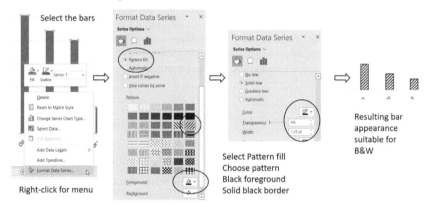

FIGURE 2.31 Reformatting the bars.

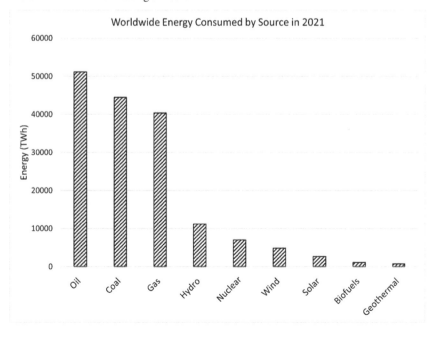

FIGURE 2.32 Pareto chart for energy consumption by source in 2021.

Alternately, we can depict categorical distributions in the form of a *pie chart*. The data from the previous example are not that well suited to a pie chart because several categories are so small that they would be hardly visible as "slices of the pie." We will use another example better suited to pie charts.

Example 2.7 Comparing Solar Energy Production by Country in 2022

Table 2.7 provides data on solar energy production in megawatts (MW) for the five highest producers. The countries are listed in alphabetical order. We would like to depict these data in a pie chart. Figure 2.33 shows the steps to create the initial chart.

After moving the chart to its own sheet, we need to replace the 1-through-5 category labels with the country names. Then, we delete the legend and replace it with data callout labels identifying the pie slices directly. A title is added, and the font sizes and color (black) are adjusted. This is described in Figure 2.34.

TABLE 2.7
Solar Energy Production by Country in 2022

Country	Solar Energy Production (MW)
China	175,018
Germany	45,930
India	26,869
Japan	55,500
U.S.	62,200

Source: https://worldpopulationreview.com/country-rankings/solar-power-by-country

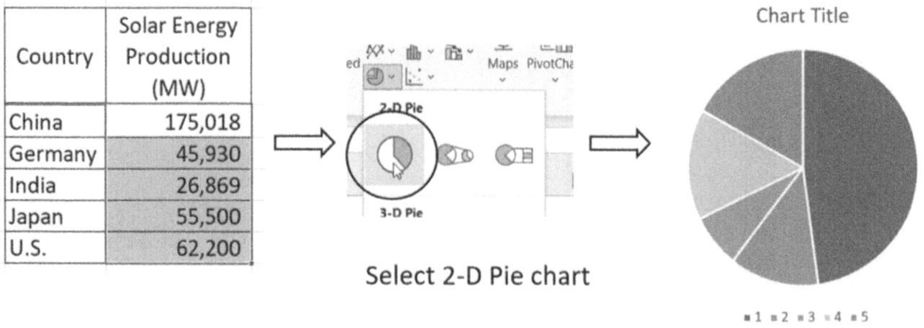

FIGURE 2.33 Creating the initial pie chart for the solar energy production data.

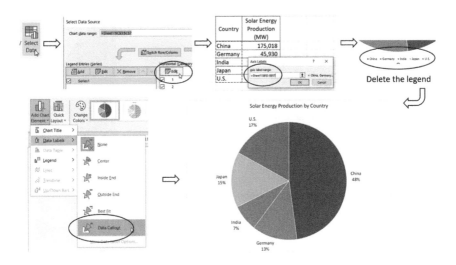

FIGURE 2.34 Modifications to the pie chart.

By default, the data callout labels show percentages. If we prefer the actual power values, we can change the display as shown in Figure 2.35 and then modify the title to indicate MW.

The last issue we deal with is that the pie chart presents different colors for the slices. That may be acceptable, but it would not look that good if the chart were printed in B&W or shades of gray. We can change the colors to provide a better-colored appearance or more contrast in B&W. We can also change the appearance of the slices to include different patterns. The latter is illustrated in Figure 2.36.

The final appearance of the pie chart is shown in Figure 2.37. Of course, there are many options that might be preferred over this presentation. The first would be to use colors instead of B&W patterns. Then, there are different styles of pie charts available in the Charts group of the Insert Ribbon. The use of a legend or data callouts provides another option. The example we use here is intended to give you a start.

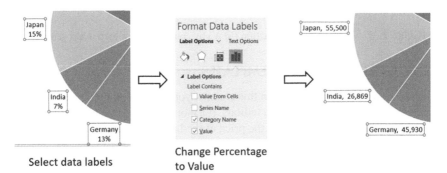

FIGURE 2.35 Changing the callout labels from percentage to power (MW).

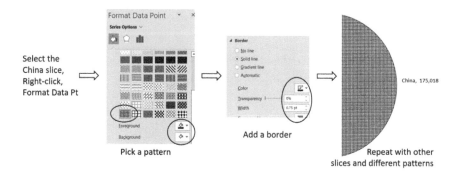

Select the
China slice,
Right-click,
Format Data Pt

Pick a pattern

Add a border

Repeat with other
slices and different patterns

FIGURE 2.36   Changing pie slice colors to B&W patterns.

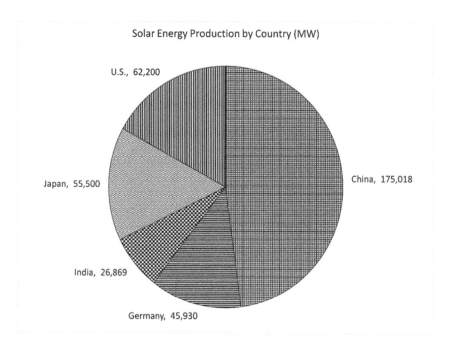

FIGURE 2.37   Solar energy production (MW) by country shown with B&W pie chart.

2.5   UTILIZING CHART TEMPLATES

If you completed the examples in this chapter, you will have experienced that the creation of charts in Excel can be tedious. If a plot is particular and is created only once, it will be necessary to pay the price of the laborious steps to create the plot. However, there are many plots that are created repetitively. Excel provides a way to create a custom chart template that completes most of the formatting of a chart, which can then be finished in a few final steps. The objective of this section is to provide what you need to know to do this.

Example 2.8   Creating a Template for a Markers-and-Lines Plot of a Single Series

We will consider creating a template for the simple X-Y chart shown in Figure 2.7. The first step is to go through the steps to create a generic version. Using the data from Table 2.1, we create the plot shown in Figure 2.38. You will note the following features in the plot:

- the lines and markers are black
- the line weight has been set at 1.5
- the marker size has been set at 8
- axis tick labels are black and 14pt
- general axis labels are created, black and 16pt
- a generic title is in place, black and 18pt

Note: No custom changes are made to the axis tick labels or axis ranges. These would be adjusted when the template is customized to a particular series. Of course, other changes could then be made, but a compromise is struck for initial formats, which are likely not to be changed.

The next step is to create the template. This is illustrated in Figure 2.39.

With the template now in place, it can be applied to a new scenario. Consider Table 2.8, which reports the freezing point of aqueous solutions of ethylene glycol (commonly called "glycol"). These solutions are typically used as antifreeze in engine cooling systems.

First, we create a scatter chart and move it to its own sheet. Figure 2.40 illustrates this. We could have chosen any scatter plot. Here, we have selected the Straight Lines and Markers type.

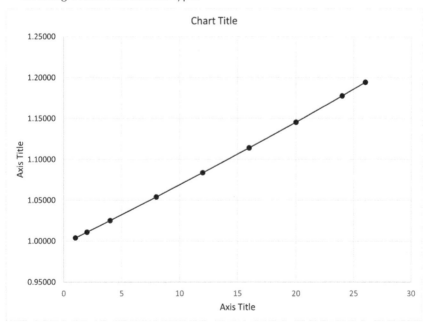

FIGURE 2.38   Generic markers-and-lines plot.

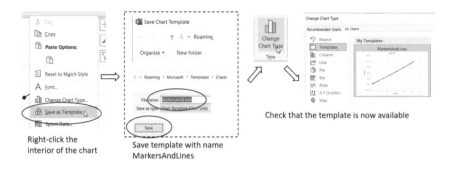

FIGURE 2.39   Creating the MarkersAndLines template.

___

TABLE 2.8
Freezing Point of Glycol-Water Solutions

Glycol (Percent by Volume)	Freezing Point (°C)
0	0.0
10	−3.4
20	−7.9
30	−13.7
40	−23.5
50	−36.8
60	−52.8
80	−46.0
90	−30.0
100	−12.8

Source: https://www.engineeringtoolbox.com/ethylene-glycol-
d_146.html

___

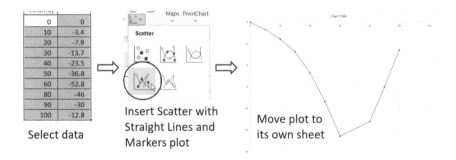

FIGURE 2.40   Inserting a standard scatter plot for the glycol data.

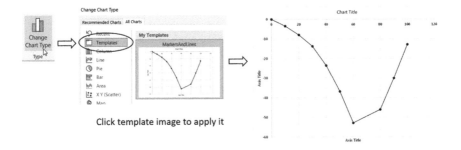

FIGURE 2.41   Applying the chart template.

Next, we select and apply the template. This is shown in Figure 2.41.
Now that the template has been applied, there are a few changes required
(or optional) to arrive at the final version of the plot:

- enter appropriate axis labels
- change to an appropriate chart title
- if desired,
  - move the crossing point of the horizontal axis
  - adjust the limits of the horizontal axis

Making these changes results in the plot presented in Figure 2.42.

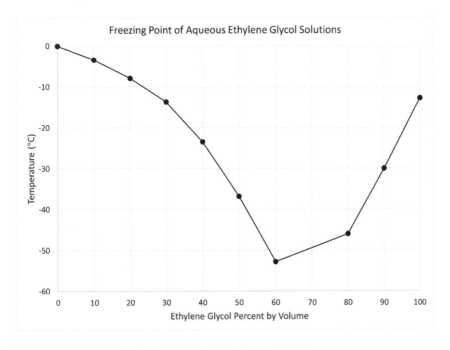

FIGURE 2.42   Final version of the glycol freezing point plot.

An observation we make from the chart is that we would not make an antifreeze solution greater than 60% glycol. Actually, because of concern for partially melted snow or ice (slush), commercial antifreeze is not generally more than 50%, providing protection to around −40°C.

The ability to create chart templates for many different scenarios can streamline chart creation and greatly improve efficiency. It is also possible to share templates within an organization or even commercially in order to provide consistency. To do that, you need to note the file path where the template (.crtx) files are stored on your computer when you save a template.

In creating plots using Excel, we are doing our best to adapt features that are often not designed directly for our purposes. We do the best we can, and, generally, can produce an acceptable product. If our needs clearly exceed Excel's capabilities, then we should consider additional software that is specialized in graphical presentations.

PROBLEMS

2.1 Table P2.1 tabulates worldwide wind power generation from 2000 to 2021. Create a well-formatted, B&W, markers-and-lines chart of these data. Describe any insights that you gain by expressing the data as an image.

TABLE P2.1
Worldwide Wind Power Generation in TWh

Year	Wind Generation (TWh)	Year	Wind Generation (TWh)
2000	31.4	2011	440.4
2001	38.4	2012	530.5
2002	52.3	2013	635.5
2003	63.3	2014	705.9
2004	85.6	2015	831.3
2005	104.6	2016	962.1
2006	133.5	2017	1140.4
2007	171.5	2018	1270.0
2008	221.4	2019	1420.5
2009	276.8	2020	1596.4
2010	346.4	2021	1861.9

Source: https://ourworldindata.org/grapher/wind-generation

2.2 Greenhouse gases are composed of various constituents including carbon dioxide. Table P2.2 provides a recent history of these emissions in four categories. Create a well-formatted, markers and lines plot of these data. Experiment with the use of a secondary axis to make the plot more presentable.

TABLE P2.2

Worldwide Atmospheric Emissions of Greenhouse Gases in Million Metric Tons (MMT) CO_2 Equivalent

	Atmospheric Emissions (MMT CO_2 equiv.)			
Year	Carbon Dioxide	Methane	Nitrous Oxide	Fluorinated Gases
2010	5675.8	692.1	455.0	168.2
2011	5540.2	666.2	445.6	175.5
2012	5338.7	658.2	416.8	172.3
2013	5474.3	654.4	463.9	172.1
2014	5522.8	651.0	474.0	177.1
2015	5371.8	651.5	468.2	179.6
2016	5248.0	642.4	450.8	179.1
2017	5207.8	648.4	446.3	180.9
2018	5375.5	655.9	459.2	180.8
2019	5255.8	659.7	457.1	185.7

2.3 In 1848, Johann Rudolph Wolf devised a method for quantifying solar activity by counting the number of individual dark spots or groups of spots on the sun's surface. He computed a quantity, now called the *Wolf sunspot number*, by adding ten times the number of groups plus the total count of individual spots in a year.

The data set for annual counts extends back to 1700! These data are available in .csv format for download from http://www.sidc.be/silso/datafiles, as filename SN_y_tot_V2.0.csv.

Create a well-formatted, straight-lines-only plot of these data. Describe any insights that you gain by expressing the data as an image.

2.4 Experimental data of a response variable versus time have been collected and are displayed in Table P2.4. Create a plot of the data using only markers. Create a second series on the worksheet for time intervals of 0.1 from 0 to 3.8 based on the polynomial model

$$y = 0.559 + 185.6t - 53.88t^2 + 4.351t^3$$

Add this series to the plot using the straight lines-only style. Employ Select Data ⇨ Add. Comment on whether the model appears to be an adequate representation of the data.

TABLE P2.4
Experimental Response versus Time

t	y
0.09	15.1
0.32	57.3
0.69	103.3
1.51	174.6
2.29	191.5
3.06	193.2
3.39	178.7
3.63	172.3
3.77	167.5

2.5 The *compressibility factor*, Z, is included as a modification of the *ideal gas law* to account for nonideal behavior, as shown in

$$P\hat{V} = ZRT$$

where P = pressure, $\hat{V}$ = specific volume, R = gas law constant, and T = temperature.

The data in Table P2.5 represent the compressibility factor of methane (CH_4) over a wide range of temperatures at elevated pressures. Create a

TABLE P2.5
Compressibility Factors for Methane at Different Temperatures and Pressures

	Pressure (bar)				
Temperature (K)	20	40	60	80	100
100	0.0874	0.1741	0.2604	0.3459	0.4313
150	0.0708	0.1401	0.2078	0.2748	0.3405
200	0.8629	0.6858	0.3755	0.3218	0.3657
250	0.9356	0.8694	0.8035	0.7403	0.6889
300	0.9663	0.9342	0.9042	0.8773	0.8548
350	0.9821	0.9657	0.9513	0.939	0.9293
400	0.9908	0.9833	0.9771	0.9721	0.9691
450	0.9965	0.9941	0.9923	0.9917	0.9922
500	1.0003	1.0009	1.0021	1.0043	1.0068

Source: https://www.govinfo.gov/content/pkg/GOVPUB-C13-c19f48b42b2c70c3422cda30d301b8b2/pdf/GOVPUB-C13-c19f48b42b2c70c3422cda30d301b8b2.pdf

well-formatted markers-and-lines plot of the data for Z versus T with different series for each pressure. Include a legend.

2.6 For the data in Table P2.5, create well-formatted contour and surface plots. If you also solved Problem 2.5, comment on the differences in observations depending on the type of plot.

2.7 The following table reports values from a response surface model of the form

$$y = a_0 + a_1x_1 + a_2x_2 + a_{11}x_1^2 + a_{22}x_2^2 + a_{12}x_1x_2$$

Create contour and surface plots, both mesh and shaded, based on these data. Interpret the results.

					x_2			
		-5	-3	-1	0	1	3	5
	-4	-301	-160	-66	-37	-20	-20	-67
	-2	-140	-24	45	61	67	41	-31
	-1	-84	20	76	86	85	48	-37
x_1	0	-44	47	91	95	88	37	-60
	1	-20	58	89	87	74	11	-99
	2	-13	53	72	63	44	-32	-154
	4	-47	-6	-12	-33	-66	-166	-313

2.8 Wind energy capacities in 2021 by country for the top ten producers are displayed in Table P2.8. Create a color, well-formatted Pareto chart for these data. Then, convert the chart to B&W patterns instead of color.

TABLE P2.8
Wind Energy Capacity in GW by Country in 2021

	Capacity (GW)
Brazil	21.1
Canada	14.3
China	329.0
France	18.7
Germany	63.8
India	40.1
Italy	11.3
Spain	27.5
UK	27.1
USA	132.7

Source: https://en.wikipedia.org/wiki/Wind_Power_by_country. *Wind Power by Country.* Last edited: June 30, 2023. Page version ID: 1162686131

2.9 For the five countries with the highest capacity in Table P2.8, create a well-formatted pie chart using color. Convert the chart's pie slices to B&W using patterns.

2.10 The populations of the countries in Table P2.8 are listed in Table P2.10. Create a well-formatted, color Pareto bar chart based on these data. If you completed Problem 2.8, comment on the comparison between the bar chart of that problem with this one.

TABLE P2.10
Population of Countries in Millions in 2021

	Population (millions)
Brazil	214
Canada	38.2
China	1412.3
France	67.5
Germany	83.1
India	1393.4
Italy	59.1
Spain	47.3
UK	67.3
USA	331.9

Source: https://en.wikipedia.org/wiki/List_of_countries_and_dependencies_by_population. *List of Countries and Dependencies by Population.* Last edited: July 22, 2023. Page version ID: 1166651887

2.11 Create a chart template called LinesOnly for plotting a larger single series of data connecting the data points. Illustrate the use of the template with the series of Atlantic hurricane history available at http://tropical.atmos.colostate.edu/Realtime/index.php?arch&loc=northatlantic. You can download these data as a comma-delimited (.csv) file and open that file in Excel, to be saved as an Excel workbook. After applying your template, comment on the plot.

NOTES

1. https://gml.noaa.gov/webdata/ccgg/trends/co2/co2_annmean_gl.txt
2. The chart derives its name from a noted Italian economist, Vilfredo Pareto (1848–1923).

3 Engineering and Scientific Formulas

CHAPTER OBJECTIVES

- Review relative, absolute, and mixed cell addressing
- Be able to name cells and ranges of cells
- Learn the arithmetic, relational, and logical operators, and their precedence
- Review the various built-in mathematical functions common to formulas
- Learn how to create array formulas

As engineers and scientists, we implement calculations on spreadsheets using mathematical formulas based on chemistry and physics. These formulas involve the typical arithmetic operations of addition, subtraction, multiplication, division, and exponentiation, and they often include trigonometric, hyperbolic, logarithmic, and exponential functions. For example, let's say we wanted to calculate the surface area, S, of an *oblate spheroid*, like the Earth. The formula for this is

$$S = 2\pi a^2 + \pi \frac{b^2}{\varepsilon} \ln\left(\frac{1+\varepsilon}{1-\varepsilon}\right) \tag{3.1}$$

where
- a: major semiaxis (radius at the Equator)
- b: minor semiaxis (radius at the poles)
- ε: eccentricity, given by $\sqrt{a^2 - b^2}/a$

Starting with the two axes, this calculation first requires evaluation of the *eccentricity*, ε, and then the longer formula involving the constant, π, and the natural logarithm, $\ln(\bullet)$. We need to learn a reliable way to implement calculations like this on the spreadsheet. That is the purpose of this chapter. Obviously, the formulas you encounter will be different from this but will probably have similar characteristics. Much later in Chapter 10, you will learn how to create your own functions to mechanize formulas to make their use efficient and more reliable.

3.1 CELL ADDRESSING

Spreadsheet users are well familiar with the basic concept of *cell addressing* illustrated in Figure 3.1. The cell is located by its column letter and row number, and the address is noted in the Name Box. The style, such as C4, is column first, row second. This is different from the typical mathematical subscript notation for

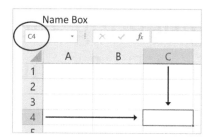

FIGURE 3.1   Cell address and Name Box.

a two-dimensional array, which is row then column, as in x_{43}.[1] As we noted previously, row numbers range from 1 to 1,048,576 (2^{20}), and column letters from A to XFD (2^{14}) yielding 2^{34} cells on the worksheet.

Spreadsheet workbooks are organized into worksheets. Often, there will be only one worksheet, but there may be many. That leads to the need to reference a cell on another worksheet. The name of the worksheet is noted on the tab at its bottom, for example, Sheet1. A cell on another worksheet is referenced using the syntax:

SheetName!CellAddress

and an example might be Sheet2!C4. Occasionally, we want to reference a cell in another spreadsheet workbook. The syntax for this is

[WorkbookName]SheetName!CellAddress

with an example [Book2.xlsx]Sheet1!B2. The other workbook file must be open for this reference to work.

In addition to referencing individual cells, we often want to reference rectangular blocks or ranges of cells. These are addressed by denoting the upper left cell and the lower right cell in the syntax:

UpperLeftAddress:LowerRightAddress

An example might be B2:E5. It should also be noted that the range reference could be a single row, like B2:E2, or a single column, B2:B5.

3.1.1 CELL ADDRESSING IN FORMULAS

With this background on cell addressing, we can consider how cell addresses are used in formulas. Perhaps the simplest example of this is the *pointer formula*, illustrated in Figure 3.2. The formula is entered with the initial equal sign, =, and followed by a cell address, here B2. When entered, as shown in the figure, the cell with the formula displays the same value as the cell pointed to. If the value in cell B2 is changed, this will be automatically updated in cell C3.

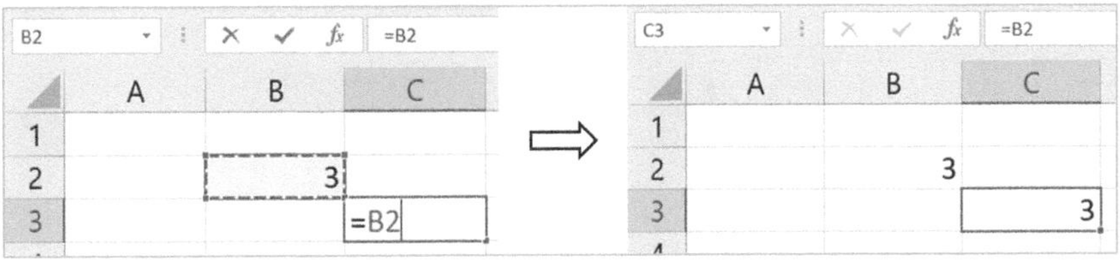

FIGURE 3.2   Pointer formula with cell address before and after entry.

As noted in Chapter 1, the content of cells can be moved or copied. It is important to be clear on what happens to formulas when these actions take place. First, what happens to the pointer formula in cell C3 when cell B2 is moved to A1? This is illustrated on the left in Figure 3.3, and on the Formula Bar we note that the formula is automatically updated to reference cell A1. Alternately, on the right in the figure, if the cell with the formula is moved from C3 to C1, the pointer reference to B2 is not changed.

A critical feature of the spreadsheet is what happens when a cell with a formula is copied to another cell. A copy down one cell of the formula shown in cell C3 of Figure 3.3 is illustrated in Figure 3.4. As demonstrated in Chapter 1, there are several techniques available to implement this copy. Perhaps the most convenient is to drag the *fill handle* in the lower right corner of the selected cell down one cell. The important observation here is that the copied formula in cell C4 now references or points to cell B3. The reference has moved down in lockstep with the

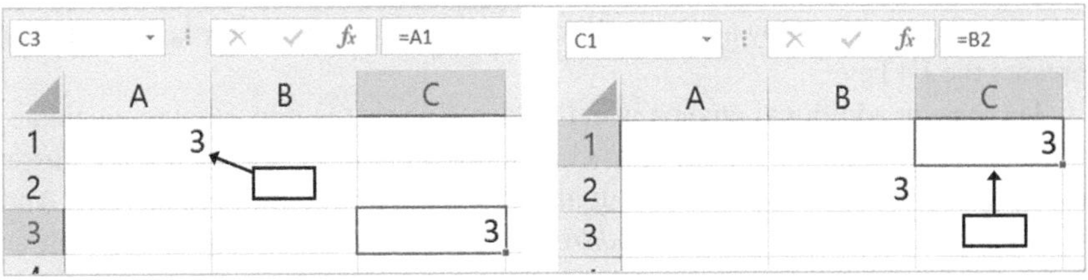

FIGURE 3.3   Moving the pointed-to cell and the formula cell.

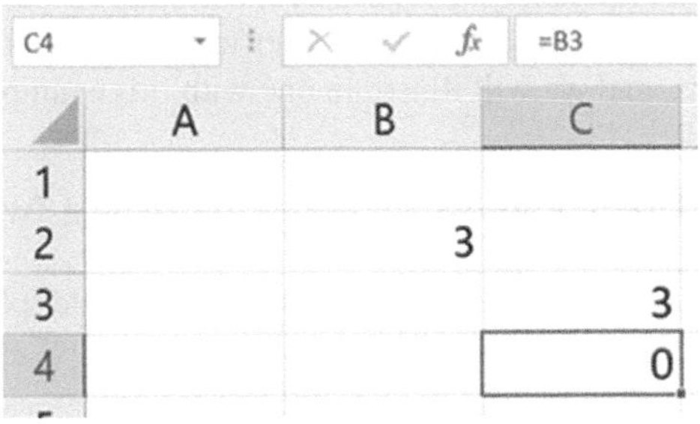

FIGURE 3.4   Cell C3 with formula copied down to cell C4.

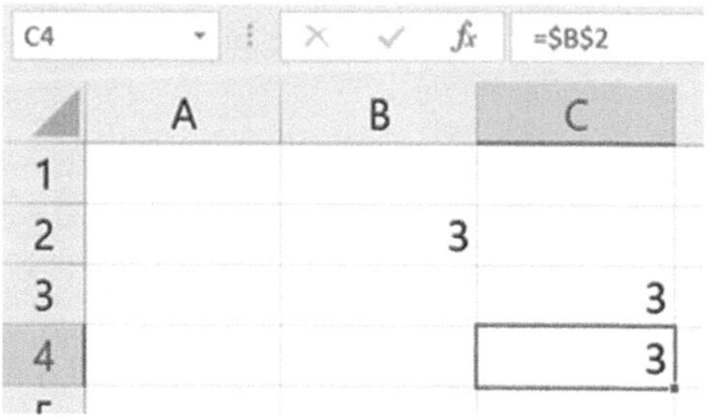

FIGURE 3.5 Copied formula with an absolute cell reference.

formula copy. For this reason, this is called a *relative copy* and the cell address is called a *relative address*. If we were to carry out the copy to a cell far away, even on another worksheet or in another workbook, the new formula would point to the cell one row above and one column to the left of the formula cell.

The example shown in Figure 3.4 raises a question. What if, when the formula is copied down, we don't want the reference to cell B2 to shift with the copy? The answer to this is to use an *absolute reference* to cell B2 in the formula using $ signs, as $B$2. When this is done, the copy scenario in Figure 3.5 results. You will note that the formula in cell C4 is the same as that in cell C3. The reference to cell B2 is preserved. In entering the original formula with an absolute reference, one can type in the $ signs. Alternately, one can type in the relative reference as B2 and then press the F4 key while the cursor is within the address or immediately adjacent. And, commonly, rather than type in a relative cell address, one selects the desired cell by a mouse click or movement of the arrow keys, and then presses the F4 key.[2]

There is an additional nuance in cell addressing that is important to elucidate. It has to do with answering the following question. How do we use a cell address that, upon copying, fixes the row but allows the column to move relatively? And vice versa. The answer is *mixed addressing*. With an absolute address, such as $B$2, the $ signs in front of the column and row indicators cause the reference to be absolute; that is, invariant in copying formulas that reference cell B2. If we use instead $B2, that provides for the column B to be locked or invariant, but row 2 shifts with the copy. Alternately, if we use B$2, that fixes row 2 but allows column B to shift with the copy. When would you ever use mixed addresses? Well, there are many occasions, and we will illustrate one with this example.

Example 3.1　Use of Addressing Modes to Create a Two-way Table

In Equation 3.1, we described the surface area of an oblate spheroid. The volume, V, of such a spheroid is given by

$$V = \frac{4}{3}\pi a^2 b \qquad (3.2)$$

⊿	A	B	C	D	E	F	G	H	I	J	K	L
1							b					
2			10	20	30	40	50	60	70	80	90	100
3		100										
4		110										
5		120										
6		130										
7		140										
8	a	150										
9		160										
10		170										
11		180										
12		190										
13		200										

FIGURE 3.6 Two-way table for oblate spheroid volume calculations.

Imagine that we want to display a table of volumes for ranges of values of the major and minor semiaxes, a and b, respectively. We can set up the spreadsheet with the ranges as shown in Figure 3.6.

In cell C3, we enter Equation 3.2 using the $ symbol to anchor the a in the second column ($B3) and the b in the second row (C$2),

$$= 4/3*PI()*\$B3 \wedge 2*C\$2$$

and copy that formula throughout the table. We can do that in two operations by dragging the fill handle across to column L and down to row 13, or initiate a copy with Ctrl-C, select the entire interior of the table and press Enter. The result is shown in Figure 3.7 with one interior cell, E7, selected to show its formula. You can see how the copied formula refers to the appropriate values of a and b for the location of the formula's cell.

We introduced the use of the F4 key to convert a relative address to absolute. Actually, there is more to it than that. As in Figure 3.8, the F4 key allows you to toggle through all possible ways of anchoring the address. Repeated F4 key entries cycle through the various addressing modes returning to the relative address on the fourth step.

⊿	A	B	C	D	E	F	G	H	I	J	K	L
1							b					
2			10	20	30	40	50	60	70	80	90	100
3		100	418879	837758	1256637	1675516	2094395	2513274	2932153	3351032	3769911	4188790
4		110	506843.6	1013687	1520531	2027374	2534218	3041062	3547905	4054749	4561593	5068436
5		120	603185.8	1206372	1809557	2412743	3015929	3619115	4222301	4825486	5428672	6031858
6		130	707905.5	1415811	2123717	2831622	3539528	4247433	4955339	5663244	6371150	7079055
7		140	821002.9	1642006	=4/3 * PI() * $B7 ^ 2 * E$2			4926017	5747020	6568023	7389026	8210029
8	a	150	942477.8	1884956	2827433	3769911	4712389	5654867	6597345	7539822	8482300	9424778
9		160	1072330	2144661	3216991	4289321	5361651	6433982	7506312	8578642	9650973	10723303
10		170	1210560	2421121	3631681	4842241	6052802	7263362	8473923	9684483	10895043	12105604
11		180	1357168	2714336	4071504	5428672	6785840	8143008	9500176	10857344	12214512	13571680
12		190	1512153	3024307	4536460	6048613	7560766	9072920	10585073	12097226	13609379	15121533
13		200	1675516	3351032	5026548	6702064	8377580	10053096	11728613	13404129	15079645	16755161

FIGURE 3.7 Completed two-way table for oblate spheroid volume calculations.

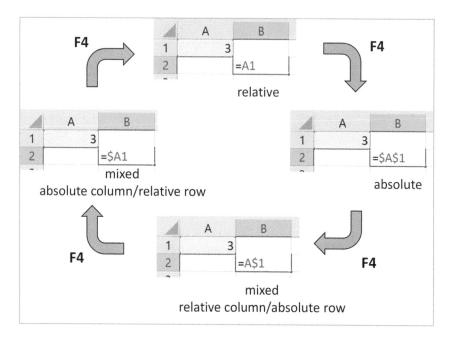

FIGURE 3.8 Use of the F4 key to cycle through the addressing modes.

It is common, when referring to a table or range of cells in a formula, to use an absolute reference because when the formula might be copied, we don't want the reference to the table to shift. This requires $ signs on all column and row indicators, as in $B$2:$E$5. Although it's possible, we don't often encounter mixed addresses for ranges. The easiest way to create the absolute reference is, again, using the F4 key. This is illustrated in Figure 3.9. Note: A pointer formula from a single cell to a range of cells *won't work*. Later, we will deal with array formulas for relating ranges to comparable ranges. Ranges can be included as arguments in some functions, e.g., AVERAGE($B$2:$E$5), as illustrated in the figure.

You can always go back and edit a cell address to change its mode from relative to absolute and beyond. To do that, the editing cursor (insertion point cursor |) must be within the address or immediately adjacent. To modify an entire range address, it must be selected before using the F4 key.

	A	B	C
1	=AVERAGE(B2:C3		
2		1	2
3		3	4

F4 ⇒

	A	B	C
1	=AVERAGE($B$2:$C$3		
2		1	2
3		3	4

FIGURE 3.9 Creating an absolute range reference with the F4 key.

3.2 NAMING CELLS

In engineering and scientific calculations, numerical quantities are stored in cells. The cells, of course, have their respective addresses, but they also represent variables in equations. This presents a problem to the user – relating the cell addresses that appear in cell formulas with the actual quantities that have meaning in the problem being solved. One common technique is to place *tag labels* in adjacent cells. For example, if we are going to implement the calculation of Equation 3.1 on a spreadsheet, the initial setup might look like Figure 3.10. The cells to the right of the tag labels would contain the values and formulas. The cells to the right of the figure in column C could contain labels of the units of the adjacent quantities in column B.

Excel provides a feature to carry cell identification one important step further. For the example in Figure 3.10, we can create *names* for the cells in the B column and choose the labels in column A as the names. One common way to do this is to select the cell in question, for example B1, and type the name into the Name Box, a, and press the Enter key. This is shown in Figure 3.11. To be sure, cell B1 can still be referenced by its cell address, but now it has an alternate identification, a relevant name. The name is an absolute reference and will not shift if a formula referencing it is copied.

An alternate method to create a name is to transfer a label from an adjacent cell. After selecting the adjacent cells that hold the name and the value, use the Ribbon commands, Formulas → Defined Names → Create from Selection.

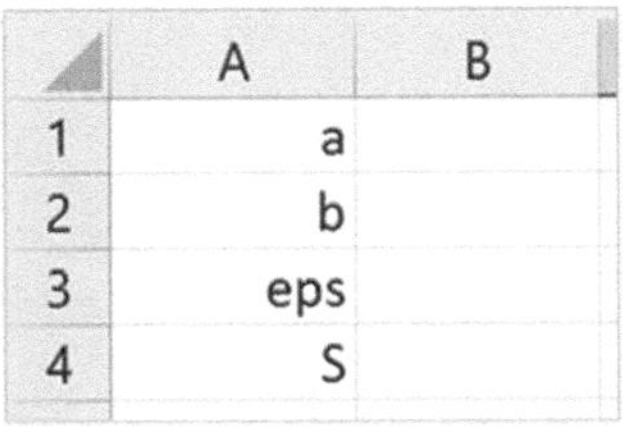

FIGURE 3.10   Initial setup for calculation of surface area of oblate spheroid.

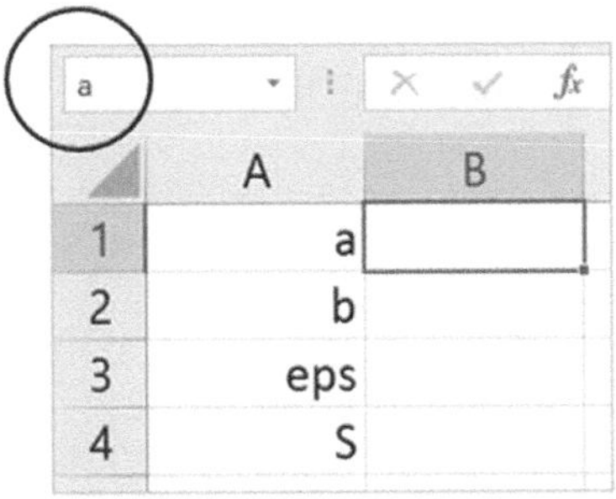

FIGURE 3.11   Using the Name Box (the circle) to create a cell name.

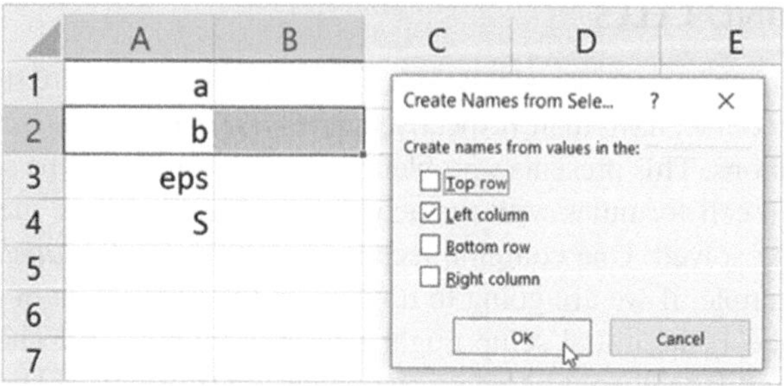

FIGURE 3.12 Creating a name from a label in an adjacent cell.

The final step is shown in Figure 3.12. Once this operation is complete, you always want to select the named cell and confirm the name appears in the Name Box. Similarly, it is possible to transfer more than one label to adjacent cells. This is shown in Figure 3.13.

Once a name is created for a cell, pointing to that cell in creating a formula in another cell will automatically enter the name and not the cell address. This is shown in brief in Figure 3.14.

The great advantage of using named cells is that formulas in other cells will immediately look familiar without having to go through the step of relating a cell address to each equation variable. Also, named cells provide for greater reliability. As you will find out in later chapters, named cells are very important in VBA programming.

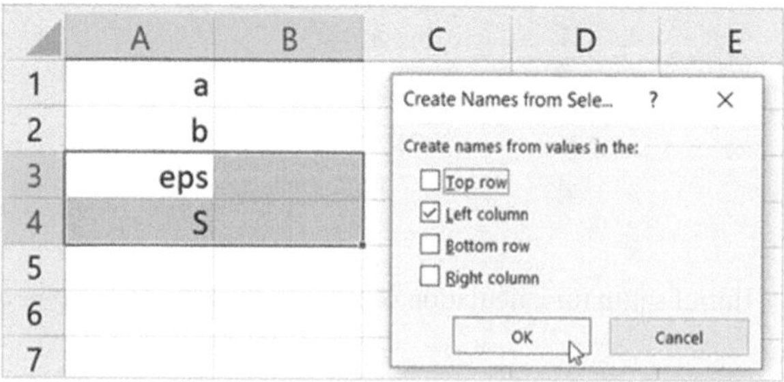

FIGURE 3.13 Creating names from labels in adjacent cells.

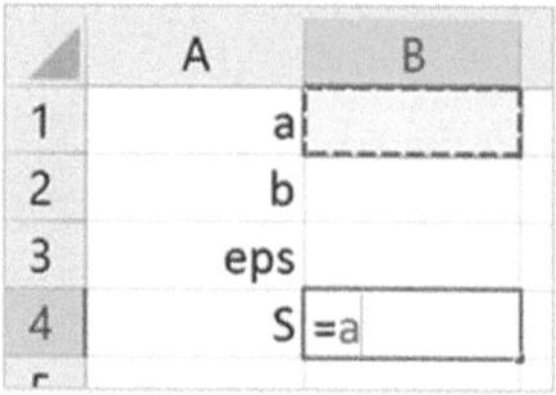

FIGURE 3.14 Pointing to a named cell.

FIGURE 3.15 Name Manager icon.

There are restrictions that come into play when creating names. First, and logically, you cannot have a name that is also a cell address. For example, you cannot name a cell x1. If x_1 is a meaningful variable to you, you can use x_1 or x1_ instead. If you attempt to enter a cell address in the Name Box, it will jump to that cell. If you attempt to transfer an x1 label to an adjacent cell as a name, Excel will modify the name to x1_. You cannot use R or C as a name nor R or C followed by any numeral and any other characters. This is because of the conflict with the legacy R1C1 method of cell addressing (see Note 1). The name must be a contiguous character string with no spaces. If you are ever concerned about whether a name is valid, you can try to transfer it from a label and observe the result.

The remaining topic regarding names is how to change them or the addresses to which they refer. You cannot use the Name Box to change or delete a name. The Defined Names group on the Formula tab of the Ribbon is used for this. The general tool is the Name Manager with the command icon shown in Figure 3.15.

Opening the Name Manager, we see the display in Figure 3.16. We see here options for creating a new name, similar to using the Name Box or Create from Selection, editing an existing name (changing its spelling or modifying its Refers to: cell address), or deleting the name.

By default, names are defined for use throughout the workbook, i.e., on all worksheets. This is described by the term *scope* and is shown in Figure 3.15 as Workbook for all names there. On occasion, we may wish to restrict the scope of

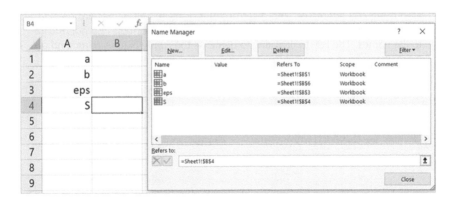

FIGURE 3.16 Name Manager.

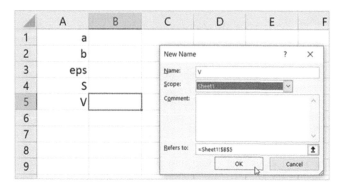

FIGURE 3.17   Creating a name with worksheet scope.

a name just to the worksheet where it is defined. That is common when the same name is to be used on multiple worksheets. To do this, you use the New… button on the Name Manager or the Define Name command in the Defined Names group on the Ribbon. Creating a name with Worksheet scope is illustrated in Figure 3.17. You will see here that the Scope has been selected from the drop-down list as Sheet1.

Example 3.2   Implementing Formulas for the Oblate Spheroid Using Names

Given the spreadsheet depicted in Figure 3.17, we can enter example values for the major and minor semiaxes, $a = 10$, $b = 8$, and then enter a formula to compute the eccentricity, ε. The resulting spreadsheet is shown in Figure 3.18.

   With the values for a, b, and ε in hand, we can now enter the formulas for surface area, S, and volume, V, based on Equations 3.1 and 3.2 These steps are captured in Figure 3.19. Notice how well the entered formulas relate to the original algebraic versions.

   A scenario that sometimes presents itself is when a formula has been created using cell addresses, and, later on, one or more cell names are created. When the names are created, the cell addresses in the formulas do not update to the names. The key to updating to names is the Apply Names command on

	A	B	C
1	a	10	
2	b		
3	eps	=SQRT(a^2-b^2)/a	
4	S		
5	V		

	A	B
1	a	10
2	b	8
3	eps	0.6
4	S	
5	V	

FIGURE 3.18   Implementing the formula for eccentricity.

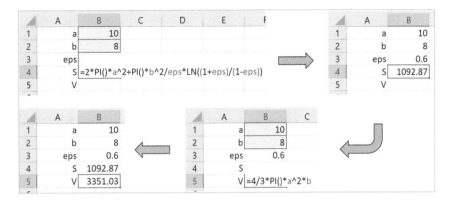

FIGURE 3.19 Completion of formulas for the surface area and volume of an oblate spheroid.

the Define Name drop-down list in the Defined Names group. An example is illustrated in Figure 3.20. It is possible to apply more than one name at a time by selecting more names from the Apply Names list.[3]

After creating several names, you will notice that, if you select the drop-down arrow on the right side of the Name Box, a list of current names is displayed. You can click a name, and its cell will be located. This is an excellent way to familiarize yourself with a spreadsheet that is new to you, created by others, or even a spreadsheet you created some time ago. Additionally, with larger spreadsheets, there may be tens, possibly hundreds, of names. Even when there are a few names, good documentation suggests that there be a nomenclature table. This can be created on its own worksheet using the Paste List command found in the Paste Name window available from the Use in Formula drop-down list in the Defined Names group of the Ribbon. Figure 3.21 illustrates the steps.

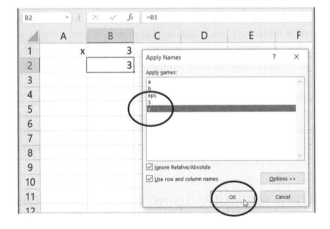

FIGURE 3.20 Applying a name to a formula to convert a cell address.

FIGURE 3.21 Creating a nomenclature table using Paste List from the Paste Name window.

	A	B	C
1	Name	Description	Location
2	a	major semiaxis	=Sheet1!$B$1
3	b	minor semiaxis	=Sheet1!$B$6
4	eps	eccentricity	=Sheet1!$B$3
5	S	surface area	=Sheet1!$B$4

Sheet1   **Nomenclature**

FIGURE 3.22 Formatted nomenclature table on its own worksheet with modified tab.

With a little work, the list can be converted into a meaningful nomenclature table, and the sheet tab can be modified to Nomenclature. This is shown in Figure 3.22. An informative nomenclature table will be very helpful to others who inherit the spreadsheet.

Experience has shown that spreadsheet applications grow in scope over time. If good organization and, in particular, the use of names are not adopted, the application may become so unwieldy that the spreadsheets are unfathomable for a new user/developer (or even you, if you haven't worked with the spreadsheet in a while). In such cases, a common alternative is to start over from scratch. Just the well-documented use of names can avoid this inefficient, uneconomical "reinvention of the wheel." You will find that we use names extensively in this text.

That does leave one lingering question. When not to use names? The common situation when cell addresses are used is when formulas are copied down adjacent to a table. Relative addressing provides an advantage using the autofill feature of Excel. Even then, the use of an array formula may be a better alternative. Array formulas are introduced later in this chapter.

3.3 CREATING FORMULAS USING EXCEL'S OPERATORS

In implementing scientific and engineering calculations on the spreadsheet, we translate algebraic formulas into Excel expressions. The algebraic formulas are

---

TABLE 3.1

Arithmetic Operators

−	negation (unary minus)
^	exponentiation
*, /	multiply, divide
+, −	addition, subtraction

---

typically two-dimensional with numerators and denominators. Arithmetic operations and inclusion of special functions (trigonometric, hyperbolic, logarithmic, exponential) are common. The result is usually numeric, but it is possible to have a true/false, Boolean result.

The *arithmetic operators* in Excel formulas are shown in Table 3.1. They are listed in order of precedence from highest to lowest. When compared to other mathematical and programming software, what is unusual here is that negation (−) precedes exponentiation (^). Also, some other systems/languages like Python and Fortran use ** for exponentiation.

Evaluation of formula expressions is generally left to right, applying the operator precedence first. This order can be overridden by the use of parentheses where expressions within (•) are evaluated first. Also, any functions, e.g., SQRT(•), are evaluated first.

As a simple example, we can write an Excel formula for Equation 3.2, assuming that appropriate names have been created for the cells containing values for a and b.

$$= 4/3*PI()*a^2*b$$

The steps for evaluating this formula are the following:

1. Evaluate the PI() function and return a value of π in its place.
2. Compute the a^2 term → result 1.
3. Compute the rest of the formula from left to right.
   $4 \div 3 \rightarrow$ result 2
   result $2 \times \pi \rightarrow$ result 3
   result $3 \times$ result $1 \rightarrow$ result 4
   result $4 \times b \rightarrow$ final result

Equation 3.1 provides a more complicated formula that might look like that depicted in Figure 3.23.

Here the PI() functions are evaluated first and return values of π, then the LN function. For the latter, given the internal parentheses, the (1+eps) term is evaluated first, then the (1-eps) term. The divide operation comes next, and finally the

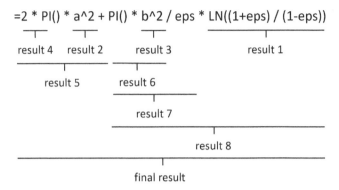

FIGURE 3.23 Formula calculation sequence.

LN function is invoked to produce result 1. After those, the remaining evaluations are generally made left-to-right following operator precedence.

1. a^2 → result 2
2. b^2 → result 3
3. $2 \times \pi$ → result 4
4. result 4 × result 2 → result 5
5. π × result 3 → result 6
6. result 6 ÷ eps → result 7
7. result 7 × result 1 → result 8
8. result 5 + result 8 → final result

There are three other simple examples that are worthwhile to illustrate the details of operator precedence. The first is the formula

=−3^2

Applying precedence rules, we see that the negation operation, −3, occurs first followed by the exponentiation, ^2, yielding a result 9. It is important to illustrate this because, in most other mathematical and programming software, the precedence is reversed and the result is −9.[4] If we want to override the order of operations in an Excel formula, parentheses would be required, as in

=−(3^2)

The next example involves repeated exponentiation. Here, we implement the algebraic expression

$$x^{y^z}$$

Our "algebraic tendencies" would lead us to evaluate y^z first and then x to the power of that result. However, given left-to-right evaluation and the Excel formula,

```
=x^y^z
```

left-to-right order has x^y computed first, followed by exponentiation by z, counter to our algebraic interpretation. To force the latter, we would use parentheses, as in

```
=x^(y^z)
```

The final example has to do with coding multiple denominators, as in the simple algebraic expression

$$\frac{x}{yz}$$

There are two common ways to implement this expression as an Excel formula. The first is

```
=x/(y*z)
```

where the denominator is set apart within parentheses. The second style, which is also valid, is

```
=x/y/z
```

This takes advantage of left-to-right order as x is first divided by y and the result is divided by z. You can use either style. Some prefer the first because it "looks like" the algebraic expression. A common error in creating Excel formulas is illustrated by this formula:

```
=x/y*z
```

This looks somewhat like the algebraic expression, but, because of left-to-right order, is equivalent to the algebraic expression,

$$\frac{xz}{y}$$

So, be careful to avoid errors like this.

A final piece of advice for creating longer Excel formulas is to "divide and conquer." That is, break up the formula into parts and compute them separately.

An illustration of this involves Equation 3.1. The design would break this equation into two parts as follows:

```
=2*PI()*a^2
=PI()*b^2/eps*LN((1+eps)/(1-eps))
```

and then sum these two to obtain the final result. This arrangement makes the formulas more readable and is helpful in debugging errors, as an error will likely show up on one of the two parts. This overall formula perhaps doesn't call for breaking it into two parts, but others are much longer, and "divide and conquer" should be considered.

3.4 USE OF BUILT-IN FUNCTIONS IN ENGINEERING AND SCIENTIFIC FORMULAS

We have seen the use of the PI() and LN functions earlier in this chapter. They are part of a family of hundreds of *built-in functions* provided by Excel. The concept of a function is illustrated in Figure 3.24. The function may have no input arguments, as in PI(), or one argument, as in LN(•). It may have multiple arguments, as in ATAN2(x,y), and may have arguments that are ranges of cells, like AVERAGE(B2:D4). The function returns a single result, either used in a longer expression or simply displayed in the cell. There is another type of function, the array function, that returns results in a cell range. That is the subject of the next section.

In the current section, we will cover the families of built-in functions that appear commonly in engineering and scientific formulas. These include trigonometric, hyperbolic, logarithmic, exponential, and a few general functions. A good way to explore and find out about functions is the Insert Function window available via the command icon to the left of the Formula Bar as shown in Figure 3.25.

FIGURE 3.24 Excel function.

FIGURE 3.25 Insert Function command icon.

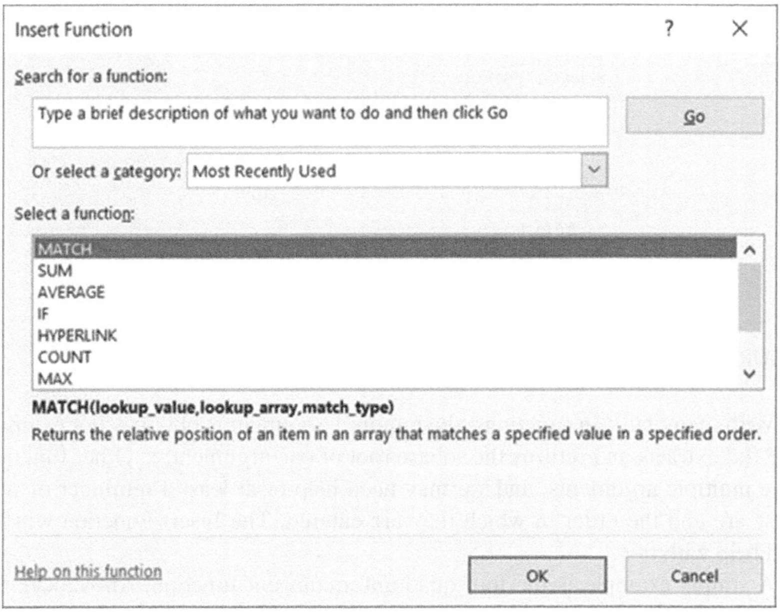

FIGURE 3.26 Initial Insert Function window.

The Insert Function window is shown in Figure 3.26. There are different ways to find a desired function. One can type a name close to the desired function in the Search for a function: field, as illustrated in Figure 3.27. Here, entering "absolute value" yields the ABS function in the Select a function: and a brief syntax for the function below. The list also provides two alternate functions. A second way is to use the Or select a category: field. For example, we can select Math & Trig from the drop-down list, and the result is shown in Figure 3.28. By scrolling through the Select a function: list, we might also find the desired function.

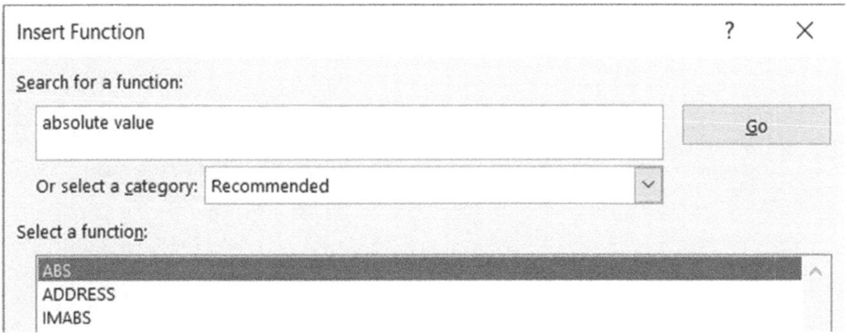

FIGURE 3.27 Search for a function by name.

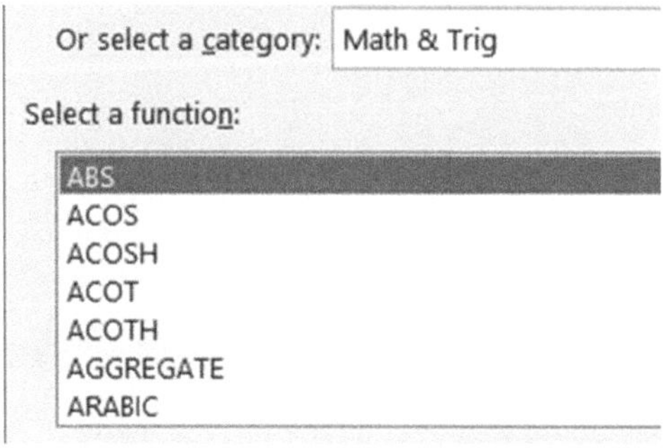

FIGURE 3.28   Function list in the Math & Trig category.

With many built-in functions, the required argument is obvious. For example, SQRT(x) extracts and returns the square root of one argument, x. Other functions have multiple arguments, and we may need help or at least a reminder of what these are and the order in which they are entered. The Insert Function window will help with this.

A simple example is the four-quadrant arctangent function, ATAN2(x,y). See Figure 3.29. The information on the function in the lower part of the window reminds us that the x-axis value comes first and the y-axis value second.

Although we represent all built-in functions in upper-case font here, you can enter them in a formula in lower-case (or any combination of upper- and lower-case), but, once the formula is entered, Excel will reformat the function name to upper-case.

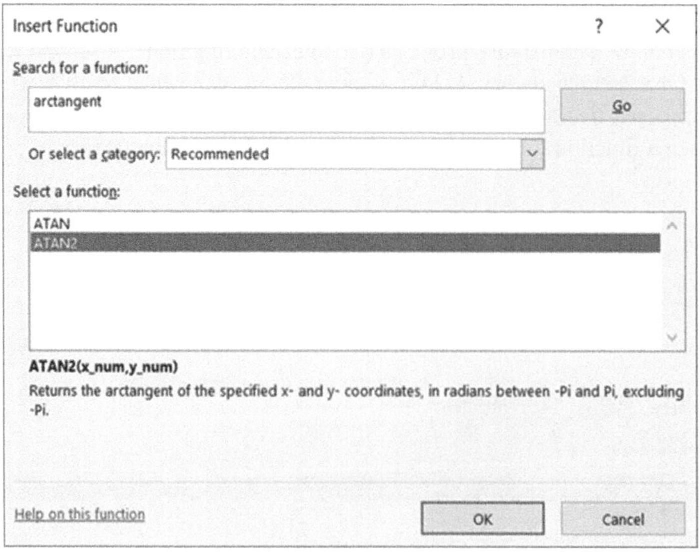

FIGURE 3.29   Details on the ATAN2 function.

3.4.1 TRIGONOMETRIC FUNCTIONS

Excel provides a complete family of *trigonometric and inverse trigonometric functions*. These are summarized in Table 3.2 and relate to the diagrams in Figure 3.30.

The ATAN function returns an angle in quadrants I and IV. For the second example in the figure, it will not return the correct angle, θ'. The ATAN2 function is called a *four-quadrant arctangent* because it will return θ' correctly in all quadrants.

The angles, θ and θ', represented in Table 3.2 and Figure 3.28 are in radians. These can be converted to degrees and back using the relationships based on the fact that $180° \equiv \pi$ radians,

$$\text{degrees} = \text{radians} \times (180/\pi) \quad \text{and} \quad \text{radians} = \text{degrees} \times (\pi/180)$$

Alternately, Excel provides functions DEGREES(•) and RADIANS(•) that carry out these conversions.

TABLE 3.2
Trigonometric and Inverse Trigonometric Functions

Sine	SIN(θ)	y/h	Cosecant	CSC(θ)	h/y	1/SIN(θ)
Cosine	COS(θ)	x/h	Secant	SEC(θ)	h/x	1/COS(θ)
Tangent	TAN(θ)	y/x	Cotangent	COT(θ)	x/y	1/TAN(θ)
Arcsin	ASIN(y/h)	θ				
Arccos	ACOS(x/h)	θ				
Arctan	ATAN(y/x)	θ				
Arctan	ATAN2(x′,y′)	θ'				

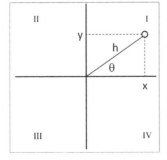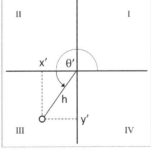

FIGURE 3.30   Trigonometric relationships in quadrants I and III.

Example 3.3   Series Approximations of Trigonometric Functions

The *Maclaurin series* representation of the sine function is

$$\sin(x) = x - \frac{x^3}{3!} + \frac{x^5}{5!} - \frac{x^7}{7!} + \cdots \tag{3.3}$$

How fast does this series converge to the true value for the sine? Let's set up a spreadsheet to test this. First, note that a general formula for the ith term of Equation 3.3 is

$$(-1)^{i+1} \frac{x^{2i-1}}{(2i-1)!}$$

The initial format of the spreadsheet provides a named cell for the x value and the use of the built-in SIN(•) function to produce the "true value." Also, there is a column for the term indices, i. This is shown in Figure 3.31. With some care, we can add a formula to the right of the term 1 index. See Figure 3.32 for this. The formula takes advantage of the built-in FACT(n) function that computes the factorial of n. When this formula is entered, it can be copied down adjacent to the rest of the terms. Table 3.3 presents the results from these cells.

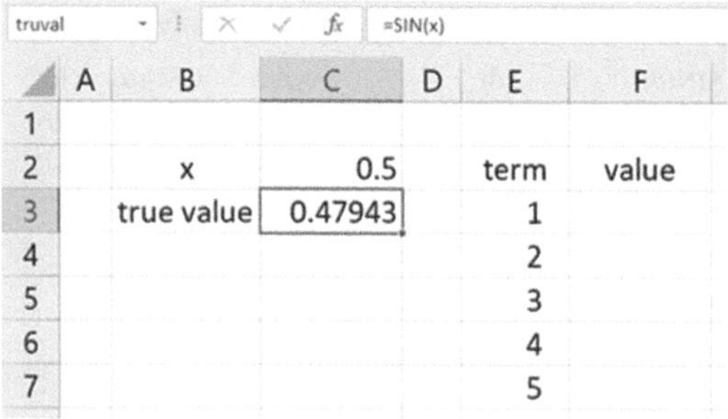

truval		▾ ⋮ ✕ ✓	f_x	=SIN(x)		
	A	B	C	D	E	F
1						
2		x	0.5		term	value
3		true value	0.47943		1	
4					2	
5					3	
6					4	
7					5	

FIGURE 3.31   Initial format of series approximation spreadsheet.

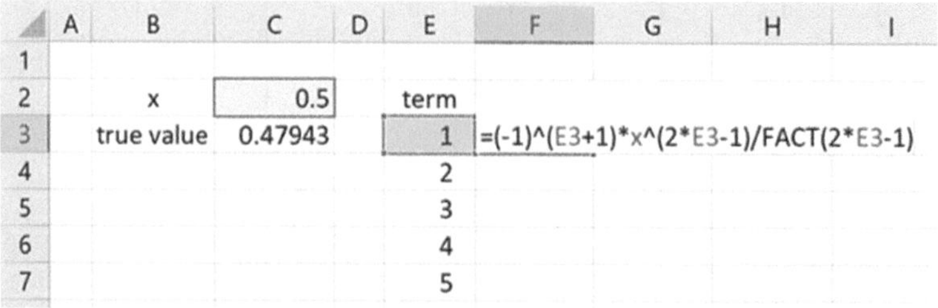

	A	B	C	D	E	F	G	H	I
1									
2		x		0.5		term			
3		true value	0.47943			1	=(-1)^(E3+1)*x^(2*E3-1)/FACT(2*E3-1)		
4						2			
5						3			
6						4			
7						5			

FIGURE 3.32   Add a formula for the general series term.

TABLE 3.3

Evaluation of Series Terms

Term	Value
1	0.5
2	−0.02083
3	0.0002604
4	−1.550E-06
5	5.382E-09

x	0.5		term	value	sum	error		
true value	0.47943		1	0.5	0.5	4.29%		
			2	-0.02083	0.47917	-0.054%		Formulas
			3	0.0002604	0.47943	0.0003%		
			4	-1.550E-06	0.47943	-0.000001%		
			5	5.382E-09	0.47943	0.000000003%		

	A	B	C	D	E	F	G	H
1								
2		x	0.5		term	value	sum	error
3		true value	=SIN(x)		1	=(-1)^(E3+1)*x^(2*E3-1)/FACT(2*E3-1)	=F3	=(G3-truval)/truval
4					2	=(-1)^(E4+1)*x^(2*E4-1)/FACT(2*E4-1)	=G3+F4	=(G4-truval)/truval
5					3	=(-1)^(E5+1)*x^(2*E5-1)/FACT(2*E5-1)	=G4+F5	=(G5-truval)/truval
6					4	=(-1)^(E6+1)*x^(2*E6-1)/FACT(2*E6-1)	=G5+F6	=(G6-truval)/truval
7					5	=(-1)^(E7+1)*x^(2*E7-1)/FACT(2*E7-1)	=G6+F7	=(G7-truval)/truval

FIGURE 3.33   Series example complete with formulas shown.

To complete the example, we add a column that computes a running sum of the term values and an error column showing the percent error compared to the true value. The final spreadsheet is presented in Figure 3.33. We observe that the series converges to the true result quickly to about one part in 10^{12} in five iterations. Speed of convergence will depend on the value of x.

As a trailing note to this example, it is pointed out that, when many terms are used in an approximation like this, the separate values of the numerator and denominator may become very large, and a better technique is to compute the next term starting with a value of the previous term.

3.4.2 HYPERBOLIC FUNCTIONS

The hyperbolic functions arise through relationships from a *hyperbola curve*, as illustrated in Figure 3.34. The primary property of interest is the hyperbolic angle, denoted POP' in the figure and u in the accompanying equations, and the *hyperbolic functions* are defined in terms of y and x coordinates of a point on

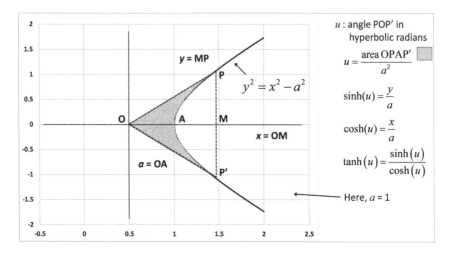

FIGURE 3.34 Hyperbolic relationships and functions – geometric interpretation.

the curve and the parameter a. Further, it is possible to derive the following definitions of the hyperbolic functions:

$$\sinh(u) = \frac{e^u - e^{-u}}{2} \qquad \cosh(u) = \frac{e^u + e^{-u}}{2} \qquad \tanh(u) = \frac{e^u - e^{-u}}{e^u + e^{-u}} \qquad (3.4)$$

and the reciprocal functions:

$$\operatorname{csch}(u) = \frac{1}{\sinh(u)} \qquad \operatorname{sech}(u) = \frac{1}{\cosh(u)} \qquad \coth(u) = \frac{1}{\tanh(u)} \qquad (3.5)$$

Excel provides a family of built-in hyperbolic functions. These are summarized in Table 3.4.

TABLE 3.4

Hyperbolic and Inverse Hyperbolic Functions

Sine	SINH(u)	Cosecant	CSCH(u)	1/SINH(u)
Cosine	COSH(u)	Secant	SECH(u)	1/COSH(u)
Tangent	TANH(u)	Cotangent	COTH(u)	1/TANH(u)
Arcsin	ASINH(x)			
Arccos	ACOSH(x)			
Arctan	ATANH(x)			

▲	A	B	C
1			
2		u	0.3
3		v	0.7
4		left side	0.38222
5		right side	0.38222

⟺

▲	A	B	C
1			
2		u	0.3
3		v	0.7
4		left side	=SINH(u)*COSH(v)
5		right side	=1/2*SINH(u+v)+1/2*SINH(u-v)

FIGURE 3.35   Validation of hyperbolic identity.

Example 3.4   Validation of a Hyperbolic Product Identity

There are many identities associated with trigonometric and hyperbolic functions. One is the *hyperbolic product rule* given by

$$\sinh(u)\cosh(v) = \frac{1}{2}\sinh(u+v) + \frac{1}{2}\sinh(u-v) \tag{3.6}$$

We set up a simple spreadsheet to test and validate this identity. First, we create named cells for the *u* and *v* values and then add formulas in adjacent cells for the left and right sides of Equation 3.6. This is illustrated in Figure 3.35, which also shows the associated formulas. Since the left- and right-side formulas yield the same result, the identity is validated.

3.4.3   LOGARITHMIC AND EXPONENTIAL FUNCTIONS

We have seen in Equation 3.1 the need to compute the natural logarithm. Excel provides the built-in LN(•) function for this. As with the trigonometric functions, the natural logarithm can be computed with an infinite series,

$$\ln(x) = 2\left[\frac{x-1}{x+1} + \frac{1}{3}\left(\frac{x-1}{x+1}\right)^3 + \frac{1}{5}\left(\frac{x-1}{x+1}\right)^5 + \cdots\right] \qquad x > 0 \tag{3.7}$$

A calculation can employ as many terms of the series as needed to converge to a result with 16 significant figures. Alternately, since the exponential function is the inverse of the natural logarithm, and can be computed from the series,

$$e^x = 1 + x + \frac{x^2}{2} + \frac{x^3}{3!} + \cdots \tag{3.8}$$

A numerical method can be constructed based on

$$y = \ln(x) \qquad \Rightarrow \qquad e^y = x \tag{3.9}$$

where *y* is adjusted until $e^y \cong x$.

TABLE 3.5
Built-in Logarithmic
Functions

	Base
LN(x)	e
LOG(x)	10
LOG10(x)	10
LOG(x,b)	b

There are alternate *logarithmic functions* listed in Table 3.5. The LOG(x) function is the LOG(x,b) function with the b left out − the default value for b is 10. Alternately, there is an explicit LOG10(x) function.

The *exponential function* provided is EXP(x).

3.4.4 GENERAL FUNCTIONS

Of the hundreds of built-in functions available in Excel, there are several that occur frequently in scientific and engineering formulas. We have summarized these in Table 3.6. There are certainly others we might have included, and we will encounter some of these in upcoming chapters.

The ABS(x) function returns a positive value, x for $x > 0$, or −x for $x < 0$, or 0 for $x = 0$. The SIGN(x) function returns +1 for $x > 0$, −1 for $x < 0$, and 0 for $x = 0$. One common application is when we want to square a quantity but carry the sign of the quantity into the result. This can be accomplished in two ways:

ABS(x)*x   or   SIGN(x)*x^2

The INT(x) function truncates or removes the fractional part of a numerical quantity. Two examples are

INT(3.99) → 3   INT(−2.5) → −2

TABLE 3.6
Common General Functions

Absolute value	ABS(x)
Sign	SIGN(x)
Truncate to integer	INT(x)
Round	ROUND(x,n)

Note in the first case that the result moves to the left on the real number line, but, in the second case, the result moves to the right.

The ROUND(x,n) function is more versatile. Setting n = 0, the function rounds to the nearest integer. Here are three examples:

$$\text{ROUND}(2.5,0) \rightarrow 3 \quad \text{ROUND}(3.5,0) \rightarrow 4 \quad \text{ROUND}(-1.5,0) \rightarrow -2$$

Notice that a fractional value of 0.5 rounds up, regardless of sign. This is different from standard rounding principle: round a fractional 0.5 value to evens. Here are examples for different values of n:

$$\text{ROUND}(\text{PI}(),5) \rightarrow 3.14159 \quad \text{ROUND}(2 \wedge 20,-3) \rightarrow 1049000$$

In the first example, the full-precision value of PI() is 3.14159265358979. For n = 5, the rounding takes place at the 10^{-5} or fifth decimal fraction place. For the second example, 2^{20} is equal to 1048576. For n = −3, the rounding takes place at the 10^3 or 1000s place.

3.5 ARRAY FORMULAS

Up to now, we have discussed formulas and functions that produce a single result. Many spreadsheet users do not know about *array formulas*. Array formulas return results to a block or range of cells, and there are array functions that return similar results. In this section, we will consider array formulas that compute results based on similarly sized arrays. In Chapter 5, you will learn how to use Data Tables, an important array-based feature of Excel. In Chapter 7, matrix calculations will be introduced using built-in array functions. And, in Chapter 10, you will learn how to create your own array functions.

Let's create a simple example to illustrate an array formula. Based on Equation 3.2, we will set up a spreadsheet to compute the volume of the oblate spheroid for a given value of the major semiaxis, *a*, and a series of values for the minor semiaxis, *b*. The initial spreadsheet is depicted in Figure 3.36. Here, we have created the first formula to compute the volume adjacent to the b = 25 value and then copied that formula down in the three cells below. This would be a typical approach.

Next, we remove the four formulas, select the four cells, and enter an array formula into the multi-cell selection. The formula is typed into the top cell and the Enter key is pressed. This is shown in Figure 3.37.

a	100	V		a	100	V
b	25	1047198		b	25	=4/3*PI()*a^2*C3
	50	2094395			50	=4/3*PI()*a^2*C4
	75	3141593			75	=4/3*PI()*a^2*C5
	100	4188790			100	=4/3*PI()*a^2*C6

FIGURE 3.36 Initial scenario for conversion to an array formula.

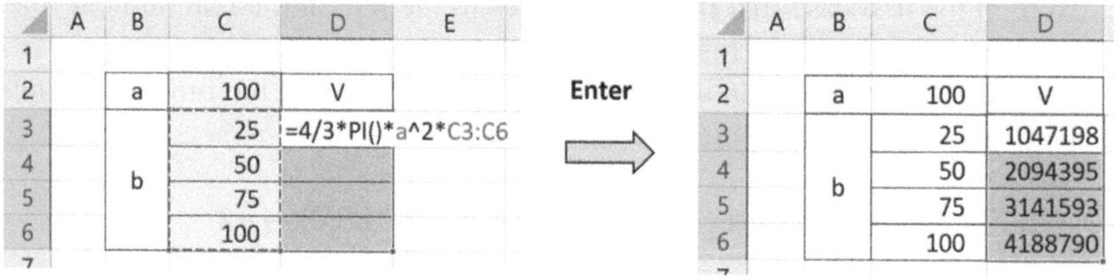

FIGURE 3.37 Creation of the array formula.

When the array formula is completed, selecting any of the four cells will display the same formula in the Formula Bar, but, except for the top cell, the formulas appear faded out. There is only one formula governing the results in cells D3:D6. The formula can only be edited with the top cell selected, either in the cell or on the Formula Bar. See Figure 3.38.

There is another type of array formula that was prevalent in earlier versions of Excel. It is now called a *legacy array formula*, and you will likely encounter it.[5] We create this type of formula by pressing the combined keys, Ctrl-Shift-Enter, instead of just the Enter key. Figure 3.39 illustrates this.

We notice that the formula appearing in each cell is the same,

$$\{= 4/3 * PI() * a \wedge 2 * C3:C6\}$$

and is surrounded by braces, {•}. This is the signal that it is a legacy array formula.

This then leaves us with a lingering question. Why use array formulas at all? The answer might be termed "one-stop shopping." With multiple formulas, as shown in Figure 3.37, if the first one is changed/edited, it must be copied down (typically by double-clicking the fill handle) to complete the change. With an array formula, there is only one formula, and any editing is completed in one place. This enhances the reliability of the spreadsheet calculation and makes it easier to comprehend. In upcoming chapters, we will also see that array formulas and functions are important in other ways.

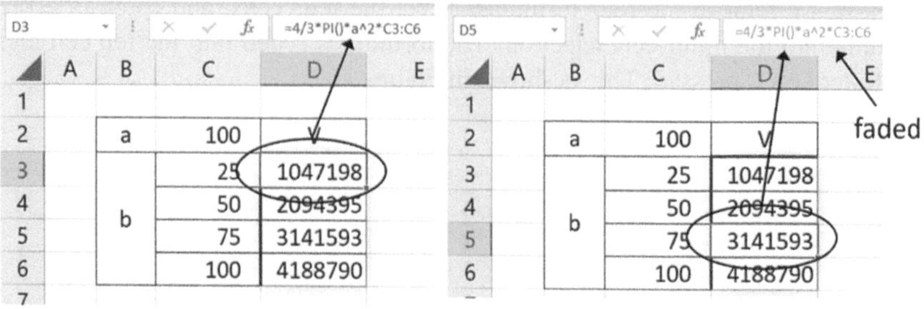

FIGURE 3.38 Array formula in place – edited only in top cell.

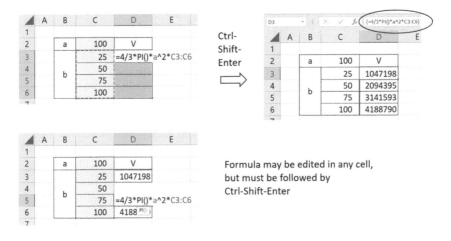

FIGURE 3.39 Entering a legacy array formula.

PROBLEMS

3.1 Open a blank spreadsheet, enter the four numbers shown in Figure P3.1 and the formula shown in cell C3.

   a. What results in cell C3 when the formula is entered? (This is the easy part!)

   b. If the formula is copied to cell D4, what formula will appear there? Predict this before doing it, and then check your understanding by examining the result. Note any errors in your prediction and what you learned by interpreting them.

   c. Following part (b), what result is displayed in cell D4? Explain in detail what happened with the copy from C3 to D4.

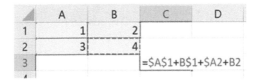

FIGURE P3.1 Entry of four numbers and an associated formula.

3.2 The Earth's major semiaxis, measured at the Equator, is 6378 km. The minor semiaxis at the poles is 6357 km. Create a spreadsheet to compute the volume and surface area of the Earth in km^3 and km^2, respectively. Add calculations based on the Earth as a sphere with radius equal to its major semiaxis. Compute the fractional errors in volume and surface area by assuming the Earth is a sphere instead of an oblate spheroid. Comment on your results.

3.3 The *segment of a circle* of radius R and segment depth h is depicted in Figure P3.3. The formula to compute the area, A, of the segment is

$$A = R^2 \cos^{-1}\left(\frac{R-h}{R}\right) - (R-h)\sqrt{2Rh - h^2}$$

   a. Set up a spreadsheet to produce a table of areas for $1 \le R \le 2$ and $0.1 \le h \le 2$. Use appropriate mixed cell addressing. Sketch the area you are computing for $R < h < 2R$.
   b. On the spreadsheet from part (**a**), add cells to store values of R and h and name those cells appropriately. In a third cell, add a formula, using the names, to compute A. Include tag labels on adjacent cells.

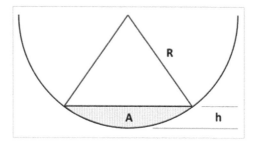

FIGURE P3.3   Segment of a circle, the shaded area.

3.4 The *trajectory of a projectile* launched at a velocity v_0 and an upward angle θ from an elevation y_0 is described by the equation,

$$y = \tan(\theta)x - \frac{g}{2v_0^2 \cos^2(\theta)}x^2 + y_0$$

Here, g is gravitational acceleration, $\cong 9.81$ m/s^2. This model assumes that air resistance is negligible, and the trajectory is parabolic. A sample trajectory is plotted in Figure P3.4.
   a. Start with a blank spreadsheet and implement the formula above to calculate y in meters given x in meters, v_0 in m/s, θ in radians, and y_0 in meters. For the angle θ, specify it first in degrees and convert it to radians in an adjacent cell for use in the formula. Use cell names for these parameters.
   b. Using your spreadsheet from part (**a**), for $v_0 = 70$ m/s, $y_0 = 10$ m, and $\theta = 12°$, experiment to find the value of x where y reaches zero.

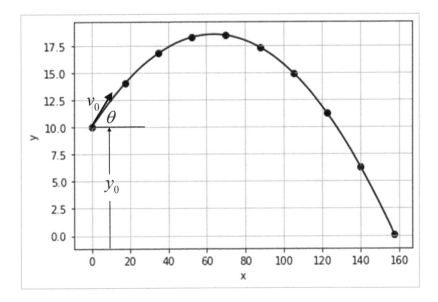

FIGURE P3.4 Trajectory of a projectile.

c. Based on your results from part (**b**), create a table of about 50 entries of y versus x for values of x from zero to the value of x that yields $y = 0$. First, enter a formula for y in the first cell of the table and copy that formula down to evaluate y for all the x values. Then, replace the set of formulas with a single array formula.

3.5 Create a spreadsheet that compares the ATAN(y/x) and ATAN2(x,y) for values in each of the four quadrants, as illustrated in Figure 3.28. Also, test values on the positive and negative portions of both the x and y axes. For each result, explain the results, especially any differences between the two function evaluations.

3.6 A sum of angles identity for the tangent is

$$\tan(\theta + \phi) = \frac{\tan(\theta) + \tan(\phi)}{1 - \tan(\theta)\tan(\phi)}$$

Set up a spreadsheet with named cells for θ and ϕ, and enter formulas for both sides of the identity to confirm it.

3.7 A situation has arisen where you want to round one or more numbers to the nearest 0.5. A suggestion to accomplish this is to multiply the number by two, round it to the nearest integer, and divide the result by two. Create a spreadsheet to test this method and validate it with several numbers, both positive and negative.

3.8 As depicted in Figure P3.8, a cable that is suspended between two points and hangs under its own weight forms a profile that is called a *catenary curve*. The nomenclature in the figure is as follows:

A   subscript identifying the left point of attachment
B   subscript identifying the right point of attachment
x lateral dimension
y vertical dimension
s cable length
y_0 cable height at minimum
W cable weight
w cable weight per unit length
T_0 lateral tension at $x = 0$ where cable is at minimum height

A mathematical derivation based on vertical and horizontal force balances yields an equation of the cable height, y, versus lateral position, x, also in terms of y_0, T_0, and w.

$$y = \frac{T_0}{w}\left[\cosh\left(\frac{w}{T_0}x\right) - 1\right] + y_0$$

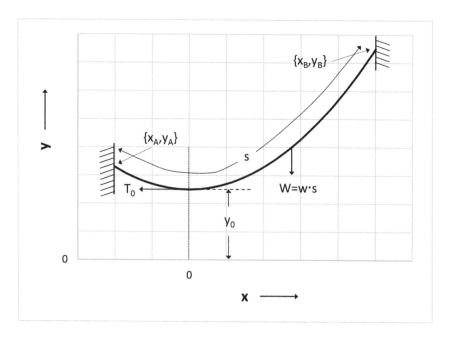

FIGURE P3.8   A catenary cable hanging under its own weight.

a. Create a spreadsheet with named cells for x_A, y_A, x_B, y_B, T_0, s, W, w, and y_0. Enter the following values:

$s = 5$ m
$W = 175$ N   Note: $W = Mg$ where M is mass of the chain and
$\quad g \cong 9.81$ m/s^2.
$y_0 = 1.5$ m
$x_A = -1.5$ m
$x_B = 2$ m
$T_0 = 200$ N

Enter formulas to compute w, y_A, and y_B.
b. Experiment with values of T_0 to obtain a value of y_B close to 1.75 m.
3.9 The *boiling point of water* changes with altitude. As altitude increases above sea level, atmospheric pressure decreases. This can be modeled with the equation,

$$P = P_0 \cdot \exp\left(-\frac{Mg}{RT_0}h\right)$$

where
P: atmospheric pressure, Pa[6]
P_0: atmospheric pressure at sea level, 101325 Pa
g: gravitational acceleration, 9.81 m/s^2
M: molar mass of dry air, 0.02897 kg/mol
T_0: sea level standard temperature, 288.2 K
R: gas law constant, 8.314 J/(mol·K)

When water is heated, its vapor pressure increases and can be modeled by the *Antoine equation*,

$$\log_{10}(P_V) = A - \frac{B}{C+T}$$

where
P_V: vapor pressure, Pa
T: temperature, °C
A, B, C: model parameters, $A = 10.183$, $B = 1724.6$, $C = 233.11$

When the vapor pressure reaches the atmospheric pressure, water boils.
a. Start with a blank spreadsheet and create named cells for P_0, g, M, T_0, R, A, B, and C. Name cells for h and P. Enter a value for the altitude, h, in meters. In the cell named P, enter the first formula to compute the pressure. Experiment with the altitude value to obtain a pressure of 80,000 Pa.

b. Name cells T and PV and enter a value of 100°C for *T*. In the P_V cell, enter the Antoine equation. Confirm that the result is close to sea level atmospheric pressure, 101325 Pa.

c. Experiment with the temperature value in the T cell until the P_V value is equal to the *P* value of 80,000 kPa. This will be the boiling point at the altitude determined in part (**a**).

d. Use your spreadsheet to determine the boiling point of water at the summit of Mt. Everest.

e. On your spreadsheet, create a column of temperatures from zero to 100°C in steps of 1°C. Name the column Tb. In the adjacent column, enter an array formula to compute the vapor pressures using the Antoine Equation. Create a chart of these data using a logarithmic vertical scale.

NOTES

1. It is worth mentioning that there is another style of cell addressing, called *R1C1*, that was used commonly decades ago. Even today, you can switch over to that via File → Options → Formulas → Working with formulas → R1C1 reference style.

2. A curiosity is that the F4 key has multiple personalities. When not in Edit mode, it can be used to transfer the formatting of a selected cell to one or more other cells. Just press F4 with the selected cell, and then select another cell and press F4.

3. Fair warning: If you select too many names, Excel may display an error message. This is a bug in Excel. If this happens, just select fewer names and repeat the Apply Names command in steps.

4. In fact, and somewhat disconcerting, this is the case in VBA, as we will illustrate in Chapter 11.

5. For example, a legacy array formula using the TABLE function is created by the built-in Data Table procedure. This is introduced in Chapter 5.

6. Pa is the abbreviation for the pascal unit of pressure. The pascal is one newton per square meter (N/m^2). It is the standard SI unit for pressure.

4 Table-based Calculations

CHAPTER OBJECTIVES

- Learn how to sort tables of data by standard and custom criteria
- Be able to determine the rank of table entries
- Know how to convert tables to lists and apply filtering
- Learn how to create an Excel Table and use Table Tools and Functions
- Explore the various lookup functions available to extract information from tables
- Learn about pivot tables and pivot charts

Over the decades, one of the most common general applications of spreadsheets has been to represent and manage tables of information. Engineering and science students do not encounter managing tables of data that often, but it is a frequent activity for practicing professionals. This chapter focuses on their needs.

Sorting tables by criteria applied to one or more of the table columns is a common activity. We introduce how to do this here for tables with or without headings. Instead of sorting, it is possible to determine the order of items in a column using the RANK function, and that will be illustrated.

Excel has three special forms of tables that can be created from a conventional table. Lists are convenient when filtering table entries is desired. Tables provide the capability to combine filtering with common sample statistical calculations. Pivot tables are convenient for tables that depend on several categorical factors as well as providing special charts. You will learn how to use all these in this chapter.

Excel has a family of lookup functions that allow the extraction of information from tables based on a search for a table entry. These are frequently used by engineers and scientists, so it is appropriate to learn these functions, and we will present them in detail in this chapter.

The example tables we use in this chapter are brief to make them manageable here. We should note that tables commonly have hundreds, even thousands, of entries. What you will learn here will scale up readily to large tables.

4.1 SORTING TABLES

We will consider a general table in Excel to be a block of cells with or without column headings. Figure 4.1 provides an example of a small table of properties of several particulate materials. The order of entries in the table is not particular, e.g., alphabetical by bed material name. In a scenario like this, we must also be

DOI: 10.1201/9781003361053-4

98 Spreadsheet Problem Solving and Programming

	A	B	C	D	E	F
1	Bed Material	Absolute Density kg/m^3	Bulk Density kg/m^3	Percent Void	Particle Diameter mm	Shape Factor
2	Wheat	1400	864.7	39.24	3.607	1.073
3	Polystyrene2	1058	592.7	43.94	3.185	1.176
4	Rice	1457	904.6	37.92	2.720	1.041
5	Millet	1180	726.8	38.44	1.989	1.070
6	Polyethylene	922	592.2	35.80	3.429	1.020
7	Corn	1342	743.1	44.60	7.257	1.500
8	Polystyrene1	1058	640.7	39.41	1.565	1.141
9	Barley	1279	724.7	43.36	3.703	1.141
10	Flaxseeds	1129	702.9	37.77	2.093	1.050

FIGURE 4.1 Properties of various particulate materials.

concerned about the possibility, after a sort, to return the table to its original order. A safe way to proceed is to add an index column to the table, usually on the left or right. Then, the table can be resorted by that column to return to the original order. The modified table with an index column to the right is shown in Figure 4.2.

To execute a sort, first we select the table. When headings are present, it is best to select them too. On the Home tab of the Ribbon in the Editing group, there is a drop-down Sort & Filter menu. This is shown in Figure 4.3 along with the selected table. The first three options on the drop-down menu are for sorting. The first two are applied to the left-most column of the table. They are shown as alphabetical sorts, but they also apply to numerical columns as ascending/descending sorts.

	A	B	C	D	E	F	G
1	Bed Material	Absolute Density kg/m^3	Bulk Density kg/m^3	Percent Void	Particle Diameter mm	Shape Factor	Index
2	Wheat	1400	864.7	39.24	3.607	1.073	1
3	Polystyrene2	1058	592.7	43.94	3.185	1.176	2
4	Rice	1457	904.6	37.92	2.720	1.041	3
5	Millet	1180	726.8	38.44	1.989	1.070	4
6	Polyethylene	922	592.2	35.80	3.429	1.020	5
7	Corn	1342	743.1	44.60	7.257	1.500	6
8	Polystyrene1	1058	640.7	39.41	1.565	1.141	7
9	Barley	1279	724.7	43.36	3.703	1.141	8
10	Flaxseeds	1129	702.9	37.77	2.093	1.050	9

FIGURE 4.2 Table with index column added.

	A	B	C	D	E	F	G
1	Bed Material	Absolute Density kg/m³	Bulk Density kg/m³	Percent Void	Particle Diameter mm	Shape Factor	Index
2	Wheat	1400	864.7	39.24	3.607	1.073	1
3	Polystyrene2	1058	592.7	43.94	3.185	1.176	2
4	Rice	1457	904.6	37.92	2.720	1.041	3
5	Millet	1180	726.8	38.44	1.989	1.070	4
6	Polyethylene	922	592.2	35.80	3.429	1.020	5
7	Corn	1342	743.1	44.60	7.257	1.500	6
8	Polystyrene1	1058	640.7	39.41	1.565	1.141	7
9	Barley	1279	724.7	43.36	3.703	1.141	8
10	Flaxseeds	1129	702.9	37.77	2.093	1.050	9

FIGURE 4.3 Selected table and Sort & Filter drop-down menu.

If we select the Sort A to Z option, the table shown in Figure 4.4 results. You will notice how the Polystyrene items are ordered by the final digit. We can reverse the sort immediately with the Undo operation (Ctrl-Z shortcut).

If we wish to reverse the sort later, it will be necessary to do so using the Index column. This procedure requires the Custom Sort option and is illustrated in Figure 4.5. Care must be taken with the My data has headers checkbox. If a selected table, not like the one here, doesn't have headers, that box should be left unchecked. We will illustrate that in an upcoming example.

Using the Custom Sort option, we can also choose a sort on a different column of the table. Let's say that we want to sort on Shape Factor from largest to smallest. Notice in Figure 4.1 that there are two materials with the same shape factor, Polystyrene1 and Barley at 1.141. We might want to apply a secondary sorting factor for duplicates, such as Particle Diameter from largest to smallest. Figure 4.6 illustrates how to do this.

	A	B	C	D	E	F	G
1	Bed Material	Absolute Density kg/m³	Bulk Density kg/m³	Percent Void	Particle Diameter mm	Shape Factor	Index
2	Barley	1279	724.7	43.36	3.703	1.141	8
3	Corn	1342	743.1	44.60	7.257	1.500	6
4	Flaxseeds	1129	702.9	37.77	2.093	1.050	9
5	Millet	1180	726.8	38.44	1.989	1.070	4
6	Polyethylene	922	592.2	35.80	3.429	1.020	5
7	Polystyrene1	1058	640.7	39.41	1.565	1.141	7
8	Polystyrene2	1058	592.7	43.94	3.185	1.176	2
9	Rice	1457	904.6	37.92	2.720	1.041	3
10	Wheat	1400	864.7	39.24	3.607	1.073	1

FIGURE 4.4 Table sorted alphabetically by Bed Material column.

FIGURE 4.5   Reversing the alphabetic sort with a sort on the Index column.

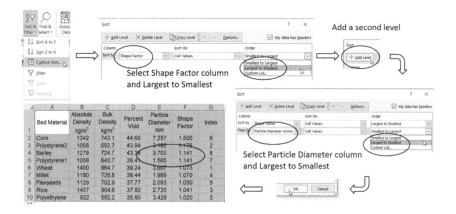

FIGURE 4.6   Completing a two-level sort on Shape Factor then on Particle Diameter.

Notice in Figure 4.6 the result of the sort where Polystyrene1 and Barley are sorted together because their shape factors are the same, but Barley is above Polystyrene1 because it has a larger particle diameter.

As a final example of sorting, we consider the bed materials table without headings. This is shown in Figure 4.7 as the last form of the table from Figure 4.6.

	A	B	C	D	E	F	G
1	Corn	1342	743.1	44.60	7.257	1.500	6
2	Polystyrene2	1058	592.7	43.94	3.185	1.176	2
3	Barley	1279	724.7	43.36	3.703	1.141	8
4	Polystyrene1	1058	640.7	39.41	1.565	1.141	7
5	Wheat	1400	864.7	39.24	3.607	1.073	1
6	Millet	1180	726.8	38.44	1.989	1.070	4
7	Flaxseeds	1129	702.9	37.77	2.093	1.050	9
8	Rice	1457	904.6	37.92	2.720	1.041	3
9	Polyethylene	922	592.2	35.80	3.429	1.020	5

FIGURE 4.7   Bed materials table without headings.

	A	B	C	D	E	F	G
1	Corn	1342	743.1	44.60	7.257	1.500	6
2	Polystyrene2	1058	592.7	43.94	3.185	1.176	2
3	Barley	1279	724.7	43.36	3.703	1.141	8
4	Polystyrene1	1058	640.7	39.41	1.565	1.141	7
5	Wheat	1400	864.7	39.24	3.607	1.073	1
6	Millet	1180	726.8	38.44	1.989	1.070	4
7	Flaxseeds	1129	702.9	37.77	2.093	1.050	9
8	Rice	1457	904.6	37.92	2.720	1.041	3
9	Polyethylene	922	592.2	35.80	3.429	1.020	5
10							
11							

FIGURE 4.8   Initiating a sort on the table without headings.

Figure 4.8 shows what happens when we initiate a sort back to the original Index order. It *may* occur, as shown in the figure, that the My data has headers box is checked and the first row of the table is excluded from the sort. It is important to note this and remove the check. Figure 4.9 then shows the completion of the sort.

Most of the tables that we encounter in engineering and science applications are organized in columns, and these are the sorting examples we have shown. It is also possible to carry out sorts on rows of a table, if the table is organized in this way. Figure 4.10 shows how to change the option from column to row for a sort, although the row-based sort is not carried out because it is inappropriate for the given table.

That covers the basics of sorting tables in Excel. We'll now consider how to extract table rankings without carrying out a sort.

FIGURE 4.9   Carrying out the sort on Index without the My data has headers box checked.

FIGURE 4.10   Changing to a row-based sort.

4.2 DETERMINING RANKS OF TABLE ENTRIES

The *rank* of a number is its position relative to other values in a list. For example, if you sorted a list of numbers, the rank of a number would be its position. Rather than sorting a range of cells or tables, we would prefer in certain applications just to know the ranks of the items either by columns or rows. Excel provides a *RANK function*[1] to accomplish this, which has the general syntax:

RANK(number, ref, [order])

where number (required) = the number whose rank you want to find, ref (required) = an array of, or a reference to, a list of numbers, and order (optional) = a number specifying how to rank the number. If order is 0 (zero) or omitted, Excel ranks number as if ref was a list sorted in descending order. If order is any nonzero value, it ranks number as if the list was sorted in ascending order.

A simple application of the function for a block of random numbers is illustrated in Figure 4.11. The first argument of the RANK function points to a cell in the table, and the second argument provides the cell address of the table as an absolute reference. The latter is because we don't anticipate moving the location of the table and may copy the formula. The result is 8, which indicates that 62.2 is 8th value in the table ranked from maximum to minimum.

Using a similar formula to that in Figure 4.11, we can create a mirror block of ranks for the entire table. This is shown in Figure 4.12. In this case, because the RANK formula is copied, the referenced range of cells must be either absolute or named.

Returning to our bed materials table, as shown, for example, in Figure 4.2, we can create a mirror table of ranks where each column is ranked individually.

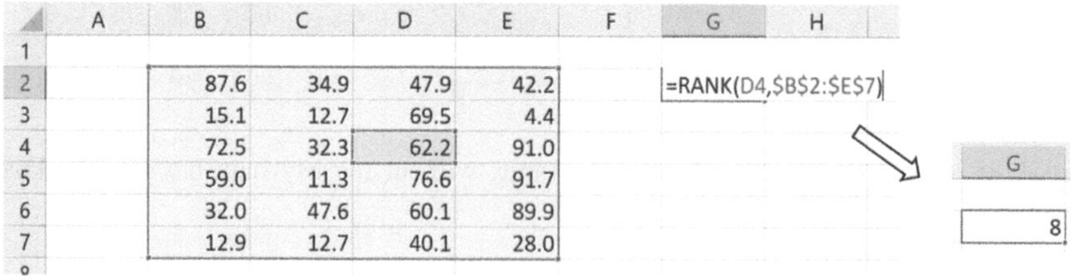

FIGURE 4.11 Use of the RANK function with a block of random numbers.

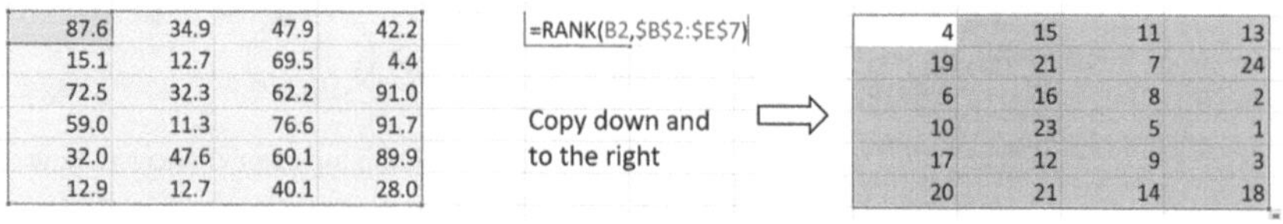

FIGURE 4.12 Creating a mirror block of ranks.

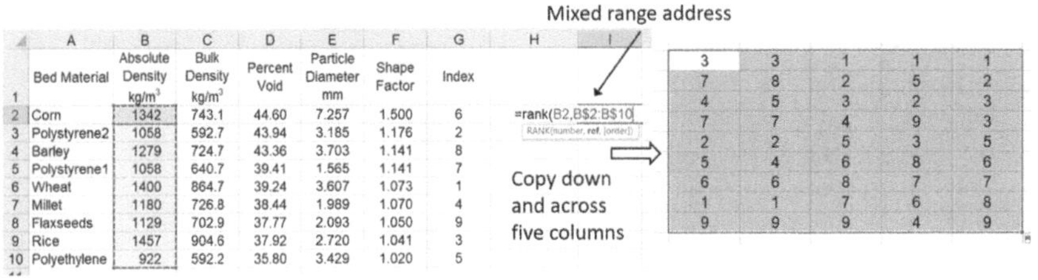

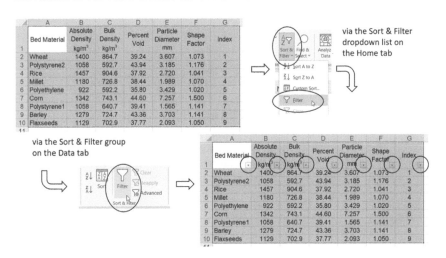

FIGURE 4.13   Creating a mirror table with ranks by columns using a mixed-range address.

This is illustrated in Figure 4.13. You will note that we create a formula using the RANK function that references the Absolute Density column values but does so with the mixed address range B$2:B$10. This will lock down rows 2 through 10 but allow the column reference to move as we copy the formula across. The rank table provides some immediate observations. First, corn has the largest values of Percent Void, Particle Diameter, and Shape Factor. Second, Propylene has the smallest values of Absolute and Bulk Density, and Percent Void.

Thus, the RANK function is a convenient way to get a picture of the order of data without having to sort them.

4.3   CONVERTING TABLES TO EXCEL LISTS

Excel provides a special form of a table called a *list* that facilitates "live" sorting by criteria applied to columns of the table. The list format also allows us to apply criteria that will alter the rows that appear on the worksheet. This "filtering" will not change the overall list information. The table can be converted to a list and then back to a table conveniently and quickly. Figure 4.14 shows two routes for conversion of our bed materials table to a list.

FIGURE 4.14   Two routes to converting a table into a list.

Bed Material	Absolute Density kg/m³	Bulk Density kg/m³	Percent Void	Particle Diameter mm	Shape Factor	Index
Sort A to Z	64.7	39.24	3.607	1.073	1	
Sort Z to A	92.7	43.94	3.185	1.176	2	
Sort by Color	104.6	37.92	2.720	1.041	3	
Sheet View	26.8	38.44	1.989	1.070	4	
Clear Filter from "Bed Material"	92.2	35.80	3.429	1.020	5	
Filter by Color	43.1	44.60	7.257	1.500	6	
Text Filters	40.7	39.41	1.565	1.141	7	
Search	24.7	43.36	3.703	1.141	8	
	02.9	37.77	2.093	1.050	9	

#	Bed Material	Absolute Density kg/m³	Bulk Density kg/m³	Percent Void	Particle Diameter mm	Shape Factor	Index
2	Barley	1279	724.7	43.36	3.703	1.141	8
3	Corn	1342	743.1	44.60	7.257	1.500	6
4	Flaxseeds	1129	702.9	37.77	2.093	1.050	9
5	Millet	1180	726.8	38.44	1.989	1.070	4
6	Polyethylene	922	592.2	35.80	3.429	1.020	5
7	Polystyrene1	1058	640.7	39.41	1.565	1.141	7
8	Polystyrene2	1058	592.7	43.94	3.185	1.176	2
9	Rice	1457	904.6	37.92	2.720	1.041	3
10	Wheat	1400	864.7	39.24	3.607	1.073	1

FIGURE 4.15   Sorting the list on Bed Material using the drop-down tab.

The evident characteristic of a list, as illustrated in Figure 4.14, is the drop-down tabs on the headings. Sorting and filtering are now enabled. Figure 4.15 depicts the use of the Bed Material drop-down menu to sort the table in alphabetical order.

A second feature of lists that we frequently use is to display only certain rows, here by selection of Bed Material items. These are the checkboxes in the drop-down menu. In Figure 4.16, we illustrate this by selecting only the polymeric materials. This filtering can be easily reversed by clicking on the Select All checkbox.

The third, and most useful feature of lists, is to apply criteria to chosen columns. For example, we might want to display only the entries with Particle Diameter greater than 2.0. Figure 4.17 shows how to do this.

One could add another numerical criterion on another column, such as Bulk Density > 700, and the list would be filtered further. We do not illustrate that here though. Another consideration has to do with formulas that apply to column data. Figure 4.18 illustrates an AVERAGE function applied to all the columns, both with the original unfiltered list and with the filtered list of Figure 4.17. The key observation here is that the formula's results are not dependent on the filtering of the list – they apply to the entire column.

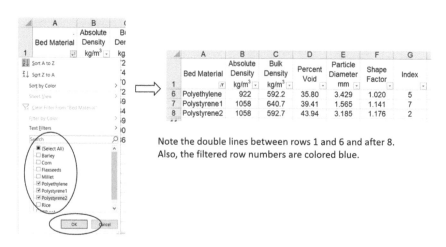

Note the double lines between rows 1 and 6 and after 8.
Also, the filtered row numbers are colored blue.

FIGURE 4.16   Using the filter to display only the polymeric bed materials.

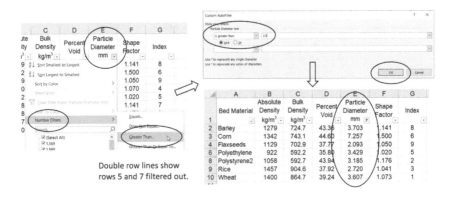

FIGURE 4.17 Applying a numerical filter: Particle Diameter > 2.0.

	A	B	C	D	E	F	G
	Bed Material	Absolute Density	Bulk Density	Percent Void	Particle Diameter	Shape Factor	Index
1		kg/m³	kg/m³		mm		
2	Barley	1279	724.7	43.36	3.703	1.141	8
3	Corn	1342	743.1	44.60	7.257	1.500	6
4	Flaxseeds	1129	702.9	37.77	2.093	1.050	9
5	Millet	1180	726.8	38.44	1.989	1.070	4
6	Polyethylene	922	592.2	35.80	3.429	1.020	5
7	Polystyrene1	1058	640.7	39.41	1.565	1.141	7
8	Polystyrene2	1058	592.7	43.94	3.185	1.176	2
9	Rice	1457	904.6	37.92	2.720	1.041	3
10	Wheat	1400	864.7	39.24	3.607	1.073	1
11							
12	Average	1203	721.4	40.05	3.283	=AVERAGE(F2:F10)	

	A	B	C	D	E	F	G
	Bed Material	Absolute Density	Bulk Density	Percent Void	Particle Diameter	Shape Factor	Index
1		kg/m³	kg/m³		mm		
2	Barley	1279	724.7	43.36	3.703	1.141	8
3	Corn	1342	743.1	44.60	7.257	1.500	6
4	Flaxseeds	1129	702.9	37.77	2.093	1.050	9
6	Polyethylene	922	592.2	35.80	3.429	1.020	5
8	Polystyrene2	1058	592.7	43.94	3.185	1.176	2
9	Rice	1457	904.6	37.92	2.720	1.041	3
10	Wheat	1400	864.7	39.24	3.607	1.073	1
11							
12	Average	1203	721.4	40.05	3.283	1.135	

The formula still applies to the entire list.

FIGURE 4.18 The AVERAGE function applied to the list's columns, both unfiltered and filtered.

Excel's list feature provides various capabilities for sorting and filtering that may be useful in various settings. Important characteristics are that sorting can be reversed conveniently and filtering doesn't alter table contents, just the appearance on the worksheet. As we now see, Excel's table feature provides different capabilities.

4.4 EXCEL TABLES

Excel tables provide more capabilities for formatting and table-based calculations. Figure 4.19 shows how to convert our unformatted, bed materials table into an Excel table. Immediately, we note that the table has a very different format including the use of color.

By using the Format as Table command in the Styles group on the Home tab, we have many choices of color combinations and intensities. Also, we can adjust column widths, and the headings will automatically format to accommodate. These are illustrated in Figure 4.20. A light intensity, gray and white scheme is selected from the many available options.

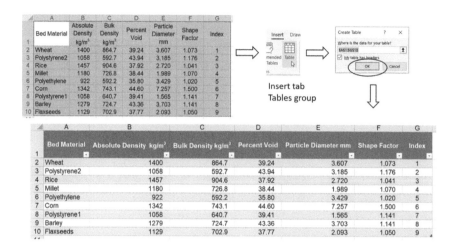

FIGURE 4.19   Creating an Excel table.

FIGURE 4.20   Changing the table color scheme and column widths.

Also, when the table is selected, a Table Design tab appears on the ribbon that includes the color styles and other options. In Figure 4.21, we select the Total Row checkbox in the Table Style Options. This provides the capability for sample statistics calculations for each column.

The initial Total Row doesn't do much for us. There is a total calculation in the Index column, but there are many more options. These can be obtained

FIGURE 4.21   Selecting the Total Row option from the Table Design tab of the Ribbon.

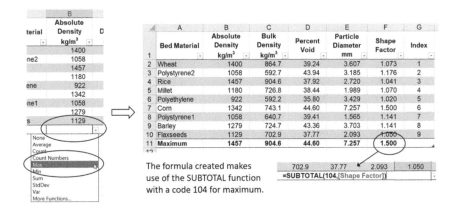

The formula created makes use of the SUBTOTAL function with a code 104 for maximum.

| | 702.9 | 37.77 | 2.093 | 1.050 |

=SUBTOTAL(104,[Shape Factor])

FIGURE 4.22 Modifying the Total Row to compute the maximum value of each column.

via the drop-down arrows in the cells in the Total Row. This is illustrated in Figure 4.22 where the maximum value in each column is displayed, and the Total label is changed to reflect this choice.

The code used in the SUBTOTAL function aligns with the drop-down list starting with 101 for the average. An important feature of the function is that it *does* adjust to the filtered items in the table. This is shown in Figure 4.23 for the case where we filter to display only the polymeric bed materials. Note: If the code 1, not 101, were used, there would be no adjustment to the filtered items – the maximum value for the entire unfiltered column would still be displayed.

It is also possible to add rows below the first Total Row using the SUBTOTAL function. This is depicted in Figure 4.24. The formula for Absolute Density can be created and copied across the remaining four columns. Note that the code 105 is used for minimum, and the Table 1 name refers to the original table and doesn't include the Maximum row.

The remaining task when it comes to Excel tables may be to return a table to a standard range. This is accomplished via the Table Design tab, Tools group,

	A	B	C	D	E	F	G
	Bed Material	Absolute Density kg/m³	Bulk Density kg/m³	Percent Void	Particle Diameter mm	Shape Factor	Index
1							
3	Polystyrene2	1058	592.7	43.94	3.185	1.176	2
6	Polyethylene	922	592.2	35.80	3.429	1.020	5
8	Polystyrene1	1058	640.7	39.41	1.565	1.141	7
11	Maximum	1058	640.7	43.94	3.429	1.176	
12							

FIGURE 4.23 Filtered table with maximum values adjusted to those items displayed.

	A	B	C	D	E	F	G
1	Bed Material	Absolute Density kg/m³	Bulk Density kg/m³	Percent Void	Particle Diameter mm	Shape Factor	Index
2	Wheat	1400	864.7	39.24	3.607	1.073	1
3	Polystyrene2	1058	592.7	43.94	3.185	1.176	2
4	Rice	1457	904.6	37.92	2.720	1.041	3
5	Millet	1180	726.8	38.44	1.989	1.070	4
6	Polyethylene	922	592.2	35.80	3.429	1.020	5
7	Corn	1342	743.1	44.60	7.257	1.500	6
8	Polystyrene1	1058	640.7	39.41	1.565	1.141	7
9	Barley	1279	724.7	43.36	3.703	1.141	8
10	Flaxseeds	1129	702.9	37.77	2.093	1.050	9
11	Maximum	1457	904.6	44.60	7.257	1.500	
12	Minimum	922	592.2	35.80	1.565	1.020	

SUBTOTAL function uses the 105 code for minimum

102.9	37.77	2.093	1.050
904.6	44.60	7.257	1.500
=SUBTOTAL(105,Table1[Shape Factor])			

FIGURE 4.24 Adding a Minimum row to the table with the SUBTOTAL function, Table 1 designation, and code 105.

Convert to Range command. Figure 4.25 illustrates this. Notice that the SUBTOTAL functions are retained, although table headings and the Table1 name are removed. This indicates that the SUBTOTAL function can be used independently of an Excel Table. The codes used in the SUBTOTAL function are summarized in Table 4.1.

At this point, we have reviewed tables, lists, and Excel tables including calculations made with table columns, filtered or not. A frequent need, when working with tables, is to extract information. That is the topic of the next section.

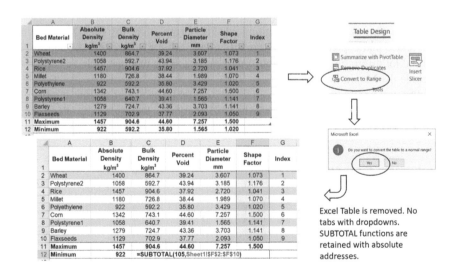

FIGURE 4.25 Returning an Excel Table to its original range format.

TABLE 4.1
Numerical Codes for the SUBTOTAL Function

	Number		
Function	Ignores Hidden Rows	Includes Hidden Rows	Description
AVERAGE	101	1	average
COUNT	102	2	count numbers
COUNTA	103	3	count all
MAX	104	4	maximum
MIN	105	5	minimum
PRODUCT	106	6	item-by-item product
STDEV.S	107	7	sample standard deviation
STDEV.P	108	8	population standard deviation
SUM	109	9	sum
VAR.S	110	10	sample variance
VAR.P	111	11	population variance

4.5 LOOKUP FUNCTIONS

Excel provides a family of lookup functions that offer various ways for extracting data from tables. The functions we will consider in this section are summarized in Table 4.2. The "X" versions of the LOOKUP and MATCH functions have been added to the most recent version of Excel but are not available in Excel 2019, 2016, or earlier versions. Eventually, the former functions may be removed. We will present both versions here for the sake of completeness.

TABLE 4.2
Excel's Lookup Function Family

LOOKUP	look for a value in a column or row of a range and return a value from a corresponding location in a column or row	XLOOKUP	a more versatile version of LOOKUP
VLOOKUP	look for a value down the left column of a range of cells and return a value in the same row but from any column of the same range of cells	MATCH	look for a value in a column or row of a range and return the index (position) of that value
HLOOKUP	look for a value across the top row of a range of cells and return a value in the same column but from any row of the same range of cells	XMATCH	a more versatile version of MATCH
INDEX	extract a value from a range of cells given a column and row index		

4.5.1 THE **LOOKUP** FUNCTION

We will start with the LOOKUP function which has the syntax

LOOKUP(lookup_value,lookup_range,return_range)

Figure 4.26 illustrates how the function works. The lookup_value is usually a reference to a cell, but it can be a literal value, number, or string. The items in the lookup_range are typically ordered either as ascending numbers or alphabetically ordered strings. An important "feature" of LOOKUP is what happens when the search through the lookup_range encounters a value that is not a match but less than the lookup_value. "Greater than" means alphabetically "after" in the case of a lookup_range with text (string) entries. The search then backs up one cell and locates that position for the item in the return_range. This includes a lookup_value that is greater than all items in the lookup_range. In this case, the last item is located and selected from the return_range. These idiosyncrasies are best illustrated through examples.

We will start with an example where the lookup_value and lookup_range are text values. A conventional application of the LOOKUP function is shown in Figure 4.27. Here we use names for the lookup_value, lookup_range, and result_range, but one can certainly use cell addresses. Also, we could enter "cork" directly as the first argument of the function. You can see that "cork" is matched in the Material range, and the value in the corresponding location of the Density range is returned by the function. There are two points worth mentioning here. First, the lookup and return ranges do not have to be adjacent columns, although that is commonly the case. Second, those ranges must be the same size so that there is an item in the return range for each in the lookup range.

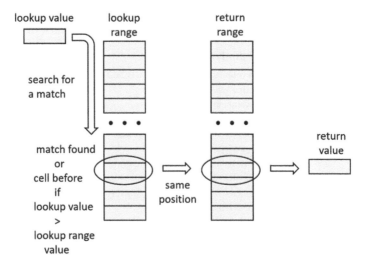

FIGURE 4.26 Scheme for the LOOKUP function.

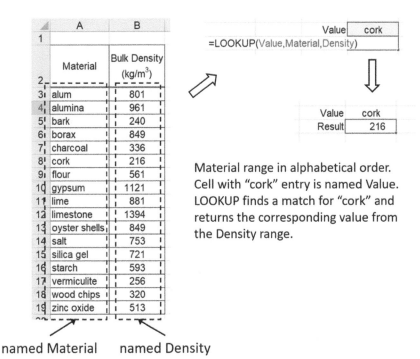

FIGURE 4.27 Table of bulk density of various particulate materials with LOOKUP function.

The LOOKUP function in this example is vulnerable to errors in the lookup_value. Three of these are illustrated in Figure 4.28.

This raises the question of how to deal with the idiosyncrasies shown in Figure 4.28. There are two approaches: (1) use a different function, or (2) restrict the lookup values to acceptable entries only. We will deal with the first of these later, but it is appropriate to introduce Data Validation as a tool for the second.

Value	gipsum
Result	561

The LOOKUP encounters "gypsum" which is after "gipsum", a typographical error, backs up one cell to "flour" and returns its bulk density.

Value	zinc sulfate
Result	513

"zinc sulfate" is beyond the last element of the lookup range, and so the bulk density of that last element, "zinc oxide" is returned.

Value	alim
Result	#N/A

A typographical error, "alim" instead of "alum", occurs before the first item in the lookup range, so a "#N/A" error code is returned.

FIGURE 4.28 Three errors involving the LOOKUP function.

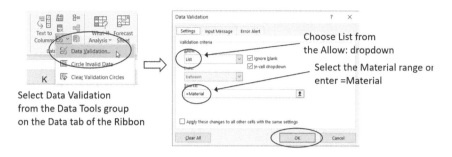

FIGURE 4.29 Setting up Data Validation for the Value cell to restrict input to the members of the Material list.

Data Validation allows us to restrict the values that can be entered in a cell. It is particularly useful with spreadsheets that will be used by others. In this example, we would like to restrict entries in the Value cell to be only those in the Material range. Figure 4.29 shows how to arrange this. The Value cell is selected, and Data Validation is selected from the Data Tools group of the Data tab of the Ribbon. There are various restrictions available from the Allow: drop-down list, and here we select the List option. In the Source: field, we can select the Material range from the spreadsheet or type in the name as =Material. Note: the = sign is required.

When the Value cell is selected, a drop-down button appears to the right of the cell. Clicking the button displays the list of materials, and one can be selected. It is still possible to type in a material in the Value cell, and when an item is entered that is not in the list, an error is displayed. These actions are illustrated in Figure 4.30.

The LOOKUP function can also be used with lookup values and ranges that are numerical. Figure 4.31 illustrates this with a table of nominal pipe diameters (formatted as fractions) and corresponding outside diameters. Again, the lookup and return ranges are named, here Nom and OD respectively. The lookup value cell is named Nominal. After entering the LOOKUP function formula, as shown in the figure, Data Validation is added to restrict entries to the values in the Nom range.

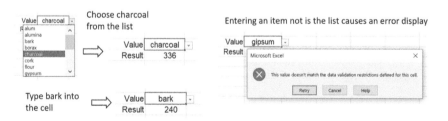

FIGURE 4.30 Entering items in the Value cell by list selection, typing in, and including an erroneous input.

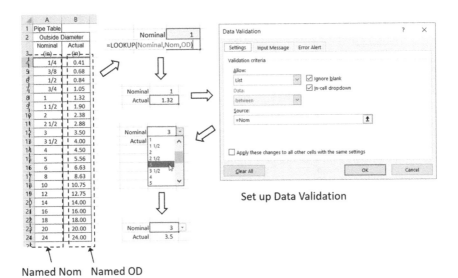

Named Nom   Named OD

FIGURE 4.31   Use of the LOOKUP function with a table of nominal and outside pipe diameters.

A lingering question is whether there would ever be a situation where we do not want to restrict the lookup values to match the entries in the lookup range but rather use the features of mismatch to our advantage. We will see that this is useful in Section 4.5.5.

4.5.2   THE **VLOOKUP** AND **HLOOKUP** FUNCTIONS

We can gain additional capability over the LOOKUP function by considering the VLOOKUP and HLOOKUP functions. Their syntax is the same, although they operate differently. For VLOOKUP,

VLOOKUP(lookup_value,table_range,column_number,match_choice)

Figure 4.32 illustrates how the VLOOKUP function works. The search for a match to the lookup value proceeds down the leftmost column of the table range. There is no separate return range. The return value is located at the same position as the left-column match in the column specified by the column number. The column number is counted 1,2,3,... from the left. There is a similar strategy in handling a lookup value that is greater than a table range value as the downward search takes place. A distinction from the LOOKUP function is that its lookup range and result range can be located separately on the worksheet, whereas VLOOKUP's table range is a contiguous block of cells. The final match choice argument is optional. There are two Boolean values possible, TRUE and FALSE, and the default value is TRUE. The default is the same match procedure as LOOKUP. Setting FALSE requires an exact match; otherwise, an error is returned.

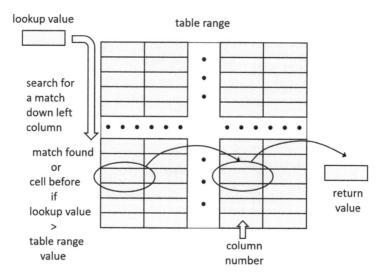

FIGURE 4.32   Scheme for the VLOOKUP function.

In a way, setting the match choice to FALSE obviates the need for Data Validation, although it can still be used for convenience, if appropriate. If we require an exact match, the items in the lookup range do *not* have to be in ascending order.

An example application of the VLOOKUP function is shown in Figure 4.33. This is a table of the nominal and inside diameters of pipe. The inside diameter depends on the Schedule number of the pipe. The Schedule number is approximated from the formula

$$\text{Schedule} \cong 1000 \frac{\text{Internal Pressure}}{\text{Maximum Allowable Stress}}$$

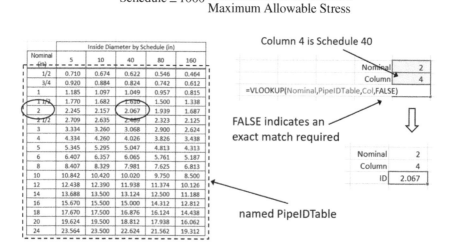

FIGURE 4.33   Use of VLOOKUP with a table of pipe nominal and inside diameters.

Nominal (in)	Inside Diameter by Schedule (in)				
	5	10	40	80	160
1/2	0.710	0.674	0.622	0.546	0.464
3/4	0.920	0.884	0.824	0.742	0.612
1	1.185	1.097	1.049	0.957	0.815
1 1/2	1.770	1.682	1.610	1.500	1.338
2	2.245	2.157	2.067	1.939	1.687
2 1/2	2.709	2.635	2.469	2.323	2.125
3	3.334	3.260	3.068	2.900	2.624
4	4.334	4.260	4.026	3.826	3.438
5	5.345	5.295	5.047	4.813	4.313
6	6.407	6.357	6.065	5.761	5.187
8	8.407	8.329	7.981	7.625	6.813
10	10.842	10.420	10.020	9.750	8.500
12	12.438	12.390	11.938	11.374	10.126
14	13.688	13.500	13.124	12.500	11.188
16	15.670	15.500	15.000	14.312	12.812
18	17.670	17.500	16.876	16.124	14.438
20	19.624	19.500	18.812	17.938	16.062
24	23.564	23.500	22.624	21.562	19.312
1	2	3	4	5	6	column number

Use HLOOKUP with Schedule no., expanded table, result from the bottom row, and exact match.

Nominal	2
Schedule	40

=HLOOKUP(M6,C4:G23,20,FALSE)

ID 2.067

Nominal	2
Schedule	40
Column	4
ID	2.067

Row of column numbers added to the bottom

FIGURE 4.34 Use of the HLOOKUP function to automate Schedule number lookup.

The higher the Schedule, the thicker the pipe wall. Schedule 40 is the most common. The example shows the VLOOKUP function searching for a 2-inch nominal pipe and returning the inside diameter for a Schedule 40 pipe. Since we use the FALSE argument, any value in the Nominal cell that isn't in the left column of the table will cause an error.

To get a result from the VLOOKUP formula in Figure 4.33, we have to translate our desired Schedule number into a column number. A way around that is to have another lookup function that finds the column number given the Schedule number. This can be accomplished by adding an index row below the table and using an HLOOKUP function. Figure 4.34 shows how this is done. The HLOOKUP function has syntax similar to the VLOOKUP function. The top row of the table range is searched for a match with the lookup value, and the column number in row 20 of the table is returned. With the match choice argument set to FALSE, an exact match is required.

HLOOKUP(lookup_value,table_range,row_number,match_choice)

We could always add Data Validation to the Nominal and Schedule cells to provide the drop-down lists and error protection.

4.5.3 THE MATCH AND INDEX FUNCTIONS

The difference between the MATCH function and the prior lookup functions is that MATCH returns the location of the lookup value in the lookup range rather than a value from the return range or table range. The syntax of the function is

MATCH(lookup_value,lookup_range,match_type)

The first two arguments are as before, and the match_type argument is given by

1   Finds the largest value in the lookup range that is ≤ the lookup value. Values in the lookup range must be in ascending/alphabetical order. This is the default if match type is left out.

0   Finds an exact match or returns an error. Just as with the VLOOKUP and HLOOKUP functions, items in the lookup range can be in any order.

−1   Finds the smallest value in the lookup range that is ≥ lookup value. Values in the lookup range must be in descending/reverse-alphabetical order.

Figure 4.35 illustrates how the MATCH function can be used with our previous example. This uses the 0 value for the match type to require an exact match. The Nominal value of 3 is located at the 9th position in the Nom range.

How can we use the result of the MATCH function? That is where the INDEX function comes in. The syntax of the INDEX function is

INDEX(table_range,row_number,column_number)

	A	B	C	D	E
1	Pipe Table				
2	Outside Diameter				
3	Nominal (in)	Actual (in)			
4	1/4	0.41		Nominal	3
5	3/8	0.68		Actual	3.5
6	1/2	0.84			
7	3/4	1.05	=MATCH(Nominal,Nom,0)		
8	1	1.32			
9	1 1/2	1.90		⇩	
10	2	2.38			
11	2 1/2	2.88			
12	3	3.50		Nominal	3
13	3 1/2	4.00		Actual	3.5
14	4	4.50			
15	5	5.56		Match	9
16	6	6.63			
17	8	8.63			
18	10	10.75			
19	12	12.75			
20	14	14.00			
21	16	16.00			
22	18	18.00			
23	20	20.00			
24	24	24.00			

FIGURE 4.35 Use of the MATCH function to locate the position of the lookup value in the lookup range.

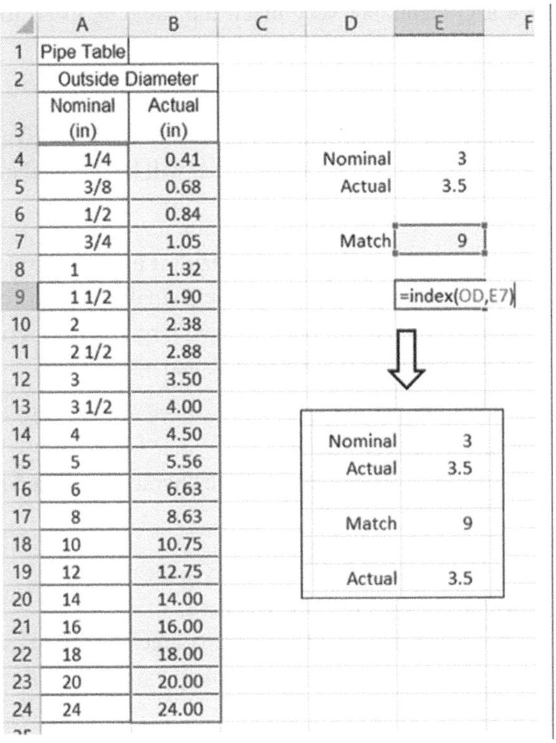

◢	A	B	C	D	E	F
1	Pipe Table					
2	Outside Diameter					
3	Nominal (in)	Actual (in)				
4	1/4	0.41		Nominal	3	
5	3/8	0.68		Actual	3.5	
6	1/2	0.84				
7	3/4	1.05		Match	9	
8	1	1.32				
9	1 1/2	1.90			=index(OD,E7)	
10	2	2.38				
11	2 1/2	2.88				
12	3	3.50				
13	3 1/2	4.00				
14	4	4.50		Nominal	3	
15	5	5.56		Actual	3.5	
16	6	6.63				
17	8	8.63		Match	9	
18	10	10.75				
19	12	12.75		Actual	3.5	
20	14	14.00				
21	16	16.00				
22	18	18.00				
23	20	20.00				
24	24	24.00				

FIGURE 4.36   Using the INDEX function to find values based on the MATCH function.

The column number argument is optional. If it is left out, only the left-most column of the table range, the only column, is used. The function returns the value in the row/column intersection of the table range. Adding this to Figure 4.35, you can see how this is accomplished in Figure 4.36. An obvious point here is that this could be accomplished with one application of the LOOKUP or VLOOKUP function. And that raises the question, what use is the MATCH function? We will see that in Section 4.5.5.

4.5.4   XLOOKUP AND XMATCH

In the most recent versions of Excel (not 2019 or before), two new functions have been introduced to extend the capabilities of LOOKUP and MATCH. Eventually, these new functions, XLOOKUP and XMATCH, may replace the former functions, but, for now, all of them are available. The point of this section is to explain what the new functions "bring to the table."

The syntax for the XLOOKUP function is

XLOOKUP(lookup_value,lookup_range,return_range,if_not_found,match_type,search_mode)

The first three arguments are very much the same as those for the LOOKUP function. The remaining arguments require some explanation:

if_not_found

When a match for the lookup value is not found in the lookup range, this argument can be a text string, e.g., "not found," that will be returned instead of #N/A.

match_type

0: Exact match is required. This is the default if match type is left out.
–1: Exact match or, if none is found, the next smaller item is returned.
1: Exact match or, if none is found, the next larger item is returned.
2: Allows for a wildcard match using *, ?, and ~.

*: Allows any number of characters.
?: Allows any single character.
~: Used before *, ?, or ~ to allow those characters in the lookup value.

search_mode

1: Search starting with the first item. This is the default.
–1: Reverse search starting with the last item.
2: Use a binary search method. The lookup range must be in ascending order.
–2: Use a binary search method. The lookup range must be in descending order.

The XLOOKUP function can perform on either a column or row basis. We will present an interesting example using the table in Figure 4.33. This is presented in Figure 4.37. There is one XLOOKUP function imbedded in another. The inner

	Inside Diameter by Schedule (in)				
Nominal (in)	5	10	40	80	160
1/2	0.710	0.674	0.622	0.546	0.464
3/4	0.920	0.884	0.824	0.742	0.612
1	1.185	1.097	1.049	0.957	0.815
1 1/2	1.770	1.682	1.610	1.500	1.338
2	2.245	2.157	2.067	1.939	1.687
2 1/2	2.709	2.635	2.469	=XLOOKUP(Nominal,$B$5:$B$22,XLOOKUP(Schedule,$C$4:$G$4,$C$5:$G$22))	
3	3.334	3.260	3.068	2.900	2.624
4	4.334	4.260	4.026	3.826	3.438
5	5.345	5.295	5.047	4.813	4.313
6	6.407	6.357	6.065	5.761	5.187
8	8.407	8.329	7.981	7.625	6.813
10	10.842	10.420	10.020	9.750	8.500
12	12.438	12.390	11.938	11.374	10.126
14	13.688	13.500	13.124	12.500	11.188
16	15.670	15.500	15.000	14.312	12.812
18	17.670	17.500	16.876	16.124	14.438
20	19.624	19.500	18.812	17.938	16.062
24	23.564	23.500	22.624	21.562	19.312

Nominal	2
Schedule	40
Column	4
ID	2.067

Nominal	2
Schedule	40
Column	4
ID	2.067
	2.067

FIGURE 4.37 Carrying out a two-way table lookup using an imbedded XLOOKUP function.

function implements a horizontal search on the row of pipe schedules to locate the match with the Schedule value. Once found, instead of returning an individual value from a single row, it returns an entire column from the absolute table range $C$5:$G$22. This entire column is used as the return range for the outer XLOOKUP function. It uses the Nominal value to search down the values in column B for a match and then returns a value from the column produced by the inner XLOOKUP. It is worth mentioning that in the previous section, we could have imbedded the HLOOKUP function in the VLOOKUP function to determine the ID value with one formula. That did include adding the column number row at the bottom of the table.

Although we describe the various options for the XLOOKUP function, we do not take the time here to illustrate them. They are not the "bread and butter" of engineering and scientific spreadsheets.

The XMATCH function is the new, extended version of the MATCH function. Its syntax is

XMATCH(lookup_value,lookup_range,match_type,search_mode)

The match type and search mode arguments are the same as those for XLOOKUP. In Figure 4.38, we illustrate the use of XMATCH along with the INDEX function for a similar imbedded application to that of Figure 4.37, except here we use two XMATCH functions, one for the row index and a second for the column index. The formula returns the same value as our previous examples with the lookup functions.

=INDEX($C$5:$G$22,XMATCH(Nominal,$B$5:$B$22,),XMATCH(Schedule,$C$4:$G$4))

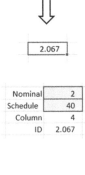

	A	B	C	D	E	F	G
4		Nominal (in)	5	10	40	80	160
5		1/2	0.710	0.674	0.622	0.546	0.464
6		3/4	0.920	0.884	0.824	0.742	0.612
7		1	1.185	1.097	1.049	0.957	0.815
8		1 1/2	1.770	1.682	1.610	1.500	1.338
9		2	2.245	2.157	2.067	1.939	1.687
10		2 1/2	2.709	2.635	2.469	2.323	2.125
11		3	3.334	3.260	3.068	2.900	2.624
12		4	4.334	4.260	4.026	3.826	3.438
13		5	5.345	5.295	5.047	4.813	4.313
14		6	6.407	6.357	6.065	5.761	5.187
15		8	8.407	8.329	7.981	7.625	6.813
16		10	10.842	10.420	10.020	9.750	8.500
17		12	12.438	12.390	11.938	11.374	10.126
18		14	13.688	13.500	13.124	12.500	11.188
19		16	15.670	15.500	15.000	14.312	12.812
20		18	17.670	17.500	16.876	16.124	14.438
21		20	19.624	19.500	18.812	17.938	16.062
22		24	23.564	23.500	22.624	21.562	19.312

2.067

Nominal	2
Schedule	40
Column	4
ID	2.067

FIGURE 4.38 Use of the INDEX function and two imbedded XMATCH functions to extract a value from the table.

This completes our presentation of the LOOKUP, MATCH, INDEX functions and their variations. We will now implement these for a typical engineering/scientific application.

4.5.5 PROPERTY TABLE INTERPOLATION

In conducting engineering and scientific calculations on spreadsheets, it is common to acquire values for physical or chemical properties from tables. We will use as an example a table of the density of liquid water in kg/m^3 at temperatures from 0°C to 100°C at atmospheric pressure, as shown in Figure 4.39. It is always useful to present such data in graphical form, so a plot is included in the figure. Interestingly, there is a maximum in the curve at 4°C. This can lead to the phenomenon of pipes breaking during freeze/thaw cycles.

We will illustrate here the use of the linear interpolation method to approximate densities at temperatures within the range of the table, but not at table values. This simple method is illustrated in Figure 4.40. One considers a straight line between a pair of table entries and approximates the density, as $\hat{\rho}$, at an intermediate temperature, T, using that line. The key formula is shown in the figure.

We now face the task of creating formulas on the spreadsheet depicted in Figure 4.39 to implement the method shown in Figure 4.40. There, we see the necessity to find the pair of adjacent table entries that bracket the desired interpolation. This is where our lookup functions come into play. There are several ways to accomplish this, and we will only illustrate one here in Figure 4.41 using MATCH and INDEX.

Let's explain these formulas in some detail starting with

=MATCH(TempIn,Temperature)

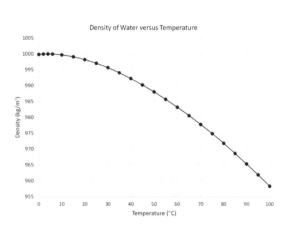

FIGURE 4.39 Density of water versus temperature from 0°C to 100°C.

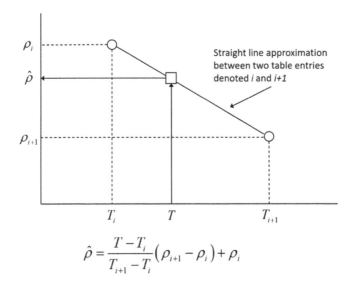

$$\hat{\rho} = \frac{T - T_i}{T_{i+1} - T_i}\left(\rho_{i+1} - \rho_i\right) + \rho_i$$

FIGURE 4.40 Linear interpolation (the square) between two table entries (the circles).

The cell with the input temperature, for which an interpolated density is desired, is named TempIn. The table columns are named Temperature and Density. The tag labels LowRow, HiRow, LowTemp, HiTemp, LowDens, and HiDens are also the names of the adjacent cells.

The MATCH function searches down the Temperature range for the TempIn value. If the TempIn value is between two Temperature values, it returns the index of the lower value to the LowRow cell. The HiRow cell is just 1 added to the LowRow cell.

With the LowRow and HiRow values in place, the temperatures and densities that bracket the TempIn value can be extracted from the table using the INDEX function. Finally, the interpolated density is computed using a formula patterned after that in Figure 4.40.

An important aspect of any spreadsheet problem solution, such as the one here, is testing whether it is "bullet-proof." Figure 4.42 shows several tests with different TempIn values. These tests include the extremes of the table, 0°C and 200°C, an intermediate matched value, 60°C, and two values outside the range of the table,

Temperature 37		°C		Temperature	37	°C
LowRow =MATCH(TempIn,Temperature)				LowRow	10	
HiRow =LowRow+1				HiRow	11	
LowTemp =INDEX(Temperature,LowRow)		°C		LowTemp	35	°C
HiTemp =INDEX(Temperature,HiRow)		°C		HiTemp	40	°C
LowDens =INDEX(Density,LowRow)		kg/m³		LowDens	994.06	kg/m³
HiDens =INDEX(Density,HiRow)		kg/m³		HiDens	992.25	kg/m³
Density =(TempIn-LowTemp)/(HiTemp-LowTemp)*(HiDens-LowDens)+LowDens		kg/m³		Density	993.34	kg/m³

FIGURE 4.41 Formulas to implement linear interpolation.

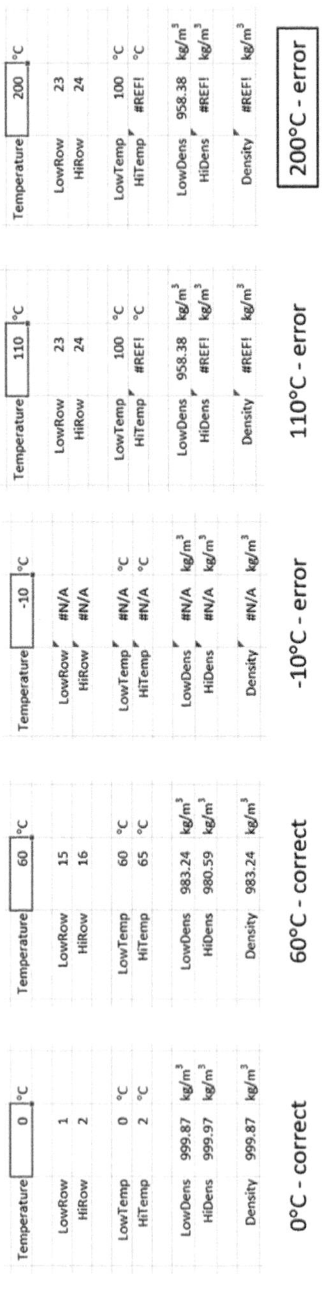

Temperature	0	°C
LowRow	1	
HiRow	2	
LowTemp	0	°C
HiTemp	2	°C
LowDens	999.87	kg/m³
HiDens	999.97	kg/m³
Density	999.87	kg/m³

0°C - correct

Temperature	60	°C
LowRow	15	
HiRow	16	
LowTemp	60	°C
HiTemp	65	°C
LowDens	983.24	kg/m³
HiDens	980.59	kg/m³
Density	983.24	kg/m³

60°C - correct

Temperature	-10	°C
LowRow	#N/A	
HiRow	#N/A	
LowTemp	#N/A	°C
HiTemp	#N/A	°C
LowDens	#N/A	kg/m³
HiDens	#N/A	kg/m³
Density	#N/A	kg/m³

-10°C - error

Temperature	110	°C
LowRow	23	
HiRow	24	
LowTemp	100	°C
HiTemp	#REF!	°C
LowDens	958.38	kg/m³
HiDens	#REF!	kg/m³
Density	#REF!	kg/m³

110°C - error

Temperature	200	°C
LowRow	23	
HiRow	24	
LowTemp	100	°C
HiTemp	#REF!	°C
LowDens	958.38	kg/m³
HiDens	#REF!	kg/m³
Density	#REF!	kg/m³

200°C - error

FIGURE 4.42 Testing the linear interpolation spreadsheet calculation.

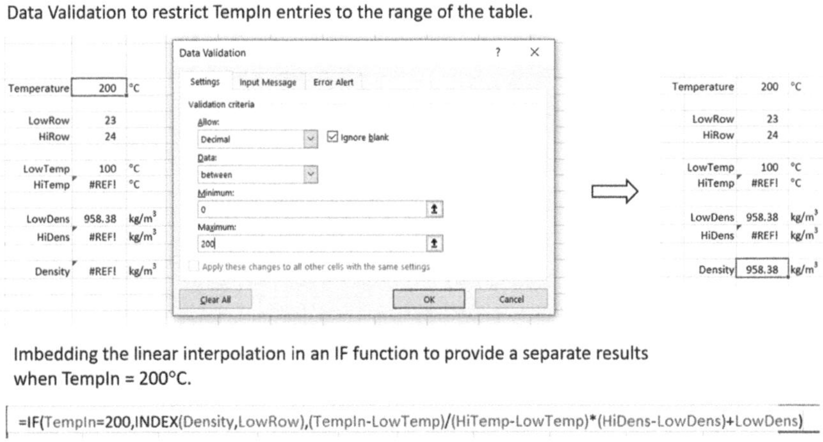

Data Validation to restrict TempIn entries to the range of the table.

Imbedding the linear interpolation in an IF function to provide a separate results when TempIn = 200°C.

=IF(TempIn=200,INDEX(Density,LowRow),(TempIn-LowTemp)/(HiTemp-LowTemp)*(HiDens-LowDens)+LowDens)

FIGURE 4.43 Adding Data Validation and an IF function to accommodate 200°C.

−10°C and 210°C. With one exception, the results are as expected. That one is highlighted on the right side of the figure, where 200°C generates an error because the scheme looks for the next row which is outside the table's range.

There are two enhancements we consider at this point to complete this example:

1. Correcting the error produced by the 200°C TempIn value
2. Restricting the TempIn values to the range of the table

Figure 4.43 shows the changes that are made.

Even though the Density result for a TempIn value of 200°C is now correct, it is disconcerting that two #REF! errors still appear on the spreadsheet. Although the calculation now performs acceptably, it is limiting in a way because, in the case of multiple property tables in an overall spreadsheet, we would have to create a dedicated interpolation set of formulas for each table. This would even be required if we needed to repeat interpolations of the same table at different locations on the spreadsheet. As we will see in Chapter 11, it will be possible to "elevate" this calculation into a flexible user-defined INTERP function using VBA.

4.6 PIVOT TABLES AND PIVOT CHARTS

Pivot tables and charts are a special feature of Excel that have proven useful to many engineers and scientists. They were designed primarily for business/financial applications, but, as with many other features of Excel, engineers and scientists have adapted them for their use.

The typical scenario for employing a pivot table is a collection of information that may consist of text and numerical data involving many factors including categorical input data and numerical results. An example is illustrated in Figure 4.44 that represents an expansion of our former pipe dimension tables.[2] In this case, there are columns of different quantities. This is different from the arrangement

	A	B	C	D	E	F	G	H	I	J	K
1								SI Units			
2		Nominal (in)	Outside Diameter (in)	Schedule	Inside Diameter (in)	Wall Thickness (in)	Flow Cross-sectional Area (ft²)	Outside Diameter (m)	Inside Diameter (m)	Wall Thickness (m)	Flow Cross-sectional Area (m²)
3		1/2	0.84	5	0.710	0.065	2.75E-03	2.13E-02	1.80E-02	1.65E-03	2.55E-04
4		1/2	0.84	10	0.674	0.083	2.48E-03	2.13E-02	1.71E-02	2.11E-03	2.30E-04
5		1/2	0.84	40	0.622	0.109	2.11E-03	2.13E-02	1.58E-02	2.77E-03	1.96E-04
6		1/2	0.84	80	0.546	0.147	1.63E-03	2.13E-02	1.39E-02	3.73E-03	1.51E-04
7		1/2	0.84	160	0.464	0.188	1.17E-03	2.13E-02	1.18E-02	4.78E-03	1.09E-04
8		3/4	1.05	5	0.920	0.065	4.62E-03	2.67E-02	2.34E-02	1.65E-03	4.29E-04
9		3/4	1.05	10	0.884	0.083	4.26E-03	2.67E-02	2.25E-02	2.11E-03	3.96E-04
10		3/4	1.05	40	0.824	0.113	3.70E-03	2.67E-02	2.09E-02	2.87E-03	3.44E-04
11		3/4	1.05	80	0.742	0.154	3.00E-03	2.67E-02	1.88E-02	3.91E-03	2.79E-04
12		3/4	1.05	160	0.612	0.219	2.04E-03	2.67E-02	1.55E-02	5.56E-03	1.90E-04
13		1	1.32	5	1.185	0.068	7.66E-03	3.35E-02	3.01E-02	1.71E-03	7.12E-04
14		1	1.32	10	1.097	0.112	6.56E-03	3.35E-02	2.79E-02	2.83E-03	6.10E-04
15		1	1.32	40	1.049	0.136	6.00E-03	3.35E-02	2.66E-02	3.44E-03	5.58E-04
16		1	1.32	80	0.957	0.182	5.00E-03	3.35E-02	2.43E-02	4.61E-03	4.64E-04
17		1	1.32	160	0.815	0.253	3.62E-03	3.35E-02	2.07E-02	6.41E-03	3.37E-04
18		1 1/2	1.90	5	1.770	0.065	1.71E-02	4.83E-02	4.50E-02	1.65E-03	1.59E-03

	A	B	C	D	E	F	G	H	I	J	K
82		18	18.00	160	14.438	1.781	1.14E+00	4.57E-01	3.67E-01	4.52E-02	1.06E-01
83		20	20.00	5	19.624	0.188	2.10E+00	5.08E-01	4.98E-01	4.78E-03	1.95E-01
84		20	20.00	10	19.500	0.250	2.07E+00	5.08E-01	4.95E-01	6.35E-03	1.93E-01
85		20	20.00	40	18.812	0.594	1.93E+00	5.08E-01	4.78E-01	1.51E-02	1.79E-01
86		20	20.00	80	17.938	1.031	1.75E+00	5.08E-01	4.56E-01	2.62E-02	1.63E-01
87		20	20.00	160	16.062	1.969	1.41E+00	5.08E-01	4.08E-01	5.00E-02	1.31E-01
88		24	24.00	5	23.564	0.218	3.03E+00	6.10E-01	5.99E-01	5.54E-03	2.81E-01
89		24	24.00	10	23.500	0.250	3.01E+00	6.10E-01	5.97E-01	6.35E-03	2.80E-01
90		24	24.00	40	22.624	0.688	2.79E+00	6.10E-01	5.75E-01	1.75E-02	2.59E-01
91		24	24.00	80	21.562	1.219	2.54E+00	6.10E-01	5.48E-01	3.10E-02	2.36E-01
92		24	24.00	160	19.312	2.344	2.03E+00	6.10E-01	4.91E-01	5.95E-02	1.89E-01

FIGURE 4.44   Table of pipe dimensions in both U.S. customary and SI units.[3]

of Figures 4.37 and 4.38 where the columns represent different levels of the same quantity, Schedule. Here, it is called *database format*. Since there are several Schedule rows for each Nominal diameter, the table is deeper, extending to row 92.

In addition to Nominal diameter, Schedule, and Inside Diameter in inches, the table includes these quantities in SI units (m). Inside cross-sectional area for flow in both ft² and m² are included. The data in this table are comprehensive, but they may be difficult to manage and interpret. This is where the pivot table feature of Excel may help.

The right-most six columns of the table in Figure 4.44 are created with the following formulas, shown for row 3:

Column	Formula
F	=(C3-E3)/2
G	=PI()*E3^2/4/144
H	=C3*0.0254
I	=E3*0.0254
J	=(H3-I3)/2
K	=PI()*E3^2*6.4516E-4

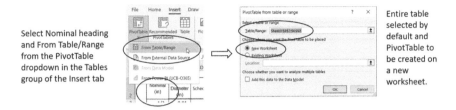

Select Nominal heading and From Table/Range from the PivotTable dropdown in the Tables group of the Insert tab

Entire table selected by default and PivotTable to be created on a new worksheet.

FIGURE 4.45   Creation of a pivot table from the pipe dimensions table.

These formulas can then be copied down using autofill (double-click on the fill handle) to complete the table.

Figure 4.45 shows how to create a pivot table from the basic table of Figure 4.44. This option is available from the Insert tab of the Ribbon in the Tables group. By default, the pivot table is created on a separate worksheet.

The initial view of the pivot table is shown in Figure 4.46. The table is empty and is awaiting selections made in the PivotTable Fields panel. The resulting display is a selective, formatted report of information selected from the complete table, now treated as a database.

In Figure 4.47, we show what happens when the Nominal category is dragged to the Rows field, the Schedule category to the Columns field, and the Cross-sectional Area category to the Values field. You will note that, automatically, a Sum of Nominal Values category is established. This doesn't make much sense for this table and can be changed to other quantities, such as average, by right-clicking the Sum label in the upper lefthand corner of the table and clicking

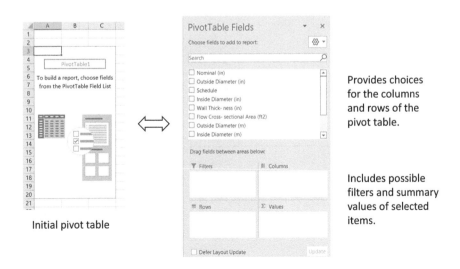

Provides choices for the columns and rows of the pivot table.

Includes possible filters and summary values of selected items.

Initial pivot table

FIGURE 4.46   Initial view of pivot table with PivotTable Fields panel.

Pivot table with initial selections for Row, Column and Value entries

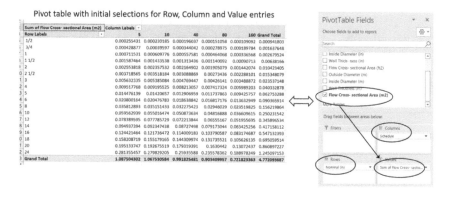

FIGURE 4.47 Creating the pivot table with Row, Column, and Value selections.

Value Field Settings. If there were a third categorical, input variable, we could use that in the Filters field to restrict the display based on particular values of that variable.

If you would rather not see the Grand Total row and column, you can right click each label and select Remove Grand Total. That setting is also available by selecting PivotTable Options by right-clicking anywhere in the selected table. Figure 4.48 shows the removal, and the numerical entries have been reformatted.

Excel also provides a charting facility that goes along with pivot tables. To illustrate that, we introduce another example showing worldwide greenhouse gas emissions by source for the years 2010–2019. This initial table is converted into

3	Sum of Flow Cross- sectional Area (m2)	Column Labels				
4	Row Labels		5	10	40	80	160
5	1/2		2.554E-04	2.302E-04	1.960E-04	1.511E-04	1.091E-04
6	3/4		4.289E-04	3.960E-04	3.440E-04	2.790E-04	1.898E-04
7	1		7.115E-04	6.098E-04	5.576E-04	4.641E-04	3.366E-04
8	1 1/2		1.587E-03	1.434E-03	1.313E-03	1.140E-03	9.071E-04
9	2		2.554E-03	2.358E-03	2.165E-03	1.905E-03	1.442E-03
10	2 1/2		3.719E-03	3.518E-03	3.089E-03	2.734E-03	2.288E-03
11	3		5.632E-03	5.385E-03	4.769E-03	4.261E-03	3.489E-03
12	4		9.518E-03	9.196E-03	8.213E-03	7.417E-03	5.989E-03
13	5		1.448E-02	1.421E-02	1.291E-02	1.174E-02	9.426E-03
14	6		2.080E-02	2.048E-02	1.864E-02	1.682E-02	1.363E-02
15	8		3.581E-02	3.515E-02	3.228E-02	2.946E-02	2.352E-02
16	10		5.956E-02	5.502E-02	5.087E-02	4.817E-02	3.661E-02
17	12		7.839E-02	7.779E-02	7.221E-02	6.555E-02	5.196E-02
18	14		9.494E-02	9.235E-02	8.727E-02	7.917E-02	6.343E-02
19	16		1.244E-01	1.217E-01	1.140E-01	1.038E-01	8.317E-02
20	18		1.582E-01	1.552E-01	1.443E-01	1.317E-01	1.056E-01
21	20		1.951E-01	1.927E-01	1.793E-01	1.630E-01	1.307E-01
22	24		2.814E-01	2.798E-01	2.594E-01	2.356E-01	1.890E-01

FIGURE 4.48 Pivot table reformatted and without Grand Total row and column.

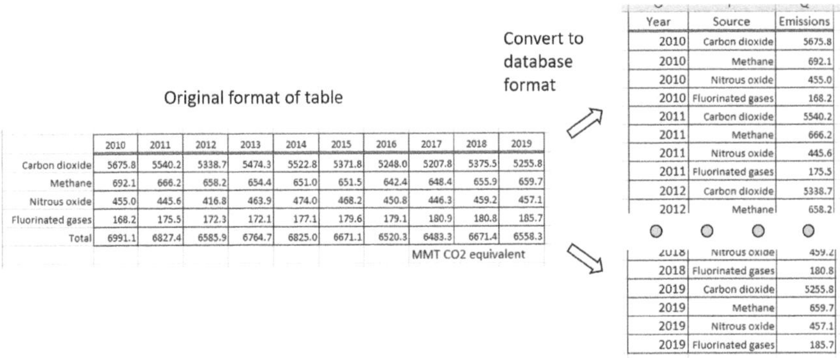

Original format of table

Convert to
database
format

MMT CO2 equivalent

FIGURE 4.49 Creation of database format for greenhouse gas emissions data. (Data from the EPA. https://cfpub.epa.gov/ghgdata/inventoryexplorer/#allsectors/allsectors/allgas/gas/all.)

database form, as shown in Figure 4.49. In creating the database form, the total values are left out because the pivot table will provide those automatically.

When the pivot table is created, it can be completed by choosing the Year for the Rows, the Source for the Columns, and the Emissions for the Values field. This will yield the appearance as shown in Figure 4.50, where the interior values have been reformatted to scientific notation.

Let's say we wanted to limit the pivot table display to years 2015–2019 and leave Carbon dioxide out of the row display. Figure 4.51 shows how this is done. We have flexibility provided in adjusting the displayed table.

Pivot charts are provided to accompany pivot tables. With the pivot chart selected, from the Insert tab of the Ribbon in the Charts group, the PivotChart command is selected, and the PivotChart Fields panel can be arranged as shown in Figure 4.52, and the accompanying bar chart is displayed. The bars are colored but could be changed to black-and-white patterns. The PivotChart Fields and PivotTable Fields panels give us the opportunity to make changes in the pivot chart, and any changes are visible immediately.

Sum of Emissions Column Labels ⊤			
Row Labels ⊤	Fluorinated gases	Methane	Nitrous oxide	Grand Total
2013	1.72E+02	6.54E+02	4.64E+02	1.29E+03
2014	1.77E+02	6.51E+02	4.74E+02	1.30E+03
2015	1.80E+02	6.52E+02	4.68E+02	1.30E+03
2016	1.79E+02	6.42E+02	4.51E+02	1.27E+03
2017	1.81E+02	6.48E+02	4.46E+02	1.28E+03
2018	1.81E+02	6.56E+02	4.59E+02	1.30E+03
2019	1.86E+02	6.60E+02	4.57E+02	1.30E+03
Grand Total	1.26E+03	4.56E+03	3.22E+03	9.04E+03

FIGURE 4.50 Pivot table for greenhouse emissions data.

Row Labels	2015	2016	2017	2018	2019	Grand Total
Fluorinated gases	179.6	179.1	180.9	180.8	185.7	906.1
Methane	651.5	642.4	648.4	655.9	659.7	3258.0
Nitrous oxide	468.2	450.8	446.3	459.2	457.1	2281.7
Grand Total	1299.3	1272.3	1275.5	1296.0	1302.5	6445.7

FIGURE 4.51   Restricting the pivot table display, removing years 2010–2014 and carbon dioxide.

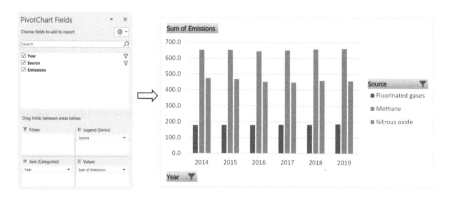

FIGURE 4.52   Pivot chart from the pivot table in Figure 4.51 created with the PivotChart Fields panel.

Pivot tables and charts are most attractive for scenarios where there are multiple categorical variables and one or more numerical response variables. With some practice, one can use these tables and charts for presentations and make adjustments on the fly in response to audience requests. On the other hand, for many, most, engineering and science applications, pivot tables and charts are not the best choices. When we need to extract and interpolate information from numerical tables and make typical engineering plots, we should stick with conventional tables. Lists and Excel tables also find use when we need to sort, filter, and make column or row calculations with tables of data.

PROBLEMS

4.1 Sort the table in Figure 4.44 first by Schedule and then by Nominal diameter. Document your sorted table.

4.2 Create a list from the table in Figure 4.44 and use it to create the sort requested in Problem 4.1.

4.3 For the data in Figure 4.39, create an adjacent column that displays the ranks of the density values.

4.4 Figure P4.4 presents a table of enthalpy values (kJ/kg) of aqueous sodium hydroxide (NaOH) solutions at different temperatures and different concentrations of NaOH (by mass fraction) (Olsson et al., 1997).

   a. Create a linear interpolation scheme for values of mass fraction given a temperature that matches a column of the table.

   b. Create a linear interpolation scheme for values of temperature for a given mass fraction that matches a row of the table.

   c. Challenge problem: Create a two-way interpolation scheme that approximates an enthalpy value for values of mass fraction and temperature within the ranges of the table.

   Hint: Use lookup functions to determine the four table values ("interpolation square") that surround the mass fraction – temperature location. Next, interpolate the two vertical pairs by mass fraction, and then the resulting two horizontal values by temperature.

	A	B	C	D	E	F	G	H	I
1						Temperature			
2		°F	50.00	90.00	150.00	200.00	250.00	300.00	350.00
3		°C	10.0	32.2	65.6	93.3	121.1	148.9	176.7
4		0.00	41.9	134.9	279.1	395.4	518.7	616.4	744.3
5		0.05	41.9	127.9	267.5	372.2	488.5	597.8	711.8
6		0.10	39.5	121.0	251.2	353.6	465.2	562.9	674.5
7		0.15	37.2	114.0	237.3	341.9	446.6	553.6	651.3
8		0.20	39.5	111.6	230.3	330.3	435.0	546.6	644.3
9		0.25	46.5	114.0	237.3	332.6	437.3	548.9	642.0
10		0.30	51.2	130.3	251.2	346.6	444.3	553.6	649.0
11		0.35	88.4	158.2	279.1	376.8	469.9	569.9	655.9
12		0.40	148.9	190.7	318.7	411.7	511.7	602.4	697.8
13		0.45	172.1	260.5	367.5	455.9	546.6	644.3	739.7
14		0.50	260.5	302.4	418.7	511.7	611.7	690.8	786.2
15								Entries: enthalpy in kJ/kg	

(Left side label for rows 6–12, column A: Mass Fraction NaOH)

FIGURE P4.4 Enthalpy data for NaOH solutions.

4.5 For the data in Figure P4.4, create an Excel Table that displays in the row below the table the average of the individual column values.

4.6 Table P4.6 presents data of the viscosity of water at different temperatures. Create a spreadsheet scheme to provide a linear interpolation of viscosity based on a temperature input value within the range of the table.

4.7 Given the table in Figure P4.4, create first a database table then a pivot table with rows of mass fraction and columns of temperature.

   a. Select for display only mass fractions at even intervals of 0.10 and only temperatures between 50°C and 150°C.

   b. Based on the display from part (a), add a pivot chart that displays a bar chart of enthalpy versus mass fraction with bars grouped by temperature.

TABLE P4.6

Viscosity of Water (Pa·s)[4] versus Temperature (°C)

Temperature (°C)	Viscosity (Pa·s)
0.0	1.79E-03
4.4	1.55E-03
10.0	1.31E-03
15.6	1.13E-03
21.1	9.82E-04
26.7	8.62E-04
32.2	7.64E-04
37.8	6.82E-04
48.9	5.59E-04
60.0	4.70E-04
71.1	4.01E-04
82.2	3.47E-04
93.3	3.05E-04

4.8 Pivot table and chart project.

Via an Internet search, acquire data on annual energy generation by source (oil, natural gas, coal, nuclear, renewable) for the countries below and for a recent five-year period. Create a database table of these data and then a pivot table. Demonstrate various selective displays and create more than one pivot chart to depict the data.

Australia	India
Brazil	Japan
Canada	Sweden
China	United Kingdom
France	United States

4.9 Because of its geography and meteorology, tornadoes are more common in the United States than in any other country. As depicted in Figure P4.9, most tornadoes in the United States occur in *Tornado Alley*, a colloquial term for this area that is particularly prone to tornadoes. Table P4.9 lists the average number of tornadoes per state by month from 1989 through 2013. Using concepts from this chapter:

a. Sort the table by columns in descending order based on the number/ area row.

b. Then, sort the rows in descending order based on the values in the TOTAL columns.

c. Based on the final sorted table and the map in Figure P4.9, speculate how rising air temperatures might influence tornadoes in the coming decades.

FIGURE P4.9 Map of the contiguous United States; the shaded area is Tornado Alley along with the individual states and their abbreviations. This is a public site that facilitates downloading of data and images without restrictions. (From https://worldpopulationreview.com/state-rankings/tornado-alley-states.)

TABLE P4.9
The Average Number of Tornadoes per State by Month from 1989 through 2013

	IA	KA	ND	NE	OK	SD	TX	TOTAL
Jan	0	0	0	0	0	0	5	5
Feb	0	1	0	0	0	0	3	4
Mar	2	5	0	2	5	0	11	25
Apr	8	13	0	5	14	0	29	69
May	11	36	3	16	27	7	43	143
Jun	15	20	12	20	7	14	24	112
Jul	6	7	10	6	1	6	3	39
Aug	3	3	4	3	1	3	4	21
Sep	2	3	0	1	2	1	5	14
Oct	1	3	0	3	3	1	8	19
Nov	1	1	0	0	1	0	7	10
Dec	0	0	0	0	0	0	4	4
TOTAL (#)	49	92	29	56	61	32	146	465
Area	206.7	54.2	64.6	60.7	59.9	44.1	45.1	490
#/area	0.24	1.70	0.45	0.92	1.02	0.73	3.24	0.95

Source: https://worldpopulationreview.com/state-rankings/tornado-alley-states

NOTES

1. In more recent versions of Excel, there are two additional functions, RANK.AVG and RANK.EQ. These are suggested to replace the earlier RANK function, which is still available. RANK.AVG returns the average rank for duplicate entries, and RANK.EQ returns the top rank for duplicate entries. The RANK function assigns the same value for duplicate entries, but subsequent rank values will be skipped.
2. Downloadable pipe dimension data are available from numerous sources on the Internet. Most are not protected by copyright.
3. Standard pipe sizes outside the U.S. are different. They are typically classified by nominal diameters in millimeters. Canada still uses U.S. pipe conventions, calling them "Imperial." So does Mexico.
4. The pascal (Pa) is the SI unit for pressure equivalent to newtons per square meter (N/m^2). A common unit in the United States for viscosity is the poise or centipoise (cP). The relationship is 1 cP = 0.001 Pa·s.

5 Case Studies and Targeting

CHAPTER OBJECTIVES

- Learn how to create an Excel data table and to interpret the TABLE array function
- Be able to conduct one-way case studies using data tables including line plots
- Learn how to extend data tables to two-way case studies including surface and contour plots
- Explore the use of the Goal Seek method to solve targeting scenarios
- Learn how to use Excel's Solver to solve targeting problems, including multivariable scenarios, constraints, and optimization

In this chapter, we introduce two essential methods of problem solving with Excel. Although engineers and scientists who have used spreadsheets have often used targeting techniques, also called back-solving, many are unfamiliar with data tables. Introducing the latter will make an important contribution to their Excel skills.

Excel's data tables, which use the TABLE array function, provide an extension to Excel's calculation capabilities that allow us to carry out sensitivity or case studies in a surprisingly efficient fashion. This facilitates our ability to test spreadsheet calculations and share with others, often in response to their questions, table results that elucidate the dependency of results on variations in key input variables.

Here, we will introduce data tables and then show their application in one-way, or one-variable case studies, and then two-way studies. Usually, we find it convenient to depict the results of case studies graphically, so plotting results will be included.

One of the time-tested features of spreadsheets is their "live" calculations. In such *what if?* applications, we can change the values in cells and observe the effects of the change instantly. A key feature here is that the results are visible, but the formulas are underlying. A common practice is to adjust one or more input cell values in a trial-and-error manner to obtain a desired result in one or more output cells. We would term this manual targeting. In this chapter, we will introduce automated targeting using the Goal Seek and Solver features of Excel. The Solver is more versatile than Goal Seek including the ability to manipulate multiple inputs, accommodate constraints, and seek maxima or minima instead

DOI: 10.1201/9781003361053-5

of a target value. Targeting can also be seen in the context of numerical methods for solving systems of equations and curve-fitting (nonlinear regression). These will be dealt with in Chapters 7 and 8.

5.1 DATA TABLES AND THE TABLE FUNCTION – ONE-WAY CASE STUDIES

The concept of an Excel data table is illustrated in Figure 5.1. The spreadsheet calculation referenced in the figure could be a single formula or an extensive calculation involving many formulas. We pick one input to the calculation and establish a range of values we would like to use in that calculation to produce a corresponding column of output values. The data table operation is "live," meaning that any changes in the spreadsheet calculation, such as changing relevant parameter values, will be reflected automatically in the output range and any corresponding plots.

We will use a simple, initial example to show how a data table works. The basis for the calculation is depicted in Figure 5.2, a formula to compute the volume of

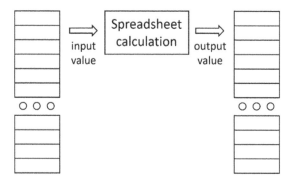

FIGURE 5.1   Data table concept.

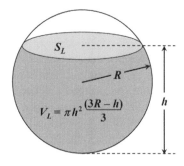

$$V_L = \pi h^2 \frac{(3R - h)}{3}$$

FIGURE 5.2   Volume of liquid in a spherical tank.

	A	B	C	D
1				
2		Radius	5.0 m	
3		Depth	1.5 m	
4		Volume	31.8 m³	

⟺

	A	B	C	D
1				
2		Radius 5		m
3		Depth 1.5		m
4		Volume =PI()*Depth^2*(3*Radius-Depth)/3		m³

FIGURE 5.3 Example calculation of liquid volume in spherical tank.

liquid, V_L, in a spherical tank of inside radius, R, and centerline liquid depth, h. We would like to set up a data table to compute the liquid volumes for a range of depths from 0 to $2R$.

The first step is to set up an example calculation of volume on the spreadsheet. This is shown in Figure 5.3 including the formula to compute the volume.

Next, we set up a table with a column of candidate values for the depth of liquid and, in the adjacent column, a formula that points to the Volume cell on the spreadsheet. This is shown in Figure 5.4. An Index column is created with values from 0 to 100. Adjacent to it, there is a Depth column that has formulas to compute a fractional value of 2*Radius so that we have 101 values of Depth, including zero. To the right of the Depth column, there is a Volume column that contains a formula pointing to the result Volume cell. This formula is located in the row *just above* the Index and Depth columns.

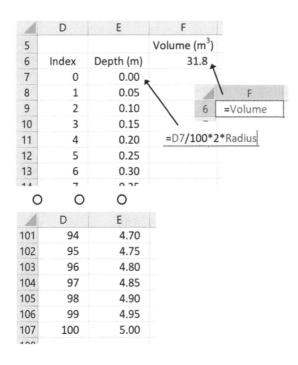

FIGURE 5.4 Setting up the data table for depths and volumes.

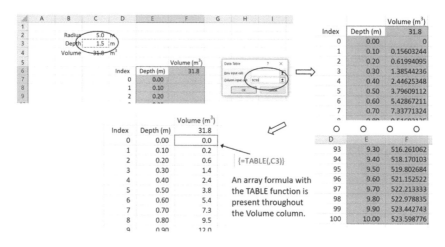

FIGURE 5.5   Completion of the data table, illustrating the resulting TABLE array formula.

To complete the data table, we select the **Depth** and **Volume** columns including the "rule" or pointer formula below the Volume (m³) heading. From the Data tab of the Ribbon, in the Forecast group, and the What-If Analysis drop-down menu, we select the Data Table… command. In the Column input cell: field, we select the **Depth** cell (or $C$3) and click OK. This procedure and the results are shown in Figure 5.5.

There is an array formula, {=TABLE(,C3)}, that is present throughout the column of Volume results. This formula is generated automatically by the Data Table command and cannot be edited other than by deleting it. The Volume values in the final table displayed in Figure 5.5 have been reformatted to show only one decimal place.

The data table is "live." If we change the Radius value from 5 to 6.5, the results update immediately. This is illustrated in Figure 5.6.

We can expand this one-way data table to include another calculation related to the spherical tank. One possibility would be to include the surface area of the tank associated with the liquid segment, S_L. The formula for this is

$$S_L = 2\pi R h \tag{5.1}$$

If we add this to the base calculation of the spreadsheet, we can re-establish the data table by including a pointer formula in a second column and following the procedure to create the data table, now including the second column. This is illustrated in Figure 5.7.

Commonly, the results of a data table are depicted in plots. Figure 5.8 shows a plot of the results from the previous figure using a secondary axis for the surface area. This follows the plotting techniques introduced in Chapter 2. Notice that the Liquid Surface Area is linearly related to Liquid Depth whereas Liquid

Radius	6.5	m
Depth	1.5	m
Volume	42.4	m^3

			Volume (m^3)
	Index	Depth (m)	42.4
	0	0.00	0.0
	1	0.13	0.3
	2	0.26	1.4
	3	0.39	3.0
	4	0.52	5.4
	5	0.65	8.3
	6	0.78	11.0
	○	○	○
	93	12.09	1134.2
	94	12.22	1138.4
	95	12.35	1142.0
	96	12.48	1145.0
	97	12.61	1147.3
	98	12.74	1149.0
	99	12.87	1150.0
	100	13.00	1150.3

FIGURE 5.6 Data table with modified tank radius.

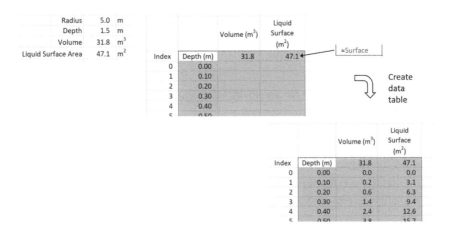

FIGURE 5.7 Expansion of the data table to two columns: Volume and Liquid Surface Area.

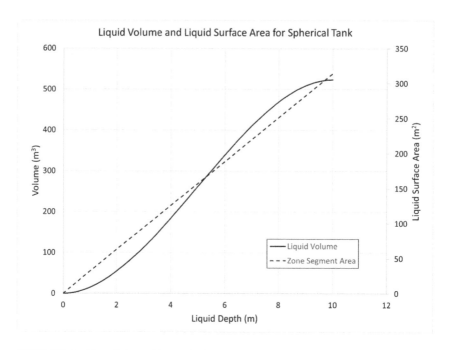

FIGURE 5.8   Plot of data table results for spherical tank.

Volume is related to depth with an S-curve. Although the former is apparent from Equation 5.2, whereas the latter might not be obvious from a cursory examination of the volume formula in Figure 5.1. Such is an advantage of data tables with plots.

There are other possibilities for our one-way data table. We could include columns in our study with different tank radii by placing a radius value at the top of the column and adjusting the pointer formula to a tank formula including that value. For now, we will move on to expanded data tables in terms of two variables – two-way case studies.

5.2   DATA TABLES FOR TWO-WAY CASE STUDIES

The strategy of the two-way data table is illustrated in Figure 5.9. There are ranges of two input variables, one in column format and the other as a row of values. These do not have to be the same length. The data table computes all intersections or combinations of the two inputs and places the results in a rectangular table below and to the right of the inputs.

We will use an example from chemistry and the behavior of gases to illustrate a two-variable data table. From your background, you are aware of the ideal gas law,

$$P = \frac{RT}{\hat{V}} \tag{5.2}$$

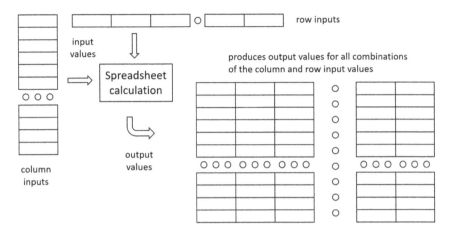

FIGURE 5.9 Two-variable data table strategy.

where P = pressure, $\hat{V}$ = specific volume, R = the gas law constant, and T = the absolute temperature. Typical units might be

$$P[=]\text{kilopascals}(\text{kPa}) \quad \hat{V}[=]\text{liters/mole of gas}(\text{L/mol}) \quad T[=]\text{kelvins}(\text{K})$$

for which $R \cong 8.3145 \dfrac{\text{L} \cdot \text{kPa}}{\text{mol} \cdot \text{K}}$.

When the conditions are extreme, particularly at high pressures, the ideal gas law becomes inaccurate, and modifications are required. One of the classic modified gas laws is the *van der Waals equation,*

$$P = \frac{RT}{\hat{V} - b} - \frac{a}{\hat{V}^2} \tag{5.3}$$

where a and b are parameters, whose values depend on the gas under study.

Here, the values for a and b for carbon dioxide, CO_2, are:

$$a = 365.4 \ \text{L}^2 \cdot \text{kPa/mol}^2, \ b = 0.04280 \ \text{L/mol}$$

We would like to investigate the behavior of pressure, P, for different values of temperature, T, and specific volume, $\hat{V}$. First, we establish the calculation of a single pressure value for example values of temperature and specific volume. This is shown in Figure 5.10. The formula that computes the pressure is included along with an imbedded typeset equation to enhance the documentation.

Next, we decide to create a case study for ranges of temperature from 200 to 500 K and specific volume from 1 to 5 L/mol. The arrangement of the two-way table is illustrated in Figure 5.11. Here, we have chosen temperature for the column values and specific volume for the row values. The rule governing the data table is to be located in the upper left corner, as shown.

▲	A	B	C
1			
2	Rgas	8.3145	L·kPa/mol·K
3	a	365.4	L^2·kPa/mol^2
4	b	0.0428	L/mol
5			
6	T	300	K
7	Vhat	2.5	L/mol
8			=Rgas*T/(Vhat-b)-a/Vhat^2
9	P	956.7	kPa
10		$P = \dfrac{RT}{\hat{V}-b} - \dfrac{a}{\hat{V}^2}$	
11			
12			

FIGURE 5.10   Spreadsheet example calculation of van der Waals equation.

		Specific Volume (L/mol)								
		1.0	1.5	2.0	2.5	3.0	3.5	4.0	4.5	5.0
Temperature (K)	200									
	225									
	250	Place "rule",								
	275	pointer formula								
	300	to P here.								
	325									
	350									
	375									
	400									
	425									
	450									
	475									
	500									

FIGURE 5.11   Arrangement for two-way data table.

Since we have our test calculation resulting in the pressure in the cell named P, the rule formula will be simply =P. Following that, we select the entire table and initiate the data table command, as before. The procedure is shown in Figure 5.12. This time, in the Data Table dialog box, we enter Vhat for the Row input cell: and T and the Column input cell:. Upon clicking the OK button, the resulting live data table is presented in Figure 5.13.

If we changed the *a* and *b* parameters to different values for a different gas, the data table would update immediately. Just as with one-way data tables, we want to get a picture of the data graphically. We can produce a two-dimensional plot with separate curves for each column (or row) of the table, but often a three-dimensional rendering is preferred, either in the form of a surface or contour plot with colors or mesh style. As we illustrated in Chapter 2, surface and contour plots are designed for categorical independent variables, but, with some work, we can produce acceptable results for the data table in Figure 5.13. Both mesh surface and contour plots are presented in Figure 5.14.

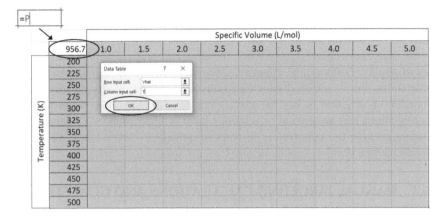

FIGURE 5.12 Procedure to create the two-way data table.

956.7	Specific Volume (L/mol)								
	1.0	1.5	2.0	2.5	3.0	3.5	4.0	4.5	5.0
200	1371.9	978.8	758.3	618.3	521.7	451.2	397.4	355.0	320.8
225	1589.0	1121.4	864.5	702.9	592.0	511.3	449.9	401.7	362.8
250	1806.2	1264.1	970.7	787.5	662.3	571.4	502.4	448.3	404.7
275	2023.3	1406.7	1076.9	872.1	732.6	631.5	555.0	494.9	446.6
300	2240.5	1549.3	1183.1	956.7	802.9	691.7	607.5	541.6	488.6
325	2457.6	1692.0	1289.3	1041.2	873.2	751.8	660.0	588.2	530.5
350	2674.8	1834.6	1395.5	1125.8	943.5	811.9	712.5	634.8	572.4
375	2892.0	1977.3	1501.7	1210.4	1013.8	872.0	765.1	681.5	614.4
400	3109.1	2119.9	1607.9	1295.0	1084.0	932.2	817.6	728.1	656.3
425	3326.3	2262.6	1714.1	1379.6	1154.3	992.3	870.1	774.8	698.2
450	3543.4	2405.2	1820.3	1464.2	1224.6	1052.4	922.7	821.4	740.1
475	3760.6	2547.9	1926.5	1548.8	1294.9	1112.5	975.2	868.0	782.1
500	3977.7	2690.5	2032.7	1633.4	1365.2	1172.7	1027.7	914.7	824.0

(Temperature (K) labels the rows.)

FIGURE 5.13 Completed two-way data table.

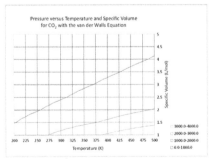
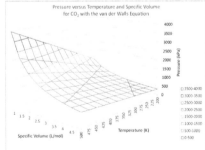

FIGURE 5.14 Contour and surface plots of data table interior.

These plots are best seen in color and are difficult to interpret, other than qualitatively, when presented in black and white. Plots associated with data tables are often used in presentations, either live with Excel or with PowerPoint slides.

In analyzing engineering projects, the ability to prepare multiple case studies to display results can be effective, especially when including economic considerations. For example, displaying results such as profitability measures as a function of variation in key parameters can be extremely effective in making presentations to management or venture capital funding sources. Dynamic case studies provide the opportunity to respond quickly to queries such as "What would happen if...?"

5.3 MANUAL TARGETING CALCULATIONS AND GOAL SEEK

The general concept of a single-input, single-output, targeting calculation on a spreadsheet is that the calculation leads to an output value, and we identify an input value to the calculation. We would like to adjust the input value to achieve a desired output called a target value. This is an iterative process as illustrated in Figure 5.15.

We will illustrate three approaches to solving targeting calculations:

1. Manual targeting
2. Goal Seek
3. Solver

The latter two are built-in features of Excel. Each of these does not provide a "live" result; in other words, each time we want the targeted result, we need to perform a procedure. In Chapter 11, we will illustrate how to use VBA to create "live" targeting solutions.

One approach that might be considered is to restructure the spreadsheet calculation so that the output value becomes the input value and vice versa. This requires solving for the input in terms of the output. It may be possible to accomplish this analytically in some cases, but not often. Otherwise, it will be required to employ numerical methods to solve one or more equations, often nonlinear. In the "spreadsheet approach" to problem solving, we generally try to use the layout of the calculation that is natural and use targeting to achieve the goal.

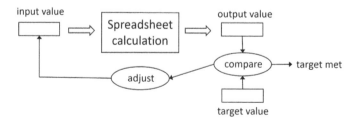

FIGURE 5.15 Targeting scheme for single-input, single-output spreadsheet calculation.

There are variations of the simple scheme in Figure 5.15 that we will consider. One that is common is to replace the target value with a goal such as maximization or minimization. Another is to consider varying more than one input variable, and thirdly, there might be constraints on input, intermediate, and output values that must be enforced. As we will see, as we move toward more complicated targeting scenarios, the Solver becomes the tool of choice.

If we need to conduct a targeting calculation once or infrequently, there may be no need to use a fancy tool for this. We can use a simple trial-and-error approach. An example to illustrate this is the *van der Waals equation* where we specify the temperature and vary the specific volume to achieve a target pressure of 1,000 kPa. Figure 5.16 illustrates this.

You will note that with four "educated guesses" we have obtained the target within one part in 10,000 – usually enough precision to meet our needs. A byproduct of the trial-and-error procedure is that we develop a "feel" for the relationship between the input and the output. Is the relationship of change direct or reverse? Here, it is reverse. How sensitive is the output value to changes in the input value? Here, changing the specific volume by 0.1 causes a change of about 40 kPa in the pressure.

Manual targeting is an excellent on-the-spot method and is effective, as long as great precision is not required and the calculation doesn't have to be repeated frequently. If this is not the case, we can move on to the other targeting tools provided by Excel.

Goal Seek is available on the Data tab of the Ribbon in the Forecast group and on the What-If Analysis drop-down menu (same location as Data Table). The Goal Seek dialog box is illustrated in Figure 5.17.

Note in Figure 5.17 that the final value of Vhat is 2.386804, and our value from manual targeting was 2.387 are identical to four significant figures. This is not a "live" calculation. If we were to change the temperature to 400 K, we would have to repeat the procedure and implement the Goal Seek dialog box from scratch. That is an unfortunate "feature" of Goal Seek. We have no control over the Goal Seek procedure, and there is no understanding of its method; in other words, Goal Seek is a "black box."

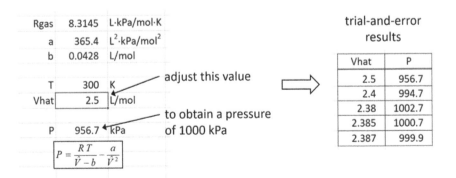

FIGURE 5.16   Manual targeting of the van der Waals equation.

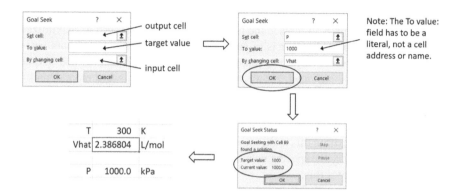

FIGURE 5.17 Completing the Goal Seek dialog box and carrying out the targeting procedure.

The Solver provides more flexibility than Goal Seek, and its methods and parameters are adjustable. It also remembers the last scenario solved, which is convenient when compared to Goal Seek. Along with this comes some complexity. We dedicate the next section to the Solver.

5.4 THE SOLVER

Excel's Solver application is an add-in provided by *Frontline Systems*. Extensive information on the Solver, in addition to the company's other products, is available at www.solver.com. To use the Solver, you must activate it as an add-in. This is done via the Add-Ins dialog box, which is available via File ⇨ Options ⇨ Add-ins ⇨ Go... or the shortcut Alt-T-I.[1] An example dialog box is shown in Figure 5.18 with Solver checked. The other items in the Add-Ins list will vary from one Excel installation to another. We will use the two Analysis Toolpak add-ins in Chapters 8 and 11. You complete the operation by clicking

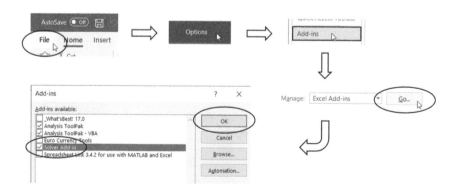

FIGURE 5.18 Activating the Solver add-in.

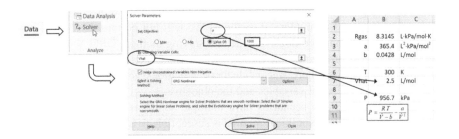

FIGURE 5.19 Setting up the Solver for the van der Waals targeting example.

the OK button. This is confirmed by the Solver command icon appearing in the Analyze group on the Data tab of the Ribbon.

As a first example of using the Solver, we can repeat the *van der Waals equation* targeting that we illustrated manually and with Goal Seek. Figure 5.19 shows this. The Solver Parameters dialog box has many options. Here, we enter P in the Set Objective: field. For the To: field, the Value Of: option is selected, and a literal value of 1,000 is entered. The By Changing Variable Cells: field is set to Vhat. We have a choice of three numerical methods for solution. The default is GRG Nonlinear (Generalized Reduced Gradient), and there are alternates: Simplex LP and Evolutionary. We have accepted the default here. There are other settings behind the Options button that we are not considering yet. To execute the Solver, we click the Solve button.

In this case, the solution result is shown in Figure 5.20. The Solver Results dialog box claims that a solution has been found, and we can confirm that by examining the spreadsheet. The pressure value is indeed 1000 kPa, and the Vhat value has been adjusted accordingly to meet that target. We can accept that solution by clicking OK. In the case the solution is unacceptable (and, yes, that happens), there is the option to Restore Original Values.

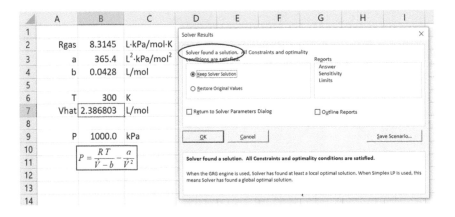

FIGURE 5.20 Solver Results dialog box and solution displayed on the spreadsheet.

In choosing the solving method and setting the Options, we need to consider some details regarding numerical methods. We will not go into these here, just touch on several considerations. The default solving method, GRG Nonlinear, is suitable for most problems. The Simplex LP method is designed for linear programming scenarios where linear, algebraic equations govern the relationship between the independent variables and the objective.

Certain optimization scenarios have multiple solutions, whether minima or maxima, and the GRG Nonlinear method is susceptible to finding a local solution that is not the true or global optimum. In such cases, the Evolutionary method, also called a genetic algorithm, may be preferred. The following example illustrates this.

Example 5.1   Finding the Maximum of the "humps" Function

A specific version of the so-called "humps" function is

$$f(x) = \frac{1}{(x-0.3)^2 + 0.01} + \frac{1}{(x-0.9)^2 + 0.04} - 6 \qquad (5.4)$$

If this function is plotted over the domain $0 \le x \le 1.2$, Figure 5.21 results. It is evident that there are two maxima in the domain with the global maximum at around $x = 0.3$.

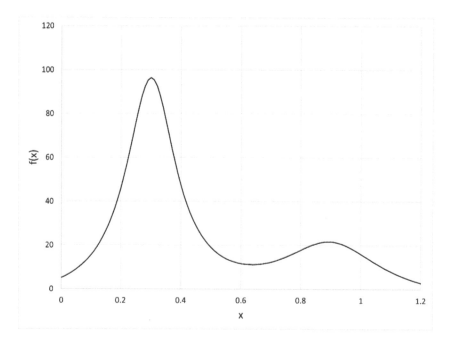

FIGURE 5.21   Plot of a version of the "humps" function.

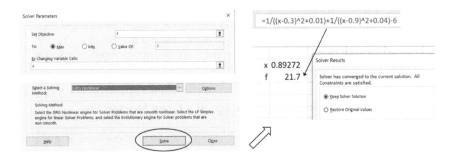

FIGURE 5.22   Solving for a maximum of the "humps" function with the GRG Nonlinear method.

In a first attempt to find the true maximum, we set up the Solver using the GRG Nonlinear method with an initial estimate $x = 0.7$. This is illustrated in Figure 5.22. We can see that the method finds a maximum of 21.7 at $x \cong 0.9$. This is the local maximum to the right in Figure 5.21. If we had used an initial estimate of say $x = 0.5$, the method would find the global maximum at $x \cong 0.3$. Nevertheless, this sensitivity to initial guesses can be problematic.

To employ the Evolutionary method, we must institute constraints within which the method will explore. Creating the constraints is illustrated in Figure 5.23.

Solving for the maximum with the Evolutionary method is illustrated in Figure 5.24. The true maximum within the interval is determined. It is worth mentioning that, in contrast to the GRG Nonlinear method that produces its

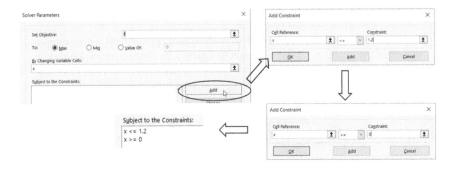

FIGURE 5.23   Establishing constraints, $0 \le x \le 1.2$.

FIGURE 5.24   Evolutionary method solution.

results almost instantaneously, the Evolutionary method is exhaustive and requires around 30 seconds to reach the result. In any case, the example illustrates its advantage when more than one local extremum is present. The following example shows how physical constraints are incorporated in a Solver scenario.

Example 5.2   Optimizing the Design of a Grain Storage Bin

Figure 5.25 depicts a grain storage bin with a cylindrical upper section and a conical base. The formulas associated with the bin volume and surface area, neglecting the top, are included in the figure. The problem to be solved is to determine, for a bin of given total volume, what are the dimensions that minimize the material (surface area) required to build the bin. In formulating the problem, there are constraints that must be incorporated:

- The total volume of the bin is specified
- The conical angle must be less than the angle of repose of the grain[2]

The first of these is an equality constraint, and the second is an inequality constraint. The reason for the second is to keep the conical wall steep enough that the grain will flow out of the bin without "hanging up" or bridging (particles binding together to form an arch above the outlet).[3]

We consider an example with the following specifications where the grain considered is dry corn:

$$V = V_{cyl} + V_{con} = 10 \text{ m}^3$$

$$\phi_{max} = 20.4°$$

To solve this optimization problem, we create a spreadsheet with example values of r, h_{cyl}, and h_{con}. This is shown in Figure 5.26. The formulas are shown in the figure, and the initial values for radius and the two heights do not meet the constraints.

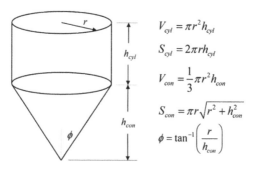

$$V_{cyl} = \pi r^2 h_{cyl}$$

$$S_{cyl} = 2\pi r h_{cyl}$$

$$V_{con} = \frac{1}{3}\pi r^2 h_{con}$$

$$S_{con} = \pi r \sqrt{r^2 + h_{con}^2}$$

$$\phi = \tan^{-1}\left(\frac{r}{h_{con}}\right)$$

FIGURE 5.25   Grain bin with dimensions and formulas.

	A	B	C	D	E	F	G	H	I	J	K	L	M
1	Grain Bin Design												
3		radius	1.000	m		Vcyl	3.142	m³	Scyl	6.283	m²	phi	0.785	radians
4		hcyl	1.000	m		Vcon	1.047	m³	Scon	4.443	m²	phid	45	degrees
5		hcon	1.000	m		V	4.189	m³	S	10.726	m²		

	A	B	C	D	E	F	G	H	I	J	K	L	M
1	Grain											
3		radius 1	m	Vcyl =PI()*radius^2*hcyl	m³	Scyl =2*PI()*radius*hcyl	m²	phi =ATAN(hcon/radius)	radians
4		hcyl 1	m	Vcon =PI()*radius^2*hcon/3	m³	Scon =PI()*radius*SQRT(radius^2+hcon^2)	m²	phid =DEGREES(phi)	degrees
5		hcon 1	m	V =SUM(F3:F4)	m³	S =SUM(I3:I4)	m²	

FIGURE 5.26 Spreadsheet calculation of grain bin volume, surface area, and angle (upper). The underlying formulas (lower).

FIGURE 5.27 The Solver Parameters settings to solve the grain bin problem.

radius	1.572	m		Vcyl	8.486	m³		Scyl	10.794	m²		phi	0.356	radians
hcyl	1.093	m		Vcon	1.514	m³		Scon	8.286	m²		phid	20.4	degrees
hcon	0.585	m		V	10.000	m³		S	19.081	m²			

Solver Results

Solver found a solution. All Constraints and optimality conditions are satisfied.

FIGURE 5.28 Optimization results for the grain bin design.

We can employ the Solver for this problem by specifying the total surface, S, as the objective, the minimization option, by manipulating the radius and the two heights, and incorporating the two constraints. The Solver Parameters dialog window is shown in Figure 5.27. Notice that the volume constraint is an equality, and the angle constraint is an inequality (≤).

When the Solve button is selected, the solution appears quickly. This is illustrated in Figure 5.28. You will notice that both constraints are met. Even though the angle constraint is an inequality, the solution is at the constraint boundary.

As a last consideration for the Solver, we will look at the Options dialog window that allows further adjustments to the solving methods. After the Options

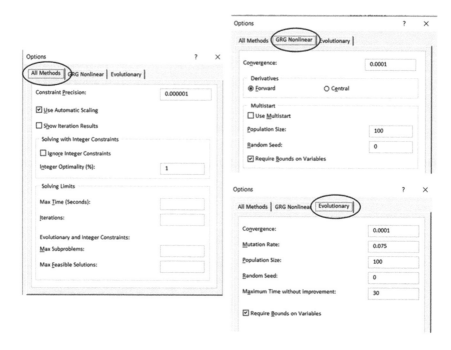

FIGURE 5.29 The Solver Options window tabs.

button is clicked in the Solver Parameters dialog window, the Options window appears with three tabs, as shown in Figure 5.29. We highlight the following:

On the All Methods tab:
- the Constraint Precision field is set by default at one part in a million. This specifies how closely constraints will be met, and
- the Use Automatic Scaling option is set by default and provides scaling of multiple independent variables (By Changing Variable Cells:) that assists in making methods more efficient.

On the GRG Nonlinear tab:
- the Convergence field is set by default at 0.0001, one part in 10,000, and may often be adjusted to a smaller value to require tighter convergence to the solution, and
- the Multistart frame allows for multiple attempts at the solution that may assist in finding a true optimum, such as in Example 5.1. See Problem 5.6.

Further study of the Solver requires delving into the nature of its numerical methods. The solver.com website is helpful here and provides references to the methods used.

It is possible to automate operation of Goal Seek and the Solver with Excel's Visual Basic for Applications (VBA) programming language. We will consider that for Goal Seek in Chapter 9 and as Problem 12.2 in Chapter 12.

PROBLEMS

5.1 Create the following data tables to carry out case studies of the grain bin scenario from Example 5.2.

a. For a radius of 1.6 m and a cylinder height of 1.1 m, vary the conical height from 0.4 to 0.8 m in steps of 0.01 m and display columns of the total surface area, total volume, and angle in degrees. Provide plots of the three columns versus conical height. Combine plots where feasible and use secondary vertical axes if appropriate.

b. For a cylinder height of 1 m, create a two-variable data table of surface area for conical height from 0.4 to 0.8 m in steps of 0.05 m and radius from 1 to 2 m in steps of 0.1 m. Create a surface mesh plot of the interior of the table with appropriately labeled axes and tick labels.

5.2 An interesting comparison is that of the *van der Waals equation* of state and the ideal gas law. Create a data table to compare pressure predicted by these in two columns with the following parameters:

- carbon dioxide, CO_2
- **temperature:** 400 K
- **specific volume:** 3 to 0.1 L/mol in steps of −0.1 L/mol

Provide a plot of pressure versus specific volume with a curve for the *van der Waals equation* of state and another for the ideal gas law. The plot should include a legend.

5.3 The *Haaland equation* is used to predict the *Moody friction factor*, f_M, for turbulent fluid flow in pipes.

$$f_M = \frac{1}{\left(-1.8\log_{10}\left[\left(\frac{\varepsilon}{3.7D}\right)^{1.1} + \frac{6.9}{Re}\right]\right)^2}$$

where
ε: pipe roughness, m
D: pipe inside diameter, m
Re: Reynolds number for the fluid and flow rate, dimensionless

The Reynolds number is computed from

$$Re = \frac{\rho u D}{\mu}$$

where

ρ: fluid density, kg/m^3

u: average fluid velocity, m/s

μ: fluid viscosity, Pa·s

The Reynolds number has a typical range of 10,000 to 1,000,000.

Create a two-variable data table to predict f_M for a 6-inch, Schedule 40 pipe with $D = 6.065$ inches $\cong 0.1541$ m. Employ the following values for ε and Re:

ε:	carbon steel	4.5×10^{-5} m
	drawn tubing	1.5×10^{-6} m
	cast iron	2.6×10^{-4} m
	galvanized iron	1.5×10^{-4} m
	glass, plastic	0 m
Re:	10,000 to 100,000 in steps of 10,000	

Provide a line plot with curves for each material.

5.4 The volume of a liquid stored in a spherical tank of radius, R, is given in terms of the center-line depth of the liquid, h, by

$$V = \pi h^2 \frac{(3R - h)}{3}$$

For a tank of radius, $R = 5$ m, use Goal Seek to determine the liquid depth, h, for a volume, $V = 200$ m^3.

5.5 Consider the modified *humps function*:

$$f(x) = \frac{1}{(x - 0.3)^2 + 0.01} + \frac{1}{(x - 0.9)^2 + 0..04} - 26$$

a.   Create a plot of the function similar to that in Figure 5.21. Identify approximate locations of roots of the function.

b.   Demonstrate how you can use Goal Seek with appropriate initial estimates of x to solve for all the roots of $f(x)$.

5.6 In Example 5.1, we demonstrated how the Evolutionary method of the Solver could find the true maximum of the "humps" function over the given domain of x. The GRG Nonlinear method, with an initial estimate of $x = 1.7$, failed to do this. Experiment with the Multistart option of the GRG Nonlinear method to see whether it can find the true maximum from the same initial estimate. If it is successful, compare its execution time with that of the Evolutionary method.

5.7 In purchasing an item like an automobile, there are financing issues. Imagine you want to acquire a Hyundai Ioniq 5 electric vehicle with a selling price of $43,680. You have $10,000 for a down payment and must finance the balance with a bank loan with a term of five years. Your budget allows a monthly payment of $750, so you need to find a source of funding with an interest rate that will make this possible. The formula for loan payment is

$$A = P\frac{i(1+i)^n}{(1+i)^n - 1}$$

where

A: loan payment, typ. monthly
P: amount borrowed
i: interest rate as a fraction, typ. monthly – annual percentage rate/12
n: number of payments, typ. monthly, term of loan in years × 12

Use Goal Seek to determine the monthly interest rate you need to find, and then multiply that by 12 to find the annual rate. Given today's economy, do you think you will be able to find loans at this rate?

5.8 Reconsider a more realistic version of the grain bin design from Example 5.2. Include the lid on the top of the cylindrical segment with a 30-cm concentric hole in the center for attachment of a grain feed duct. Also, consider that a similar duct with a slide valve is connected to the bottom of the conical segment, effectively removing the point from that segment. Formulate the problem and solve for the minimum surface area with the same volume and angle constraints as found in the example.

5.9 Figure P5.9 depicts the cross-section of a trapezoidal, open channel that carries a stream of water for irrigation. For a cross-sectional area of 25 m², use the Solver to determine the dimensions of the channel that minimize the wetted perimeter (both sides and the bottom). Can you generalize your result?

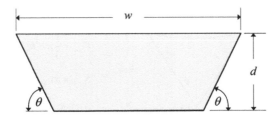

FIGURE P5.9 Cross-section of trapezoidal, open channel.

5.10 Figure P5.10 illustrates a mechanical linkage of four rigid bars with the two lower vertices being fixed. As this linkage is rotated back and forth, the two vertical members and the upper horizontal member move, but their movement is constrained by the linkage.

Following the diagram, as the angle θ_1 is increased from 0 to 180°, the angle θ_2 responds according to the linkage relationship. The equation that constrains the linkage movement is

$$\overline{AD} = \overline{AB}\cos\theta_1 - \overline{CD}\cos\theta_2 + \sqrt{\overline{BC}^2 - \left(\overline{CD}\sin\theta_2 - \overline{AB}\sin\theta_1\right)^2}$$

For the following values, set up a spreadsheet to solve for θ_2 given a value of θ_1 using the Solver. Note: This will not be a "live" calculation, so it won't be possible to use a data table for a range of θ_1 values. Use $\theta_1 = 20°$ for your test case.

$$\overline{AB} = 1.5 \text{ m}, \ \overline{BC} = 6 \text{ m}, \ \overline{CD} = 3.5 \text{ m}, \text{ and } \overline{AD} = 5 \text{ m}$$

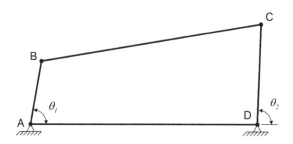

FIGURE P5.10   Four-bar linkage.

5.11 Consider the bin design depicted in Figure P5.11. It has a conical base and a hemispherical top. If the total volume of the bin is to be 5 m³, and the conical angle must be kept below 40° for wheat flour storage, what are the dimensions of the bin that minimize its surface area?

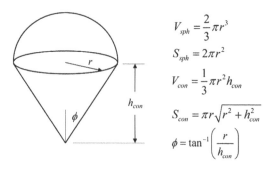

$$V_{sph} = \frac{2}{3}\pi r^3$$

$$S_{sph} = 2\pi r^2$$

$$V_{con} = \frac{1}{3}\pi r^2 h_{con}$$

$$S_{con} = \pi r\sqrt{r^2 + h_{con}^2}$$

$$\phi = \tan^{-1}\left(\frac{r}{h_{con}}\right)$$

FIGURE P5.11   Conical bin with hemispherical cap.

5.12 A production process for three different specialized electronic parts uses three materials: copper, polystyrene, and neoprene. The amounts of these materials required to produce each component are presented in Table P5.11.

    a. If 4 kg of copper, 0.8 kg of polystyrene, and 0.6 kg of neoprene have been ordered, use the Solver to determine how many components can be produced. You will need to incorporate constraints so that the number of components is either positive or zero.

    b. How would the situation change if you had ordered 100 kg of copper, 25 kg of polystyrene, and 9 kg of neoprene?

    c. Repeat parts (a) and (b) but require integer constraints on the number of components. This is one of the drop-down options in the Add Constraint dialog window. Observe and comment on the differences in your results.

TABLE P5.11
Amounts Required of Polymeric Materials

| | Amount Required per Component | | |
| | Copper | Polystyrene | Neoprene |
Component	(g)	(g)	(g)
A	15	3	1.2
B	17	4	1.5
C	19	5.5	1.9

Hint: In setting up the spreadsheet for the Solver, establish three cells for the numbers of components A, B, and C, and enter trial values. Use these and the table values to compute the amounts of copper, polystyrene, and neoprene consumed in manufacturing. Enter the amounts ordered and compute the differences between these and the amounts consumed. Compute the sum of squares of these differences, and have the Solver minimize this sum by adjusting the number of components. Add constraints to the Solver setup to require the numbers of components to be integers.

NOTES

1. Hold down the Alt key while pressing T then I.
2. The *angle of repose* of granular material is the angle of the conical side for a pile of the material poured onto a flat surface like the ground.
3. Because of other factors, such as humidity, clumping, and friction on the inner walls, this still may not be adequate, and it is common to add vibration to the bin base to maintain flow.

6 Financial Calculations

CHAPTER OBJECTIVES

- Understand Excel's basic economic functions and how to use them for financial calculations
- Appreciate a general scheme of financial flow in manufacturing operations
- Be able to set up cash flow spreadsheets to predict project economics over a number of years
- Contrast the different profitability measures and understand how to compute them on the spreadsheet

The origins and the primary users of spreadsheets have always been the business community.[1] Hence, Excel is a very powerful tool for designing, developing, and applying the types of business-oriented computations needed by companies and individuals around the world.

The primary focus of this text is engineering and scientific calculations, not those that are business oriented. However, engineers and scientists inevitably encounter economic analyses in their work, for projects that require funding, cost-benefit analyses, when developing business plans for entrepreneurial start-ups, etc. So, we dedicate a chapter to a brief and selective coverage of financial calculations that we see most often.

To start, we consider Excel's functions that deal with borrowing and lending money. These introduce basic terminology and provide an understanding of how parameter values must be specified. To put the financial functions into context, we will describe the financial flow in manufacturing operations. This will provide the basis for describing and analyzing project economics over a period of years by constructing cash flow tables.

This chapter concludes by introducing profitability measures and how to compute them with Excel's built-in functions. The properties of these measures will be discussed with advice on when to use which one in analyzing projects. Of course, further study on business and project economics is recommended for anyone who will be heavily involved in such calculations.

6.1 BASIC ECONOMIC FUNCTIONS

The concept of earning interest on an investment is fundamental to most financial calculations. In its simplest form, this is expressed as the formula

$$FV = PV + i \cdot PV = (1+i)PV \tag{6.1}$$

DOI: 10.1201/9781003361053-6

where

FV: future value of the investment ($)[2]
PV: present value of the investment ($)
i: interest rate over the period expressed as a fraction

If we carry out, or compound, this process over a number of periods, n, Equation 6.1 becomes

$$FV = (1+i)^n \, PV \qquad (6.2)$$

Equation 6.2 can be inverted to determine the present value, PV, given a future value, FV,

$$PV = \frac{FV}{(1+i)^n} \qquad (6.3)$$

It is easy to implement these formulas on the spreadsheet. Figure 6.1 illustrates these.

Now, we will expand this scenario and consider either income or payments at each period during the term of the investment. This would apply to situations like paying down loans or judging an investment where income is incremental over the term considered. The table presented in Figure 6.2 illustrates this for the latter scenario.

	A	B	C	D	E
1	PV	$10,000		FV	$50,000
2	i	5%		i	10%
3	n	10		n	15
4	FV	$16,289		PV	$11,970

	A	B	C	D	E
1	PV	10000		FV	50000
2	i	0.05		i	0.1
3	n	10		n	15
4	FV	=(1+B2)^B3*B1		PV	=E1/(1+E2)^E3

FIGURE 6.1 Simple interest formulas for future and present value.

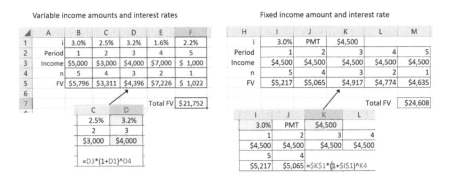

FIGURE 6.2 Future value for income distributed over the periods. Two examples.

There are two examples presented in Figure 6.2. The first allows for variable incomes and interest rates in each period, whereas the second has a fixed income (PMT) and interest rate. In each example, it is assumed that the income is accrued at the beginning of the individual period and the interest is accrued at the end. Also, in each case, the future value is computed based on the number of periods remaining, which varies with each income amount.

The second example in Figure 6.2 is common, and Excel provides a built-in *FV function* to implement the calculation. This is based on the formula

$$FV = PMT \frac{(1+i)^n - 1}{i} \tag{6.4}$$

For Equation 6.4, you will encounter the description and terminology "the future value of an annuity assuming a given discount rate." The syntax for the Excel formula is

FV(i,n,pmt,[pv],[type])

where
    i: interest rate as a fraction
    n: number of periods
    pmt: payment or income (as a negative amount) per period
    [pv]:   initial present value, can be zero
    [type]: pmt occurs at end (type = 0, default) or beginning (type = 1) of the period

In Figure 6.3, this formula is implemented to yield the same result as that calculated with the table of formulas. Notice that the income amount is negated, and the type is set to 1.

Another common application of financial calculations is the *amortization table*. It represents the payments made on an initial loan balance that leads to a zero balance at the end of the term. Figure 6.4 illustrates a typical table in its initial form. A payment of $1,000 has been estimated, and you will note that the ending balance after 60 periods (5 years) is *not* zero.

We can use Goal Seek to adjust the PMT value to drive the final balance to zero. Figure 6.5 shows how this is done. A payment of $1,160 is determined.

The amortization table is useful because it shows the entire profile of the loan term. One can make a plot, for example, of the declining balance versus payment number. It is also simple to compute the total amount and interest paid. In this case, those values are $69,598 and $9,598, respectively.

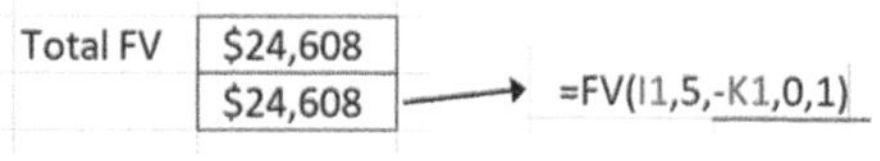

Total FV	$24,608	
	$24,608	→ =FV(I1,5,-K1,0,1)

FIGURE 6.3 Implementing Excel's FV function.

	J	K	L	M	N
1			i	0.5%		
2			PMT	$1,000		Balance
3		Number	Amount	Interest	Apply to Principal	$60,000
4		1	$1,000	$300	$700	$59,300
5		2	$1,000	$297	$704	$58,597
6		3	$1,000	$293	$707	$57,889
7		4	$1,000	$290	$711	$57,179

⟷

	J	K	L	M	N
1			i	0.005		
2			PMT	1000		Balance
3		Number	Amount	Interest	Apply to Principal	60000
4		1	=PMT	=N3*i	=K4-L4	=N3-M4
5		2	=PMT	=N4*i	=K5-L5	=N4-M5
6		3	=PMT	=N5*i	=K6-L6	=N5-M6
7		4	=PMT	=N6*i	=K7-L7	=N6-M7

	J	K	L	M	N
57	54	$1,000	$88	$912	$16,728
58	55	$1,000	$84	$916	$15,812
59	56	$1,000	$79	$921	$14,891
60	57	$1,000	$74	$926	$13,965
61	58	$1,000	$70	$930	$13,035
62	59	$1,000	$65	$935	$12,100
63	60	$1,000	$61	$939	$11,161

FIGURE 6.4 Initial amortization table.

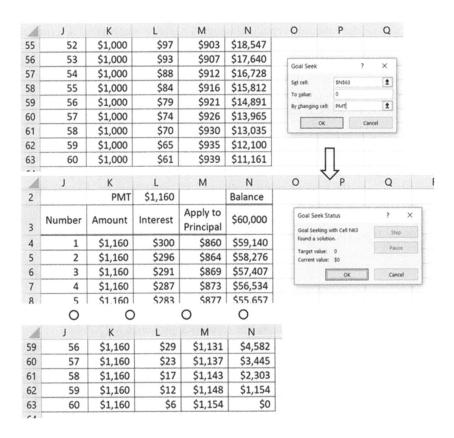

FIGURE 6.5 Using Goal Seek to determine the loan payment.

Since this is a common scenario, Excel provides a *PMT function* that automatically determines the required payment for a given initial loan balance, interest rate, and number of payments. There is a shortcut formula that is the basis for this function

$$PMT = L\frac{i(i+1)^n}{(i+1)^n - 1} \tag{6.5}$$

where
PMT:	loan payment each period
L:	original loan amount
i:	interest rate per period
n:	number of periods

The syntax for the PMT function is

=PMT(i,n,L)

and it will produce the same result (although a negative number) we determined using Goal Seek in Figure 6.5.

As a point of clarification, the interest rates we have used are per loan period. If the period is a month, that would require an adjustment to the annual interest rate. When a loan is compounded more frequently, such as monthly instead of annually, more interest is paid. A formula that describes this is

$$(i+1)^n - 1 \tag{6.6}$$

As a typical example, if we consider a monthly interest rate of 1% over 12 periods, a year, the formula result is about 12.7%, or 0.7% more than taking an interest rate of 12% for a period of one year. A question is what happens when we keep the overall term, increase the number of periods and express the interest rate as 12%/n. If we consider this in the limit, it becomes

$$\lim_{n\to\infty}\left(1+\frac{i_y}{n}\right)^n - 1 = e^{i_y} - 1 \tag{6.7}$$

Equation 6.7 is called *continuous compounding* and represents the maximum interest available. For our example, the result is approximately 12.75%.

Some confusion can result in quoting annual interest rates when the compounding period is less, typically monthly. There is a term in common use, the annual percentage rate or APR. The APR is not the true annual rate but rather is used to determine the monthly rate by dividing it by 12. So, concluding this example, an APR of 12% represents a monthly rate of 1% and a true annual rate of about 12.7%.

Excel has many financial functions, and we will not introduce all of them here. However, we choose to illustrate the RATE and NPER functions, which are complementary to the PMT function. The *RATE function* determines the interest rate required for a loan to be paid down with a given payment over a given number of periods. This function performs an iterative calculation and so requires an initial estimate of the interest rate. The *NPER function* determines the number of periods required to pay down a loan with a given payment and interest rate. The syntax for these two functions is

RATE(n,pmt,pv,fv,type,guess)
NPER(i,pmt,pv,fv,type)

where the arguments are as before with the addition of the guess argument. Note also that the functions provide the flexibility of specifying a present value, pv, or initial balance and a final value, fv, or final balance. For the loan scenario, the pv argument is the initial loan balance, and the fv is typically zero. The type argument is, as before, to describe whether payments are at the beginning or end of the period. For a loan calculation, these are usually at the end, i.e., type = 0, the default.

Using the example from Figure 6.5,

=RATE(60,-1200,60000) $\Rightarrow$ 0.32% (per month, APR $\cong$ 3.8%)
=NPER(5%,1000,60000) $\Rightarrow$ $\cong$ −28.4 (about 29 periods required)

This concludes our description of the basic financial functions. Later, in considering profitability measures, we will introduce additional built-in functions.

6.2 FINANCIAL FLOW IN MANUFACTURING OPERATIONS

Businesses operate based on cash, and this includes the manufacturing/process industries. Figure 6.6 depicts in general terms the flow of the *cash/production cycle* (Higgins, 2007). A primary path from the cash reserve is to convert cash into product

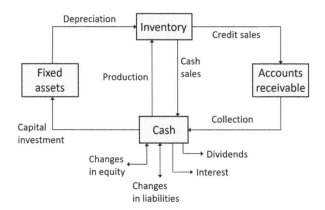

FIGURE 6.6 The cash flow/production cycle.

inventory via production. This occurs via operating and fixed costs that include acquisition of raw materials. Cash is also used to make capital investments and obtain fixed assets. A key activity that often involves engineers and scientists is determining the financial viability of these investments, and we will concentrate on that later in this chapter.

The company converts the inventory back into cash either by direct or credit sales. Outstanding invoices based on sales are noted in *accounts receivable*. There is a separate detail for credit purchases, such as for raw materials and equipment, which is called *accounts payable*. Also, the use of equity and debt to generate cash is an important element. The conversion of cash into inventory and return to cash via sales is called the *operating* or *working capital cycle*.

As production occurs and time passes, the value of fixed assets decreases. To obtain a balance, accountants consider this reduction or depreciation to be transferred to value in the inventory. At a minimum, cash must be invested in fixed assets to replace depreciation and maintain production capacity.

A logical goal is to return more to cash reserves than leaves. Cash is also distributed as dividends to shareholders for a publicly held corporation and expended as interest related to debt. It is important to understand that cash flow and profit are not equal. Cash flow is the lifeblood of the organization, and, if it gets out of balance, it can lead to insolvency and failure of the business.

Figure 6.7 provides more detail of the operating cycle that assists in creating a cash flow statement or table that focuses on analysis of proposed investments. The diagram depicts cash flow during a period of time, typically a year. A question that arises in preparing a cash flow table is when the transfers occur.

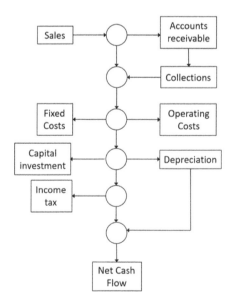

FIGURE 6.7 Cash flow cycle.

For example, sales income may be accounted for at the end of the period, and capital investment may be at the beginning of the period. In cases with more detail, this accounting may be distributed across the time period. The typical, simplified analysis has all items at the end of the time period.

Variable costs are those that scale with the production rate of the process, such as the costs of raw materials and utilities. *Fixed costs*, as the name implies, do not depend on production rate and include such factors as labor, maintenance, overhead, and insurance.

An important detail in the figure is how *depreciation* is handled. In a sense, it is a tax benefit since it is subtracted from gross income to get taxable income, and then it is added back in to get net income for net cash flow. Naively, we might expect that the company would divert cash into a depreciation fund representing savings for replacement. This is not the case. When replacement is required, this is accounted for by capital investment, and there may be an associated salvage value for the equipment replaced.

In analyzing a business project, there are variations in cash flow over the period considered, often many years. For a new manufacturing facility, the initial years may represent capital investment in construction and startup, and, consequently, the annual cash flows are negative. Later, the cash flows are positive, and the financial feasibility of the project requires that these pass the deficit at the beginning of the project. Assuming that there is debt accrued for capital investment, the cost of capital (interest rate) must be included in the analysis.

This description provides us the background to consider cash flow tables for projects and later review measures of profitability that are used in the decision to go forward with projects.

6.3 CONSTRUCTING CASH FLOW TABLES

The overall financial status of a business is represented in different ways. A common one is the *balance sheet*. It is a snapshot of the finances of a business and commonly includes two adjacent years, noting changes from one to the next. Two related presentations are the income and the cash flow statements. Although there are general standards and expectations for these statements, they will differ from company to company in the details.

The general format of the balance sheet is to list assets, liabilities, and equity in separate categories. The *assets category* will include cash, accounts receivable, inventories, property, and equipment. *Liabilities* include debt, accounts payable, expenses, and taxes. *Equity* includes retained earnings and, for a publicly held business, includes stock holdings. The sum of liabilities and equity always equals assets.

The income and cash flow statements describe the dynamics of a business' finances. The *income statement* accounts for changes in equity by partitioning those into income and expenses. It shows the difference between income and expenses as earnings or net income.

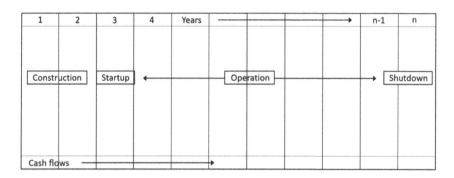

FIGURE 6.8   General format of a cash flow table for a project.

The *cash flow statement* focuses on solvency, having enough cash on hand to pay the bills. It provides a detailed description of cash flow and how it changes over time. The cash flow during each time interval, typically a year, can be used to determine the viability of a business and the basis for evaluating a new business venture. The latter involves estimates and uncertainties, which form part of the decision-making process.

Figure 6.8 provides a simplified structure for the cash flow statement. This is combined for a number of years, *n*, often considered to be the life of a project. Project life in a cash flow table is divided into four phases: construction, start-up, operation, and shutdown. This is depicted in Figure 6.8. The number of years for the phases is variable, depending on the project and circumstances. Project life varies typically between 10 and 25 years.

It is anticipated that, during the construction phase of the project, cash flows will be negative. Following that, they will be positive and may gradually increase to a full-operation stage up to the shutdown. The shutdown period will focus on the salvage value of the project, which possibly could be zero. It is typical to predict inflation during the term of the project and incorporate this into the analysis for both sales income (depends on sales price) and costs.

For a specific scenario, we can transform Figure 6.8 into a template for a cash flow table. This is shown in Figure 6.9. You will notice here that the term of the analysis is ten years with one year of construction and a final year

	A	B	C	D	E	F	G	H	I	J	K
4		Construction				Operation					Shutdown
5	Year	1	2	3	4	5	6	7	8	9	10
6	Sales										
7	Variable Cost										
8	Fixed Cost										
9	Capital Cost										
10	Depreciation										
11	Gross Income										
12	Income Tax										
13	Depreciation										
14	Net Cash Flow										

FIGURE 6.9   Cash flow table template for a specific scenario.

	A	B
4		Construction
5	Year	1
6	Sales	
7	Variable Cost	
8	Fixed Cost	
9	Capital Cost	($140,000)
10	Depreciation	
11	Gross Income	($140,000)
12	Income Tax	
13	Depreciation	
14	Net Cash Flow	($140,000)
15		

	A	B	
4		Construction	
5	Year	1	2
6	Sales		
7	Variable Cost		
8	Fixed Cost		
9	Capital Cost	-140000	
10	Depreciation		
11	Gross Income	=SUM(B6:B10)	
12	Income Tax		
13	Depreciation		
14	Net Cash Flow	=SUM(B11:B13)	

FIGURE 6.10 Completing the first column, Construction, of the cash flow table. (Recall that negative amounts can be indicated by a negative sign or by being enclosed in parentheses.)

of shutdown. It is considered that the operation will come to full production in Year 2. The cash flow categories follow the pattern of Figure 6.7.

We will now construct the table from left to right. In the construction year, there is only a capital expense for construction of the manufacturing facility. This is illustrated, along with the first formulas, in Figure 6.10. The capital expense is entered as a negative number, and sum formulas are entered for Gross Income and Net Cash Flow.

The next step is to enter the data for Year 2. This is shown in Figure 6.11. Sales income is entered as $100,000. Often, this would be computed in more detail elsewhere in the workbook and would include sales price per unit and numbers of units sold. Variable and fixed costs are entered here as numbers; however, they would likely come from other worksheets with more detailed calculations.

	A	B	C
4		Construction	
5	Year	1	2
6	Sales		$100,000
7	Variable Cost		($45,000)
8	Fixed Cost		($3,800)
9	Capital Cost	($140,000)	
10	Depreciation		($15,556)
11	Gross Income	($140,000)	$35,644
12	Income Tax		($7,485)
13	Depreciation		$15,556
14	Net Cash Flow	($140,000)	$43,715

	A	B	C
4		Construction	
5	Year	1	2
6	Sales		100000
7	Variable Cost		-45000
8	Fixed Cost		-3800
9	Capital Cost	-140000	
10	Depreciation		=$B$9/9
11	Gross Income	=SUM(B6:B10)	=SUM(C6:C10)
12	Income Tax		=-C11*IRS
13	Depreciation		=-C10
14	Net Cash Flow	=SUM(B11:B13)	=SUM(C11:C13)

IRS		×	✓	f_x	21%

	F	G	H
3		Income Tax Rate	21%

FIGURE 6.11 Cash flow table with Year 2 completed.

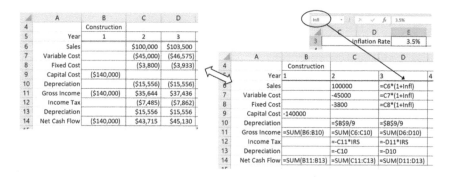

FIGURE 6.12 Cash flow table with Year 3 completed and dependent on a constant inflation rate.

Depreciation is entered here as a linearly decreasing function by year. There are different *depreciation schemes*, and Excel contains built-in functions, such as *DB* and *DDB*, to implement these. Corporate income tax is computed based on the current percentage in the U.S. (2022), and then depreciation is credited back in, as depicted in Figure 6.7. The Net Cash Flow is positive, which will help counter the negative value from the Construction year.

Moving on to Year 3, we must consider the effect of inflation. This is speculative and can be estimated to be variable over the term of the project. For example, the current inflation rate is high (~7% in 2021/22); however, we might anticipate it will be lower in coming years. For a simple analysis, a constant inflation rate can be used; otherwise, we would enter different rates for each year of the project. Figure 6.12 shows how Year 3 is established on the spreadsheet.

In this case, we choose a simple approach with a constant inflation rate of 3.5%, which is applied to the sales and cost figures. In this way, the formulas in Year 3 can be copied across the table up to Year 9. The cash flow table is then complete, except for Year 10, and is shown in Figure 6.13.

The remaining task is to complete the Shutdown year, Year 10. A simple approach is to apply a credit for sale of the capital equipment that was acquired in Year 1. More complicated considerations would be a ramping down of production

	A	B	C	D	E	F	G	H	I	J	K
3				Inflation Rate	3.5%		Income Tax Rate	21%			
4		Construction					Operation				Shutdown
5	Year	1	2	3	4	5	6	7	8	9	10
6	Sales		$100,000	$103,500	$107,123	$110,872	$114,752	$118,769	$122,926	$127,228	
7	Variable Cost		($45,000)	($46,575)	($48,205)	($49,892)	($51,639)	($53,446)	($55,316)	($57,253)	
8	Fixed Cost		($3,800)	($3,933)	($4,071)	($4,213)	($4,361)	($4,513)	($4,671)	($4,835)	
9	Capital Cost	($140,000)									
10	Depreciation		($15,556)	($15,556)	($15,556)	($15,556)	($15,556)	($15,556)	($15,556)	($15,556)	
11	Gross Income	($140,000)	$35,644	$37,436	$39,291	$41,211	$43,198	$45,254	$47,382	$49,585	
12	Income Tax		($7,485)	($7,862)	($8,251)	($8,654)	($9,072)	($9,503)	($9,950)	($10,413)	
13	Depreciation		$15,556	$15,556	$15,556	$15,556	$15,556	$15,556	$15,556	$15,556	
14	Net Cash Flow	($140,000)	$43,715	$45,130	$46,596	$48,112	$49,682	$51,306	$52,988	$54,728	

FIGURE 6.13 Cash flow table completed through the Operation phase.

A	B	C	D	E	F	G	H	I	J	K
			Inflation Rate	3.5%		Income Tax Rate	21%			
	Construction				Operation					Shutdown
Year	1	2	3	4	5	6	7	8	9	10
Sales		$100,000	$103,500	$107,123	$110,872	$114,752	$118,769	$122,926	$127,228	
Variable Cost		($45,000)	($46,575)	($48,205)	($49,892)	($51,639)	($53,446)	($55,316)	($57,253)	
Fixed Cost		($3,800)	($3,933)	($4,071)	($4,213)	($4,361)	($4,513)	($4,671)	($4,835)	
Capital Cost	($140,000)									$15,000
Depreciation		($15,556)	($15,556)	($15,556)	($15,556)	($15,556)	($15,556)	($15,556)	($15,556)	
Gross Income	($140,000)	$35,644	$37,436	$39,291	$41,211	$43,198	$45,254	$47,382	$49,585	$15,000
Income Tax		($7,485)	($7,862)	($8,251)	($8,654)	($9,072)	($9,503)	($9,950)	($10,413)	($3,150)
Depreciation		$15,556	$15,556	$15,556	$15,556	$15,556	$15,556	$15,556	$15,556	
Net Cash Flow	($140,000)	$43,715	$45,130	$46,596	$48,112	$49,682	$51,306	$52,988	$54,728	$11,850

FIGURE 6.14 Completed cash flow table with salvage credit in the Shutdown Year 10.

along with decommissioning of equipment, and, even simpler, would be not to consider salvage value. In considering depreciation, using a built-in function like DB, salvage value is taken into account. As shown in Figure 6.14, we apply a nominal salvage credit of $15,000 and consider no production in Year 10. This then shows the completed cash flow table.

This is a highly simplified cash flow analysis. Details would typically be added in the sales and cost categories. Construction would often take more than one year. Depreciation would be calculated by another scheme. However, an analysis like this is often used to provide guidance on whether or not to pursue a project into further phases, such as manufacturing process design, where a greater financial commitment is required. This is often called *Venture Guidance Analysis (VGA)*. At this level, management may consider alternate investments and need preliminary information for decision-making.

Adjoining the cash flow analysis would often be the uncertainties associated with the market, the cost factors, and even the tax rates. These can be handled by running many cash flow scenarios varying from optimistic to pessimistic forecasts. Sophisticated approaches may run hundreds or thousands of cases with random variations to provide a distribution of outcomes from least to most probable. But for now, we turn to developing key measures of profitability that quantify the merit of a proposed project. That is the objective of the next section.

6.4 PROFITABILITY MEASURES

As we study the cash flow table presented in Figure 6.14, it is evident that, after the initial investment in capital, the following years produce net positive cash flow that can be applied toward that investment. One way of interpreting that impact is a *breakeven chart* as shown in Figure 6.15. Here we see that the accumulated cash flow becomes positive in Year 5. By interpolating, we judge there is a breakeven point early in Year 4. After that, the accumulated cash flow continues to grow in the positive direction.

The chart paints a rosy picture. A traditional measurement that quantifies this is *Return on Investment (ROI)*, which, for our simplified presentation here,

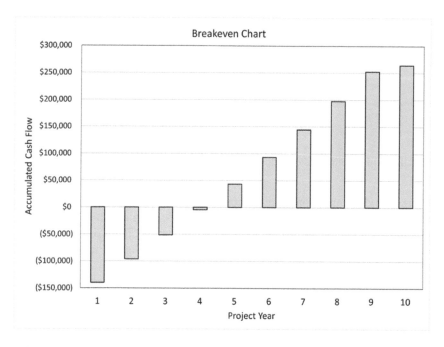

FIGURE 6.15 Breakeven chart for accumulated cash flow.

is Net Cash Flow/Initial Capital Investment. If we compute that for our cash flow table, these are the results:

Year	2	3	4	5	6	7	8	9	10
ROI	31.2%	32.2%	33.3%	34.4%	35.5%	36.6%	37.8%	39.1%	8.5%

If we project the first ROI value, that calculates as

$$\text{Breakeven} = 1/31.2\% \cong 3.2 \text{ years}$$

or 3.2 years after the production start. This is then 4.2 years after project initiation, which agrees with our analysis from Figure 6.14.

There are variations on the simple ROI measure. These include Return on Assets (ROA), Return on Equity (ROE), and Return on Invested Capital (ROIC). We do not choose to elaborate on those here for the sake of simplicity.

A major fault of the ROI measure is that it doesn't consider the time value of money. An amount earned in Year 2 is worth more than the same amount earned in later years. Another way to describe that is, if we took the amount earned in Year 2 and invested it; it will have gained value by a later year. This all depends on the interest rate at which we can invest positive cash flows and the interest rate at which we would have to borrow to finance our construction phase. This interest rate is termed the *cost of capital*.

Year	1	2	3	4	5	6	7	8	9	10
Net Cash Flow	($140,000)	$43,715	$45,130	$46,596	$48,112	$49,682	$51,306	$52,988	$54,728	$11,850
Present Value	($140,000)	$40,477	$38,692	$36,989	$35,364	$33,813	$32,332	$30,918	$29,568	$5,928

=C14/(1+Cap)^(C5-1)

| Cap | | | | | 8% |

	F	G	H
2		Cost of Capital	8%

FIGURE 6.16   Converting future cash flows into present values.

A common way to treat the cash flows in our table is to consider each a future value and compute its present value using the cost of capital interest rate. We can use our PV formula from Section 6.1 for this. The calculation is added to the cash flow table and depicted in Figure 6.16. A slight adaptation here is that the number of years, n, from the formula is our Year number minus one.

The Present Value of the ten cash flows can be summed to yield the *Net Present Value (NPV)*, which is a common measure of merit for project evaluation. This value is $144,079. A way to interpret this is, if we took our initial financing of $140,000 and invested it at the CAP rate of 8%, and, if we paid down the loan (and beyond) with our cash flows, it would yield a balance of $288,015 at the end of the project. If we brought that figure back to current time using our PV formula, the result would be $144,079. This is the present value benefit of investing in this project.

It turns out that NPV is a common measure of profitability, and Excel provides a built-in function for it. The syntax of this function is as follows:

NPV(cost_of_capital,range_of_cash_flows)

The NPV function assumes that the cash flows occur at the end of the period.

Figure 6.17 illustrates how the NPV function can be used to obtain the same result as our sum of PV values from Figure 6.16. Notice how the initial investment, as negative, is left out of the NPV argument but appended to the formula.

A negative NPV would argue against investing in the project. The extent of a positive NPV can provide guidance in pursuing the next stages of the project, which may require a significant commitment of funds. It is difficult to use the NPVs for various projects to compare and prioritize them and make choices among them.

Another profitability measure is developed by considering the question, What Cost of Capital would yield an NPV of zero? We can determine this as a targeting

	A	B	C	D	E	F	G	H	I	J	K
14	Net Cash Flow	($140,000)	$43,715	$45,130	$46,596	$48,112	$49,682	$51,306	$52,988	$54,728	$11,850

Net Present Value (NPV) =NPV(Cap,C14:K14)+B14   ⟺   Net Present Value (NPV) $144,079

FIGURE 6.17   Employing the NPV function to determine net present value.

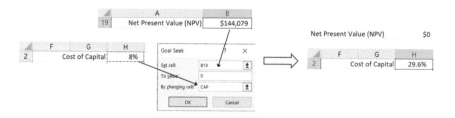

FIGURE 6.18   Using Goal Seek to determine the Cost of Capital that yields NPV = 0.

problem using Goal Seek. This is illustrated in Figure 6.18. A Cost of Capital of 29.6% is determined for an NPV of zero. This, of course, is a hypothetical interest rate for the purpose of internal evaluation. For this reason, it is called the *Internal Rate of Return (IRR)*.

An advantage of the IRR is that it is independent of the scale of the project and then perhaps better for comparing the merits of several alternative investments. Since it is a common measure of profitability, as is NPV, Excel provides a built-in IRR function with the following syntax:

IRR(range_of_cash_flows,initial_guess)

The initial_guess argument is optional. The IRR calculation is iterative, and a reasonable initial estimate may accelerate convergence to the result or provide for success in the case where the iterations fail. In the case of the IRR function, the cash flows include the initial investment. Figure 6.19 shows the implementation of this function yielding the same result as the Goal Seek targeting carried out above.

Of course, in comparing the IRR value of nearly 30% with the Cost of Capital of 8%, this project appears to be very attractive. In most situations, it would be a closer call.

After ROI, NPV, and IRR, another common profitability measure is *Benefit-Cost Ratio (BCR)*. This is the ratio of the present value of cash inflows to the present value of cash outflows. For our simple example, the latter is the initial capital investment. The BCR calculation is simply

$$BCR = \frac{\$284,079}{\$140,000} \cong 2.03$$

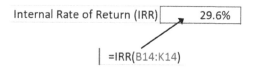

FIGURE 6.19   Application of the IRR function.

The BCR does include the time value of money since present values are used. It is also called the profitability index and, obviously, a BCR less than one indicates an investment in a project is not desirable.

One interesting case study is to consider NPV and BCR as the Cost of Capital varies. This would consider the uncertainty in the interest rates available. Figure 6.20 summarizes this case study using Excel's Data Table feature. The NPV and BCR values are positive and encouraging throughout the range of Cost of Capital values. Given the original IRR value of nearly 30%, this is expected.

A question that arises is that, with the availability of various measures of profitability, which should be chosen. The first practical answer to this is that

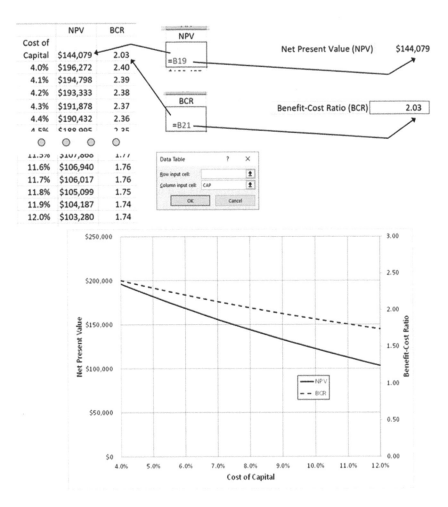

FIGURE 6.20 A Data Table and chart to explore NPV and BCR as a function of Cost of Capital.

you follow the lead of the organization considering the project. There are usually established practices. The distinction between ROI and the others is that ROI doesn't consider the time value of money. It was popular decades ago when many industrial organizations were capital rich and didn't have to depend on financing for a project. So, if those circumstances exist, ROI is still valid.

The NPV and IRR measures lead to the same conclusion regarding the financial viability of a project. An advantage, as previously stated, of IRR is that it provides a useful measure when considering various alternative projects of different scales to select one or more of the most attractive. The BCR measure has been popular in government organizations, while NPV and IRR are more so in the private sector. All consider the time value of money, so conclusions drawn in terms of financial viability are similar.

The cash flow table and associated measures of profitability provide the basis for many explorations of the viability of proposed projects for manufacturing facilities or modifications of these. Exploring variations in sales and costs, investigating projected interest and tax rates, and incorporating more detail into the entries of the cash flow table are all important activities in providing the best information to management as they make their decisions.

In certain situations, there are no opportunities for positive cash flow in ventures. This is particularly true when manufacturing changes are required to meet government regulations. In such cases, one is evaluating the project alternatives that provide the least negative NPV. In other words, it represents minimizing costs. Cash flow analysis is still viable in such situations.

The financial management of organizations is a broad topic and merits treatment in entire books, academic courses, and, in fact, advanced academic degree programs. It is impossible to consider such coverage here. We have chosen to focus on the elementary concepts and calculations associated with the time value of money and the evaluation of capital projects. The reason for our choice of the latter is that it is encountered frequently by engineers and scientists. In fact, when an engineer suggests an improvement in a manufacturing process, it is common practice to accompany that with a cash flow analysis to help determine its feasibility.

We do encourage the reader to broaden their knowledge of financial management via reading and additional training, whether as a student or practicing professional.

PROBLEMS

6.1 Develop a complete monthly amortization table for the purchase of real estate financed with a 30-year mortgage loan. The purchase price is $1.3 million, and a 20% down payment is required. The available interest rate is an APR of 5.6%. Enter an estimate of $5,000 for the monthly payment. Use Goal Seek to adjust the payment so that the final balance equals zero. What is the monthly payment?

6.2 Carry out the following calculations for the example of Problem 6.1:
   a. Use the PMT function to confirm your payment result.
   b. Conduct a case study of the monthly payment versus APR interest rate for the range of 3% to 7% in steps of 0.1%. Based on this case study, estimate the interest rate needed for a monthly payment of $5,000.
   c. Confirm your result from part (b) by using the RATE function.
6.3 For the data of Problem 6.1, if, after 10 years, it were possible to refinance the balance of the loan at 4.5% for the remaining term, 20 years, but this required a refinance payment of 2% of that balance; would it be wise to take that path? Assume that the refinance payment was not rolled into the remaining balance of the loan.
6.4 Set up the following calculations on a spreadsheet:
   a. What is the future value of an initial investment of $50,000 at an interest rate of 7.4%/year over a period of 10 years? Compounding is annual. What would be the future value with continuous compounding?
   b. If a vehicle is sold for $25,000 after five years of use, what is the present value of that amount, if the financing interest rate is 6.3%/year? Compounding is annual.
   c. Assume the vehicle in part (b) had a loan balance equal to its selling price at five years of use, and the financing interest rate was 6% APR over a six-year term. If the initial down payment was $10,000, what was the initial purchase price of the vehicle? Payments were made monthly.
6.5 In the case of variable interest rates, as shown in the table below, set up an amortization table for an initial investment of $100,000 and determine its future value at the end of the period. Using the table, expand it to determine the present value of this investment.

Year	1	2	3	4	5	6	7	8	9	10
Interest Rate	8.6%	7.2%	4.3%	5.5%	6.2%	7.7%	5.3%	4.9%	3.8%	2.9%

6.6 Based on the following information, set up a cash flow table:
   • Design and construction take place over a two-year period with an investment of $250,000 in the first year, and $15 million in the second year.
   • Cost of capital is estimated to be 9.4%.
   • The initial selling price of the product is $240/unit, and full production is 15,000 units/year.
   • In the first, start-up year, the production rate is 50% of full production.
   • Initial variable costs are $75/unit and initial fixed costs are $375,000/yr.
   • There is an estimated inflation rate of 3.9%/yr.
   • The manufacturing process is expected to be in operation for 12 years including the startup year.

- Salvage value in the shutdown year is estimated at $600,000.
- Income tax rate remains at the current 21% of gross income.
- Depreciation is computed using Excel's DB function, the fixed declining-balance method.

6.7 Determine the ROI for the project of Problem **6.6** and estimate the breakeven point.

6.8 Given a Cost of Capital of 7.8%, for the cash flow table of Problem **6.6**, determine the NPV, IRR, and BCR. Suggest whether this project should be pursued or not.

NOTES

1. They call it Microsoft "Office" for good reason!
2. As there are currently 180 currencies on the planet, we had to decide on one currency for the economic material in this book. We have decided to use dollar signs for our currency as that is where we were born. In addition, the symbol $ is used for many other currencies such as the Mexican peso and the Australian dollar. Regardless of the symbol, the approaches described in this chapter apply to any of the world's currencies that charge interest.

7 Numerical Methods

CHAPTER OBJECTIVES

- Be able to implement targeting using Goal Seek and the Solver to solve for the root of a nonlinear algebraic equation
- Learn a bracketing method, bisection, and an open method, Newton-Raphson, and how to implement these methods for "live" solutions on the spreadsheet
- Extend array formulas to include the matrix operations of transpose, multiplication, inverse, and determinant
- Learn how to create "live" solutions of sets of linear equations using matrix array formulas
- Be able to formulate sets of nonlinear equations for solution using the Solver
- Solve a differential equation that meets quadrature requirements using the trapezoidal rule method
- Understand the Euler method for solution of differential equations and how to implement it on the spreadsheet
- Learn how to approach scenarios where the result of a calculation of one or more nonlinear relationships is dependent on the result itself, yielding a circular calculation scheme

Problem solving is the "bread and butter" of engineering and science, and often these problems are represented by algebraic and differential equations that do not have analytical solutions. Consequently, one cannot set up a spreadsheet with a sequence of "straight through" calculations. We require numerical methods to solve these equations.

The topic of numerical methods (sometimes called "computer math") is a large one, and many engineering and science students take an entire academic course (or more) dedicated to this. In this chapter, we introduce elementary methods that work well on the spreadsheet and are sufficient for the solution of many of the problems that engineers and scientists face. Later, in Chapters 10 and 11, we will make the spreadsheet solution techniques more compact, versatile, and efficient by employing VBA programming.

The problem scenarios covered in this chapter are in five categories. The first is solving for the root of a single nonlinear algebraic equation, and we consider two classical methods for this purpose: bisection and Newton-Raphson. These techniques are developed for "live solution" on the spreadsheet. For these equations, we also use our knowledge of Goal Seek and the Solver from Chap. 5 to solve them. Next, we expand our knowledge of array formulas to include common matrix

DOI: 10.1201/9781003361053-7

calculations and show how these can be used to solve sets of linear algebraic equations. The third topic is the solution of sets of nonlinear equations. Details on the numerical methods for this are exhaustive, so we focus on the use of the Solver for this scenario. We then turn to ordinary differential equations and their numerical solution using the Euler method, one of the simplest methods available. This includes both single and multiple equations. Finally, a common spreadsheet scenario called a circular calculation will be described, and Excel's iterative solver will be introduced to address it.

7.1 SINGLE NONLINEAR ALGEBRAIC EQUATIONS

For most practical situations, when we have a nonlinear algebraic or transcendental equation,[1] there is one solution or root that is feasible. Of course, we can describe equations mathematically that have more than one solution, and there are equations that have no real solution but instead have complex roots. Since these are less frequent in practice, we do not focus on them here. We will illustrate a bracketing method, bisection, and an open method, Newton-Raphson in the context of an example.

7.1.1 BRACKETING METHODS – BISECTION

Example 7.1 Determining the Level in a Spherical Tank

Here, we borrow the scenario from Problem 5.4 as illustrated in Figure 7.1. As the problem stated, the volume of liquid in the tank, V, is related to the internal radius, R, and the liquid depth at the centerline of the tank, h, by the formula

$$V = \pi h^2 \frac{(3R - h)}{3} \tag{7.1}$$

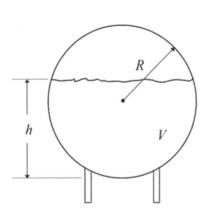

FIGURE 7.1 Spherical tank schematic drawing and example photograph. (Photograph taken by the author. It is a public view of a BASF facility in Mannheim, Germany, taken from the Rhine River.)

The problem we address here is that there is a gauge that measures the depth of the liquid, h, and we would like to calibrate that gauge directly in even increments of volume, V. In a more primitive implementation, this might be a dipstick that is inserted from the top of the tank and withdrawn to observe the wet line on its scale. It might also be an external "sight gauge" with an appended scale. Or it could be a level sensor and a computer-based lookup table. The point is that, for many intermediate values of V, we need to solve for h.

Equation 7.1 can be rearranged to provide a single algebraic expression set equal to zero. This is shown in Equation 7.2 and is a specific example of the general equation, $f(x) = 0$.

$$h^3 - 3Rh^2 + \frac{3V}{\pi} = 0 \qquad (7.2)$$

The values of h (or x for the general case) that make the function zero are formally called the function's *roots*.

Given a tank specification of R and a value of V, the goal is to find a value of h that will cause the left side of Equation 7.2 to evaluate to zero. We observe that this is a third-order polynomial in h, so it will have three roots.[2] In this case, these roots will be real numbers, and it turns out that one of them is above the tank, another is below the tank, and only one is within the range of the tank. This is typical of a practical, physical scenario where there is only one feasible solution.

First, we can view this as a targeting problem, that is, find a value of h so that the left side of Equation 7.2 meets the right-side target of zero. Figure 7.2 shows the use of Goal Seek to solve this problem for a specific set of problem specifications and values. Goal Seek adjusts h to about 4.2 m and drives the equation error from 14.986 to 0.001.

If we would like to find the depth for another value of volume, we must run Goal Seek again. And Goal Seek does not remember the last solution scheme, so we have to complete the dialog box again. (At least, the Solver does remember!). So, Goal Seek (and the Solver) do not lend themselves to our task of solving for 10s or 100s of volume values. This leads us to consider "live" numerical solution techniques.

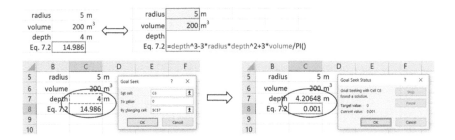

FIGURE 7.2 Solution for spherical tank depth using targeting and Goal Seek.

There are many numerical methods that can be employed to solve single nonlinear algebraic equations. These are generally grouped into two types of methods: bracketing and open. *Bracketing methods* require two initial estimates (sometimes called guesses, and we hope educated guesses) that "bracket" the root, whereas open methods only require one initial estimate (or two initial estimates that do not have to bracket the root). It is not our purpose to introduce many of these methods here. Rather, we choose to illustrate one of the simpler methods in each category and how these can be implemented on the spreadsheet as a "live" calculation.

The basis for a bracketing technique is illustrated in Figure 7.3. The general problem is to find a value of x such that $f(x) = 0$, or at least very close to 0. The $f(x)$ curve is shown in the figure, and it crosses the x axis once in the bracket interval, $x_L \le x \le x_H$. Of course, numerically, we do not know where the root crossover is located. From the figure, you can see that the signs of $f(x)$ at the two bracket values are opposite. For a well-behaved, continuous function, as are most describing physical phenomena, this implies the function crosses zero somewhere between the brackets.

The method we illustrate here is the simplest bracketing technique, *bisection*, which is based on choosing the midpoint, x_M, between x_L and x_H as the next estimate for the solution. This is shown in Figure 7.4. By observing the figure, we see that, in this instance, $f(x_M)$ is the same sign as $f(x_H)$. This implies that the solution now lies between x_L and x_M, and, to repeat the method, we would evaluate the midpoint with x_M becoming the new x_H.

Figure 7.5 illustrates the second iteration of this method. With this new midpoint, we see that it is on the same side of the solution as x_L. This means that for the third iteration, x_M will become the next x_L.

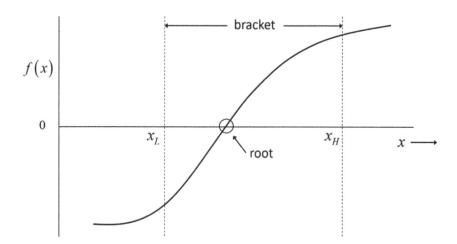

FIGURE 7.3 Bracketing scenario for finding the root of a nonlinear algebraic equation, $f(x) = 0$.

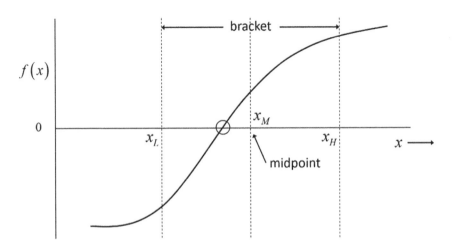

FIGURE 7.4 Bracketing method with midpoint selected.

Each time the bisection method is repeated, the interval containing the solution is cut in half. After n repetitions, in terms of the original bracket, the interval will be

$$\frac{x_H - x_L}{2^n} \qquad (7.3)$$

For 10 iterations, this will be about 1/1,000 of the original interval, and, after 20 iterations, about 1 part in a million.[3] And the solution is guaranteed to be in that final interval. For almost all practical scenarios, the latter precision is more than adequate.

Now, we will return to Example 7.1 to illustrate how to implement this method on the spreadsheet.

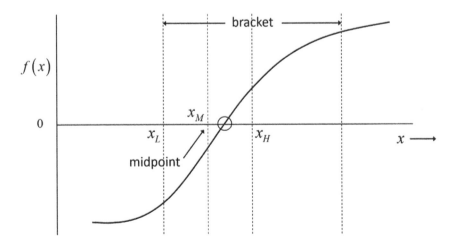

FIGURE 7.5 Bisection method with second iteration shown.

	B	C	D	E	F	G	H
2	radius	5 m					
3	volume	200 m^3					
4	depth	m					
5							
6	Bisection solution						
7							
8	Iteration	h$_L$	f(h$_L$)	h$_H$	f(h$_H$)	h$_M$	f(h$_M$)

FIGURE 7.6   Initial arrangement of the spreadsheet for bisection.

Figure 7.6 shows our initial setup of a spreadsheet using the bisection method to solve for the liquid depth in the spherical tank given the radius of the tank and the volume of the liquid. The basic data for radius and volume are entered at the top of the spreadsheet along with a blank cell for the solution. Column headings following the scheme of the previous figures are also entered.

There is a common scheme for iterative solutions on a spreadsheet. It is to provide a first row of data and results, called an *initialization row*, and then establish a second row, called the *operational row*, which can be copied down to complete the number of iterations required for an acceptable result. The establishment of the initialization row is shown in Figure 7.7. The initial values for h_L and h_H are entered as zero and $2R$, based on a bracket that, for this particular problem, will always contain the solution. Hence, because they are predicated on our knowledge of the physical system, they are what we referred to earlier as "educated" guesses.

In cell D9, a formula is entered that evaluates Equation 7.2 in terms of the cell to the left, C9, the h_L value. Since this is a relative reference, the formula is copied

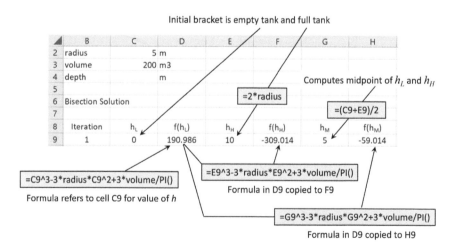

FIGURE 7.7   Initialization row for bisection solution of spherical tank example.

to the right of the h_H and h_M values for the corresponding evaluations. Cell G9 contains the simple formula to compute the midpoint of h_L and h_H, the h_M value.

By examining the $f(h_L)$ and $f(h_H)$ values in Figure 7.7, we see that they are of opposite sign. Considering Figures 7.3–7.5, this is good news. As we would expect from our physical example, there is a solution in the interval. We also note that the sign of the $f(h_M)$ value is the same as the $f(h_H)$ value. Based on our previous discussion, it appears that, for the second iteration, h_M should replace h_H.

Before we consider entering the second row of the table, the operational row, we should note that this might not always be the case; in other words, it is possible that h_M should replace h_L instead. For example, if we specified a volume of 400 m³, $f(h_M)$ and $f(h_L)$ would have the same sign, values of +132 and +382 approximately and respectively. Our second row must reflect this possibility.

If we look at the cell just below the initial value of h_L, C10, we see that it will either contain the value from the cell above, the same left side of the bracket, or the h_M value, a new left side. The latter will take place if the sign of $f(h_L)$ is the same as that of $f(h_M)$. We can implement this choice in cell C10 by using Excel's IF function as

IF(D9 * H9>0,G9,C9)

Note that the product of cells D9 and H9 will be positive if the $f(h_L)$ and $f(h_M)$ values are both either positive or negative. We can apply the same logic for the h_H, and the formula will be

IF(F9 * H9>0,G9,E9)

The formula for the $f(x)$ values and the midpoint formula in cell G9 can be copied down to row 10. The result is shown in Figure 7.8. Here you will see that we opted for a simple increment formula in cell B8 for the iteration number. The other formulas follow our discussion above. It is evident that, for the third iteration, 2.5 should replace 0 as the value for h_L.

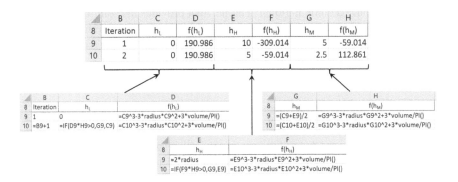

FIGURE 7.8 Implementing the first operational row of the bisection method showing results and formulas.

	B	C	D	E	F	G	H
8	Iteration	h_L	$f(h_L)$	h_H	$f(h_H)$	h_M	$f(h_M)$
9	1	0	190.986	10	-309.014	5	-59.014
10	2	0	190.986	5	-59.014	2.5	112.861
11	3	2.5	112.861	5	-59.014	3.75	32.783
12	4	3.75	32.783	5	-59.014	4.375	-12.383
13	5	3.75	32.783	4.375	-12.383	4.0625	10.474
14	6	4.0625	10.474	4.375	12.383	4.21875	0.007
23	15	4.20593	0.040	4.20654	-0.004	4.20624	0.018
24	16	4.20624	0.018	4.20654	-0.004	4.20639	0.007
25	17	4.20639	0.007	4.20654	-0.0043	4.20647	0.001
26	18	4.20647	0.001	4.20654	-0.0043	4.2065	-0.002
27	19	4.20647	0.001	4.2065	-0.0015	4.20649	0.000
28	20	4.20647	0.001	4.20649	-0.0001	4.20648	0.001

FIGURE 7.9   Complete bisection table with 20 iterations.

Now that we have a fully operational row for iteration 2, we can copy that row downward for the desired number of iterations of the bisection method. Here, based on our prior calculations, we choose 20 iterations. The results are summarized in Figure 7.9. From the table, you can see that the original equation errors are in the hundreds, and the final ones are around one thousandth. This is a reduction of five orders of magnitude. And our original bracket width of 10 has been reduced to one part in 100,000, giving us the expected reduction of one part in a million. We can use the final h_M value of 4.20468 m as our close approximation to the solution of the equation.

The finishing touch for our solution of the spherical tank equation is to enter a pointer formula to cell G28 in our depth cell, C4.

C4: =G28

It is notable that this is a *"live"* solution. What do we mean by that? By changing the volume cell to 400 m³, the depth cell displays 6.843 m automatically. Now that we have a computational "engine" that provides the solution for a given volume of liquid, we can add a case study to produce all the depth values for a set of evenly spaced volume values. Figure 7.10 summarizes this case study that employs the Data Table feature of Excel illustrated in Chapter 5.

For the given tank, the total volume is calculated as

$$V_T = \frac{4}{3}\pi R^3 \cong 523.6\,\text{m}^3 \tag{7.4}$$

Given this value, we have chosen intervals of 1 m³ which require 523 solutions of the nonlinear algebraic equation. The results on the spreadsheet could then be used to calibrate a level-measuring device, although not all values might be required. The figure also includes a plot of depth versus volume produced from the data table.

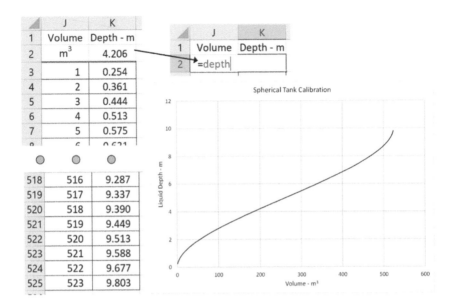

FIGURE 7.10   Case study of liquid depth versus evenly spaced values of volume.

It should be noted that, with minimal modifications, the implementation of bisection on the spreadsheet could be adapted to the solution of a different nonlinear algebraic equation. This would be accomplished by replacing one of the formulas, such as that in cell D9, with the new equation, copying this new formula throughout the table, and adjusting the initial h_L (or now x_L) and h_H (or now x_H) values. Of course, the basic data at the top of the spreadsheet and any case study would have to be replaced. This adaptation will be appropriate for certain problems at the end of this chapter.

There are other bracketing methods. One of the most common is *false position* (also known as *regula falsi*). This method uses a straight line between the points $\{x_L, f(x_L)\}$ and $\{x_H, f(x_H)\}$ with the next solution estimated where the line intersects the $f(x) = 0$ axis. This method may converge more quickly to the solution, but it may not, depending on the shape of the function between the two guesses. Additionally, the method does not provide a predictable error bound for the solution as does bisection. We choose not to illustrate false position here, although its "live" solution is a minor modification of our bisection spreadsheet technique. Rather, we include it in one of the end-of-the-chapter problems.

7.1.2   OPEN METHODS – NEWTON-RAPHSON

The *Newton-Raphson method* is the most common open technique for solving nonlinear algebraic equations. It requires only one initial solution estimate, and one needs to supply the derivative function, $f'(x)$. If that function is difficult or

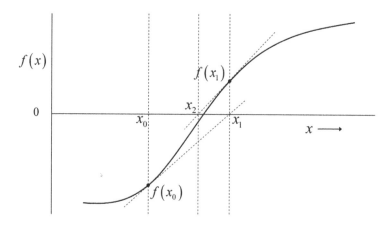

FIGURE 7.11   Schematic description of the Newton-Raphson method.

impossible to determine, there is a modification using a numerical approximation to the derivative. This is called the *modified secant method*.

The basis for the Newton-Raphson method is depicted in Figure 7.11. For an initial estimate, x_0, a tangent line to the point $\{x_0, f(x_0)\}$ is extended to the $f(x) = 0$ line to determine x_1. To do this, we need the slope of that line, $f'(x_0)$. The formula that encapsulates this from the ith to the $(i+1)^{th}$ iteration is

$$x_{i+1} = x_i - \frac{f(x_i)}{f'(x_i)} \tag{7.5}$$

The method is repeated by extending another tangent line from the point $\{x_1, f(x_1)\}$ again to the axis to determine x_2. The method is repeated until there is a judgment that there is convergence to the solution. Unlike bisection, we cannot guarantee an interval of convergence given a number of iterations. Typically, we use a relative convergence strategy. This means that we stop the process when the subsequent values of x are not changing significantly. We quantify this as an absolute relative error,

$$\varepsilon_a = \left| \frac{x_{i+1} - x_i}{x_{i+1}} \right| \overset{?}{<} \varepsilon_s \tag{7.6}$$

A typical value of the stopping criterion, ε_s, would be 1×10^{-6}, one part in a million.

Next, let's see how to implement the Newton-Raphson method on the spreadsheet. To do that, we continue Example 7.1.

Creating a "live" solution of the spherical tank equation with the Newton-Raphson method is actually simpler than for the bisection method. We first need

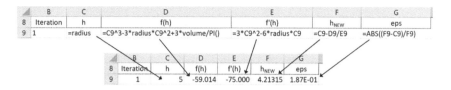

FIGURE 7.12 Initialization row for the Newton-Raphson method.

to derive the slope, $f'(x)$. For the spherical tank, this is done by differentiating Equation 7.1 to give

$$f'(h) = 3h^2 - 6Rh \qquad (7.7)$$

The initialization row of the method is shown in Figure 7.12. The initial estimate of the solution is set to the radius of the tank, in other words, at 50% full of liquid. The formula for $f(h)$ is the same as for bisection, but then a formula for $f'(h)$ is entered in cell E9. The Newton-Raphson formula is in cell F9, and the absolute relative error is computed in cell G9.

Figure 7.13 illustrates the implementation of the second, operational row and its copy down through iteration 5. The only change to the operational row is in the h cell with a pointer formula to the h_{new} value in the previous row. It is evident that the method converges rapidly with negligible error after the fourth iteration.

The case study conducted with the bisection method can now be modified by using a pointer formula to cell F13 (or F12) in the depth cell. We now see the advantage of the Newton-Raphson method here, taking only a few iterations to equal the precision of the bisection method.

However, there are two distinct disadvantages not directly evident here. First, convergence to a solution is not guaranteed for all problems, and, if the initial estimate is too far off, the method may diverge. For our example, if we pick an initial estimate very close to zero, Newton-Raphson is not successful. This is shown in Figure 7.14. A related disadvantage is that for functions with several real roots, the method can converge on different roots depending on the initial guess.

	B	C	D	E	F	G
10	=B9+1	=F9	=C10^3-3*radius*C10^2+3*volume/PI()	=3*C10^2-6*radius*C10	=C10-D10/E10	=ABS((F10-C10)/F10)

	B	C	D	E	F	G
8	Iteration	h	f(h)	f'(h)	hNEW	eps
9	1	5	-59.014	-75.000	4.21315	1.87E-01
10	2	4.21315	-0.487	-73.143	4.20649	1.58E-03
11	3	4.20649	0.000	-73.111	4.20648	3.41E-07
12	4	4.20648	0.000	-73.111	4.20648	1.60E-14
13	5	4.20648	0.000	-73.111	4.20648	0.00E+00

FIGURE 7.13 Operational row of the Newton-Raphson solution with five iterations of the method.

8	Iteration	h	f(h)	f'(h)	h_NEW	eps
9	1	0.001	190.986	-0.030	6366.83	1.00E+00
10	2	6366.83	2.57E+11	1.21E+08	4246.23	4.99E-01
11	3	4246.23	7.63E+10	5.40E+07	2832.49	4.99E-01
12	4	2832.49	2.26E+10	2.40E+07	1890	4.99E-01
13	5	1890	6.70E+09	1.07E+07	1261.67	4.98E-01

FIGURE 7.14 Newton-Raphson method is unsuccessful with initial estimate of h near zero.

A second disadvantage of the Newton-Raphson method is, if we know that it does converge to the solution, we do not know how many iterations it will take to reduce the error to an acceptable level. This will depend on the initial estimate and will change from one equation to another. We can address this disadvantage later with VBA.

In the field of numerical methods, there are more advanced techniques for solving single nonlinear algebraic equations. One of the most popular is called *Brent's method*, which is a combination of a bracketing method and an open method. Where possible, it uses the open method to achieve rapid convergence, but, if that method gets in trouble, it reverts to a stable, bracketing method until it gets in the neighborhood of the solution where it can switch back to the open method. Additional information on Brent's method and other advanced techniques can be found elsewhere (for example, Chapra and Canale 2020, Chapra and Clough 2022).

As we conclude this topic of solving single nonlinear algebraic equations, we will mention the *modified secant method* again. This method replaces the derivative, $f'(x)$, with an approximation,

$$f'(x) \cong \frac{f\big(x(1+\delta)\big) - f(x)}{\delta \cdot x} \tag{7.8}$$

where δ is a small fraction (typically on the order of 10^{-3} to 10^{-6}). You can then proceed with the Newton-Raphson formula, substituting the approximation. This is a useful modification when $f'(x)$ is difficult or impossible to derive analytically.

7.2 MATRIX OPERATIONS

This section of this chapter presumes that the reader has basic knowledge of matrix operations, such as transpose, multiplication, inverse, and computing the determinant of a square matrix. We will not expand here to provide detailed background on these. Rather, we have developed a short primer at the end of this book (Appendix A) that you can use for a review. Excel provides built-in array functions to implement these operations. We will demonstrate these through examples.

Figure 7.15 illustrates various arrays of numbers on the spreadsheet and their corresponding vector-matrix descriptions. A vertical one-dimensional array is interpreted as an $n \times 1$ column vector. A similar horizontal array is a $1 \times n$ row vector.

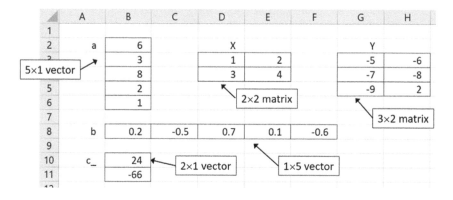

FIGURE 7.15 Examples of named cell arrays with vector-matrix descriptions.

A matrix is an array with m rows and n columns, $m \times n$. A significant number of matrices in engineering and science are square, $n \times n$.

If two vectors or matrices are to be added or subtracted, they must be of the same dimensions; that is, the same number of rows and columns. This is also considered an *array operation* because it is implemented *item-by-item*. The array operations of multiplication or division are also item-by-item. On the spreadsheet, these can be completed with individual or array formulas. The latter are illustrated with the examples in Figure 7.16.

	J	K	L	M	N
2	a	6		a·d	42
3		3			-12
4		8			40
5		2			18
6		1			-4
7					
8	d	7		a·d	-1
9		-4			7
10		5			3
11		9			-7
12		-4			5

	J	K	L	M	N
2	a	6		a·d	=a*d
3		3			
4		8			
5		2			
6		1			
7					
8	d	7		a·d	=a-d
9		-4			
10		5			
11		9			
12		-4			

	P	Q	R	S	T
2	Y			Y·Z	
3	-5	-6		-1.5	5.4
4	-7	-8		-84	-136
5	-9	2		-0.9	-6.4
6					
7	Z			Y+Z	
8	0.3	-0.9		-4.7	-6.9
9	12	17		5	9
10	0.1	-3.2		-8.9	-1.2

	P	Q	R	S	T
2	Y			Y·Z	
3	-5	-6		=Y*Z	
4	-7	-8			
5	-9	2			
6					
7	Z			Y+Z	
8	0.3	-0.9		=Y+Z	
9	12	17			
10	0.1	-3.2			

FIGURE 7.16 Example array operations with vectors and matrices.

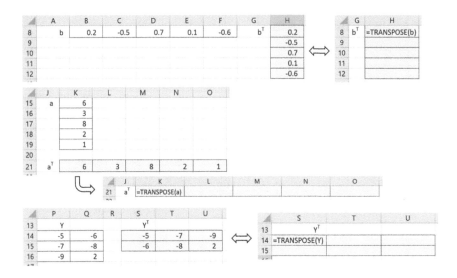

FIGURE 7.17   Examples of the TRANSPOSE function.

Matrix operations, including transpose, multiplication, and inverse, operate differently from item-by-item array operations. The *transpose* operation reverses rows and columns of a matrix and converts a column vector into a row vector, or vice versa. Thus, if $\mathbf{A}$ is an $m \times n$ matrix, then the transpose, $\mathbf{A}^T$, will be $n \times m$, and the entry in row i, column j of $\mathbf{A}$ will be in row j, column i of $\mathbf{A}^T$.

Figure 7.17 illustrates three examples of Excel's TRANSPOSE function. In each case, the destination range must be selected, and then the formula entered in the upper-left cell. The resulting range is then live, and any changes in the source array will be reflected in the transposed array immediately.

Most of us have learned how matrix multiplication works. But just in case it is new to you, the procedure is described in Figure 7.18. The figure illustrates the matrix multiplication of a general $m \times n$ matrix $\mathbf{A}$ times a general $n \times p$ matrix $\mathbf{B}$ to yield a product $m \times p$ matrix $\mathbf{C}$. Note that the inner dimensions of $\mathbf{A}$ and $\mathbf{B}$, noted as n here, must be the same.

Figure 7.19 shows several examples of matrix and vector multiplication using Excel's MMULT function. These include the inner product of $1 \times n$ row vector and $n \times 1$ column vector to produce a single, scalar result, and the outer product of an $n \times 1$ column vector and a $1 \times n$ row vector to produce an $n \times n$ matrix. Also, there is a general example of the multiplication of two matrices.

The third key matrix operation is the *matrix inverse*. It is defined for a square matrix by the following:

$$\mathbf{A} \cdot \mathbf{A}^{-1} = \mathbf{A}^{-1} \cdot \mathbf{A} = \mathbf{I} \qquad (7.9)$$

where $\mathbf{A}^{-1}$ is the inverse matrix of $\mathbf{A}$ and $\mathbf{I}$ is the *identity matrix* with ones on the diagonal and zeros off the diagonal. Excel provides a MINVERSE function

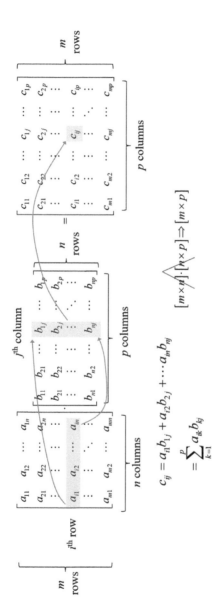

FIGURE 7.18   General scheme for matrix multiplication.

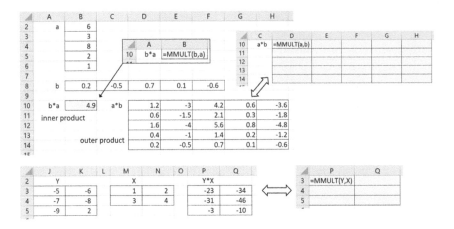

FIGURE 7.19 Examples of matrix-vector multiplication using the MMULT function.

to compute an approximate inverse matrix. Figure 7.20 shows two examples. In the first example, the MINVERSE function computes a close approximation to the inverse, and this is confirmed by multiplying the inverse by its parent matrix. The result is an identity matrix with the off-diagonal elements near zero. The second example is a matrix that does not have an inverse. The MINVERSE routine fails numerically and returns huge numbers that do not satisfy the requirements of the inverse.

The second example in Figure 7.20 is intended to be instructive. Not all square matrices have inverses. And for some, the inverse is difficult to determine numerically. This feature can be quantified by the determinant of the matrix. Analytically, the determinant is evaluated by decomposing into what are called minors. Figure 7.21 illustrates this for a 2×2 and 3×3 matrix.

	A	B	C	D	E	F	G	H	I
1	A					Ainv			
2	1	5	-7	9		-0.10016	0.07552	0.0787	0.13275
3	2	-6	4	1		-0.15501	-0.12122	-0.02107	0.18164
4	7	2	-6	-3		-0.19634	0.01312	-0.06002	0.19674
5	3	6	-4	8		0.05564	0.06916	-0.04372	0.03736
6									
7	A*Ainv								
8	1	0	0	5.6E-17	virtual				
9	1E-16	1	-2.1E-17	-2.1E-17	identity				
10	2.8E-17	-2.8E-17	1	5.6E-17	matrix				
11	5.6E-17	1.1E-16	-1.1E-16	1					

	F	G	H	I
1			Ainv	
2	=MINVERSE(A)			
3				
4				
5				

	F	G	H	I	J	K	L	
7	B					Binv		
8	1	2	3		-4.5036E+15	9E+15	-4.5E+15	inverse of B does not exist
9	4	5	6		9.0072E+15	-1.8E+16	9E+15	
10	7	8	9		-4.5036E+15	9E+15	-4.5E+15	

FIGURE 7.20 Examples of the matrix inverse with the MINVERSE function.

$$\mathbf{A} = \begin{bmatrix} a_{11} & a_{12} \\ a_{21} & a_{22} \end{bmatrix} \quad |\mathbf{A}| = a_{11}a_{22} - a_{12}a_{21}$$

sign is determined by indices of leading b_{ij} multiplier: $(-1)^{i+j}$

$$|\mathbf{B}| = \left|\begin{bmatrix} b_{11} & b_{12} & b_{13} \\ b_{21} & b_{22} & b_{23} \\ b_{31} & b_{32} & b_{33} \end{bmatrix}\right| = b_{11}\begin{vmatrix} b_{22} & b_{23} \\ b_{32} & b_{33} \end{vmatrix} - b_{12}\begin{vmatrix} b_{21} & b_{23} \\ b_{31} & b_{33} \end{vmatrix} + b_{13}\begin{vmatrix} b_{21} & b_{22} \\ b_{31} & b_{32} \end{vmatrix}$$

$$= b_{11}\left(b_{22}b_{33} - b_{23}b_{32}\right) - b_{12}\left(b_{21}b_{33} - b_{23}b_{31}\right) + b_{13}\left(b_{21}b_{32} - b_{22}b_{31}\right)$$

3x3 reduces to evaluation of three 2x2 minors
this extends upwards to 4x4 reducing to four 3x3 minors, and so on

FIGURE 7.21 Scheme for the determinant of a square matrix.

	A	B
7	|A|	-2516

⟺

	A	B
7	|A|	=MDETERM(A)

	F	G
12	|B|	6.66134E-16

⟺

	F	G
12	|B|	=MDETERM(B)

FIGURE 7.22 Application of the MDETERM function to compute determinants.

Excel provides an MDETERM function to find the determinant of a square matrix. It is based on a more efficient algorithm than the formulas depicted in Figure 7.21. Using the matrices from Figure 7.20, the results of the MDETERM function are shown in Figure 7.22. If we use the scheme in Figure 7.21, the second determinant in Figure 7.22 turns out to be identically zero. Excel's numerical method for the determinant returns a tiny number close to zero.

We can generalize our interpretation of the determinant and the MDETERM function. As the determinant of a square matrix approaches small values and eventually zero, it becomes *ill-conditioned*. This will lead to problems when, for example, we attempt to solve sets of linear equations where the matrix of coefficients yields inaccurate results related to roundoff error. Solving such sets is the topic of the next section.

7.3 SOLVING SETS OF LINEAR ALGEBRAIC EQUATIONS

Linear algebraic equations occur frequently in engineering and scientific problem solving. There are many numerical methods available for the solution of sets of these equations. Implementation of these with Excel requires coding in VBA. In this section, we will introduce a direct way for live solution of sets of linear algebraic equations utilizing the built-in matrix functions introduced in Section 7.2.

Figure 7.23 presents a specific example of three linear equations in three unknowns and a transformation that converts the system into vector-matrix format, finally with the general description $\mathbf{Ax} = \mathbf{b}$. This latter general form can be applied to sets of tens, hundreds, or even thousands of linear equations.

$$0.3x_1 + 0.52x_2 + x_3 = -0.01$$
$$0.5x_1 + x_2 + 1.9x_3 = 0.67$$
$$0.1x_1 + 0.3x_2 + 0.5x_3 = -0.44$$

$$\Longrightarrow \begin{bmatrix} 0.3 & 0.52 & 1 \\ 0.5 & 1 & 1.9 \\ 0.1 & 0.3 & 0.5 \end{bmatrix} \cdot \begin{bmatrix} x_1 \\ x_2 \\ x_3 \end{bmatrix} = \begin{bmatrix} -0.01 \\ 0.67 \\ -0.44 \end{bmatrix} \Longrightarrow \mathbf{Ax = b}$$

$$\mathbf{A} \qquad\qquad \mathbf{x} \qquad \mathbf{b}$$

FIGURE 7.23 Example set of linear equations with vector-matrix description.

Through matrix math manipulations and utilizing Equation 7.9, we can solve the general description equation for $\mathbf{x}$:

$$\left(\mathbf{A}^{-1}\mathbf{A}\right)\mathbf{x} = \mathbf{A}^{-1}\,\mathbf{b} \quad\Rightarrow\quad \mathbf{I}\,\mathbf{x} = \mathbf{A}^{-1}\,\mathbf{b} \quad\Rightarrow\quad \mathbf{x} = \mathbf{A}^{-1}\,\mathbf{b} \qquad (7.10)$$

Computing the inverse of $\mathbf{A}$ and multiplying that with $\mathbf{b}$ is not the most efficient or accurate numerical technique for the solution of $\mathbf{x}$, but it is convenient and adequately efficient in Excel because we have at hand the MINVERSE and MMULT functions. Figure 7.24 illustrates the use of these functions on the spreadsheet to solve the equations from Figure 7.23. This is a live solution, so any changes to the $\mathbf{A}$ matrix or $\mathbf{b}$ vector will affect the solution automatically.

The scheme shown in Figure 7.24 can be collapsed into a single formula for $\mathbf{x}$ by imbedding the MINVERSE function into the MMULT function. This is illustrated in Figure 7.25.

It is possible to have a set of linear equations that has no viable solution. This is caused by the fact that each equation is not unique. Rather, it can be formed from another equation or a combination of the other equations. This is revealed by

	B	C	D	E	F
1	A				b
2	0.3	0.52	1		-0.01
3	0.5	1	1.9		0.67
4	0.1	0.3	0.5		-0.44
5					
6	Ainv				x=Ainv*b
7	31.8182	-18.1818	5.45455		-14.9
8	27.2727	-22.7273	31.8182		-29.5
9	-22.7273	17.2727	-18.1818		19.8

	B	C	D	E	F
7	=MINVERSE(A)				=MMULT(Ainv,b)
8					
9					

FIGURE 7.24 Employing the MINVERSE and MMULT functions to solve the example equations.

	H
6	x
7	-14.9
8	-29.5
9	19.8

	H
6	x
7	=MMULT(MINVERSE(A),b)
8	
9	

FIGURE 7.25 Solution of the set of equations with a single formula.

the **A** matrix being singular with its determinant being zero. With the numerical methods and precision of Excel, as we saw in the previous section, the determinant may not be identically zero but very small. One way to check the validity of a set of linear equations is to compute the determinant of the **A** matrix.[4]

One could approach the solution of a set of linear equations using the Solver. In this way, one would propose estimates for the unknowns, compute the left side of the equations, subtract that from the right-side constant, and sum the squares of those remainders. The Solver would then adjust the estimates to minimize the sum of squares. Given the ability to create a live solution with the MINVERSE and MMULT functions, there is little incentive to use the Solver in this way. However, when we approach the solution of sets of nonlinear equations, using the Solver is a valid approach.

7.4   SOLVING SETS OF NONLINEAR ALGEBRAIC EQUATIONS

If they are well-conditioned, as indicated by the determinant of their **A** matrix having a reasonable value, very large sets of linear equations can be readily solved. That is not the case for sets of nonlinear algebraic equations. These may have more than one solution or no solution. They may have a solution that is difficult to find. And, as the number of nonlinear equations increases beyond just a few, solution may be challenging or even impossible. We will investigate these issues with the following example set of two nonlinear equations.

$$
\begin{aligned}
f_1(x_1, x_2) &= x_1^2 - x_2 + 1 = 0 \\
f_2(x_1, x_2) &= 2\cos(x_1) - x_2 = 0
\end{aligned}
\tag{7.11}
$$

Since we have just two equations, we can explore their solution graphically. This is a luxury we do not have with larger sets of equations. We do this by solving each equation analytically for x_2,

$$
\begin{aligned}
x_2 &= x_1^2 + 1 \\
x_2 &= 2\cos(x_1)
\end{aligned}
\tag{7.12}
$$

and then graphing the lines of the two equations. Intersections of the lines indicate solutions. Figure 7.26 presents an Excel plot for the domain $-2 \le x_1 \le 2$ and captures the two possible solutions for these equations.

If we want to find the solution for the domain $x_1 \ge 0$, our initial estimates should be in the neighborhood of the solution on the right of Figure 7.26. As an example, we might choose $x_1 = 1$, $x_2 = 3$. Figure 7.27 shows the setup on a spreadsheet to attempt a solution with the Solver. Initial estimates for x_1 and x_2 are provided in two cells, named x_1 and x_2. These are used to evaluate the left sides of the Equations 7.11 in cells named f_1 and f_2. The goal is that these values should approach zero so that the equations are satisfied. To provide a single criterion for the Solver, the SUMSQ function is used to provide the sum of squares of f_1 and f_2.

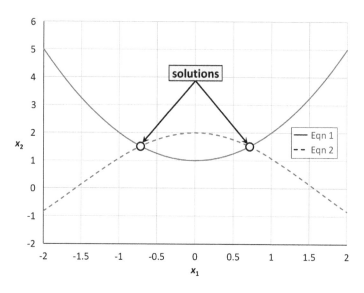

FIGURE 7.26 Plot of curves for two nonlinear equations showing solutions at their intersections.

	B	C	D	E
2	x_1	1	f_1	-1
3	x_2	3	f_2	-1.9194
4				
5			SSE	4.68408

$\Longleftrightarrow$

	B	C	D	E
2	x_1	1	f_1	=x_1^2-x_2+1
3	x_2	3	f_2	=2*COS(x_1)-x_2
4				
5			SSE	=SUMSQ(f_1,f_2)

FIGURE 7.27 Setup of two nonlinear equations for solution using the Solver.

The Solver can then be invoked to minimize the value in the SSE cell by adjusting the x_1 and x_2 cell values. This is shown in Figure 7.28. With the SSE at approximately 10^{-12}, it is clear that a close approximation to the solution has been determined in the x_1 and x_2 cells. With reference to Figure 7.26, we see that this is the intersection of the curves on the right, in the positive x direction.

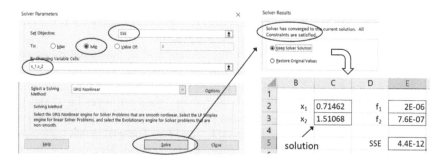

FIGURE 7.28 Utilizing the Solver to determine the solution of the two nonlinear equations.

An initial estimate closer to the solution on the left in Figure 7.26 would then converge to that location. Again, because we have only two equations, we can corroborate our results graphically. You will explore the solution of more than two simultaneous equations in the last problem at the end of this chapter.

You will note that the use of the Solver is a manual operation, thus not a live solution. It is possible to set up a live solution, and we will address that in Section 7.6.

7.5 SOLVING DIFFERENTIAL EQUATIONS

Most differential equations that arise in science and engineering do not have analytical solutions and must be solved numerically; that is, with computers. These can be single equations or sets of equations that describe rates of change with respect to an independent variable like time or position.

With reference to single equations, there are two types that lead to different numerical methods for their approximate solution. The first are equations where the rate of change of the dependent variable, dy/dt, is solely a function of the independent variable, t,

$$\frac{dy}{dt} = f_1(t) \qquad \text{where } y(t=0) = y_0 \tag{7.13}$$

The second type, which is more complex, deals with differential equations where the rate of change depends on *both* the dependent and the independent variables,

$$\frac{dy}{dt} = f_2(t, y) \qquad \text{where } y(t=0) = y_0 \tag{7.14}$$

We distinguish between these types as their solution strategies can differ. The solution to the first type can be written as

$$\int_{y_0}^{y(t)} dy = y(t) - y_0 = \int_0^t f_1(t)\,dt \tag{7.15}$$

and the technique applied is to evaluate the integral on the right by "area under the curve" or *quadrature methods*. The solution to the second type requires *numerical methods* that approximate y step by step in time. We will illustrate both solution types in this section.

As an aside, we will mention two variations on the above classification. The first is an equation of the form

$$\frac{dy}{dt} = y \cdot f(t) \quad y(0) = y_0 \tag{7.16}$$

This form is separable, and the solution can be written

$$\int\limits_{y_0}^{y(t)} \frac{dy}{y} = \ln\left(\frac{y(t)}{y_0}\right) = \int\limits_0^t f(t)dt \quad \Rightarrow \quad \frac{y(t)}{y_0} = e^{\int_0^t f(t)dt} \tag{7.17}$$

So, although Equation 7.16 is of the second type, quadrature methods can be employed to generate solutions with Equation (7.17).

The second is

$$\frac{dy}{dt} = f(t, y, F) \quad y(0) = y_0 \quad F = F(t) \tag{7.18}$$

where F is an input or *forcing function*. In the present context, it is a function that appears in the equation and is only a function of time, and not of any of the other variables. In effect, it is a constant for each value of time. For differential equations arising from physical systems, it represents how external inputs influence the system.

We will also mention that there are higher-order ordinary differential equations, such as

$$a\frac{d^2y}{dt^2} + b\frac{dy}{dt} + y = f(t) \quad y(0) = y_0 \quad \frac{dy}{dt}(0) = y_0' \tag{7.19}$$

that can be decomposed into two first-order equations. And, occasionally, one of the conditions will be at the end of the solution and the other at the beginning. This latter situation is called a two-point or split boundary value scenario.

7.5.1 Solutions by Quadrature

A simple example to illustrate a common method for quadrature on the spreadsheet is

$$\frac{dy}{dt} = f(t) = t\cos(t) \quad y(0) = 0 \tag{7.20}$$

where the solution for $y(t)$ is

$$y(t) = \int\limits_0^t t\cos(t)dt \tag{7.21}$$

The method used here is called *trapezoidal rule*, and it is illustrated in Figure 7.29. The domain of t over which the function, $f(t)$, is being integrated is divided into intervals, $\Delta t = t_{i+1} - t_i$, and the areas of the trapezoids, one of which is shown in the figure, are summed up to approximate the integral.

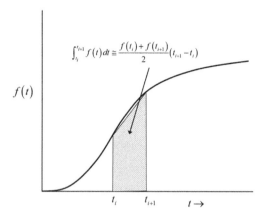

FIGURE 7.29 Trapezoidal rule scheme for quadrature.

The spreadsheet solution for this example for $0 \le t \le \pi/2$ is shown in Figure 7.30. In this case, 100 intervals are used, the trapezoidal areas are computed for each interval, and a running sum approximates the integral up to the end of each interval, including the final $t = \pi/2$. For this example, there happens to be an analytical solution for the integral,

$$y(t) = \int_0^t t\cos(t)\,dt = \cos(t) + t\sin(t) - 1$$

$$y\left(\frac{\pi}{2}\right) = \frac{\pi}{2} - 1 \cong 0.5708$$

(7.22)

and we can see that our approximate result in Figure 7.30 is accurate. If better accuracy is desired, we could divide the integration range into smaller intervals.

	B	C	D	E
1	tf	1.5708		
2	n	100		
3				
4	t	f(t)	Trapezoid	Sum
5	0	0.00000		0.00000
6	0.0157	0.01571	0.00012	0.00012
7	0.0314	0.03140	0.00037	0.00049
8	0.0471	0.04707	0.00062	0.00111
9	0.0628	0.06271	0.00086	0.00197
10	0.0785	0.07830	0.00111	0.00308
	○	○	○	○
100	1.4923	0.11708	0.00201	0.56607
101	1.5080	0.09469	0.00166	0.56773
102	1.5237	0.07177	0.00131	0.56904
103	1.5394	0.04835	0.00094	0.56998
104	1.5551	0.02443	0.00057	0.57055
105	1.5708	0.00000	0.00019	0.57074

	B	C	D	E
1	tf	=PI()/2		
2	n	100		
3				
4	t	f(t)	Trapezoid	Sum
5	0	=B5*COS(B5)		=C5
6	=B5+tf/n	=B6*COS(B6)	=(C6+C5)/2*tf/n	=D6+E5
7	=B6+tf/n	=B7*COS(B7)	=(C7+C6)/2*tf/n	=D7+E6
8	=B7+tf/n	=B8*COS(B8)	=(C8+C7)/2*tf/n	=D8+E7
9	=B8+tf/n	=B9*COS(B9)	=(C9+C8)/2*tf/n	=D9+E8
10	=B9+tf/n	=B10*COS(B10)	=(C10+C9)/2*tf/n	=D10+E9

$$\equiv \int_0^{\pi/2} t\cos(t)\,dt$$

FIGURE 7.30 Trapezoidal rule solution of Equation 7.20 for $0 \le t \le \pi/2$ with 100 intervals.

There are many numerical methods for accomplishing quadrature (see Chapra and Canale 2020). However, the trapezoidal rule has the advantage that it is simple to implement on the spreadsheet and, with intervals small enough, it provides a reasonably accurate result.

7.5.2 Solving Initial-Value Ordinary Differential Equations

The primary difference between the two Equations 7.13 and 7.14 is for the first, $f_1(t)$ is known *a priori* for any domain of the independent variable, t. That is not the case for $f_2(t)$ because it depends on the current value of the dependent variable, y. Figure 7.31 depicts the initial situation for approximate solution of an initial-value ordinary differential equation. We know the initial values of the independent and dependent variables and can then evaluate the initial rate of change from the differential equation.

The question is how to proceed from the situation in Figure 7.31 to evolve the approximate solution of the differential equation. The *Euler method*, which is the simplest approach, is well suited to implementation on a spreadsheet, and, for many real problems, is adequate. Figure 7.32 illustrates the basis of the Euler method.

From the initial scenario in Figure 7.31, a straight line based on the original derivative value is projected out to t_1, using a small time interval Δt, and the dependent variable at that point becomes the approximation to the solution, as

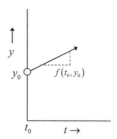

FIGURE 7.31   Initial scenario for solving a differential equation.

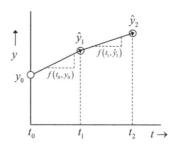

FIGURE 7.32   Illustration of the first two steps of the Euler method.

shown in Figure 7.32. This process is then repeated to a desired final t value. The general formula after the first step is shown in Equation 7.23 where the caret over the dependent variables ($\hat{y}$) denotes that they are estimates.

$$\hat{y}_{i+1} = \hat{y}_i + f(t_i, \hat{y}_i) \cdot \Delta t \tag{7.23}$$

As an example, we consider solving the following differential equation:

$$\frac{dy}{dt} = -0.2y^2 \qquad y(0) = 3 \qquad 0 \le t \le 5 \tag{7.24}$$

In a similar pattern to our live solution of single, nonlinear algebraic equations, we start on the spreadsheet with basic parameter values and an initialization row. This is shown in Figure 7.33. This represents the scenario shown in Figure 7.31 with the initial value of y set in the y_0 cell and, for $t = 0$, a pointer to y_0 and an evaluation of the first derivative value.

The second row of the table represents the operational row and can be copied downward to complete the solution. This is shown in Figure 7.34. Cell C6 is the first implementation ($i = 1$) of Equation 7.23.

	B	C	D
1	y₀	3	
2	dt	0.1	
3			
4	t	y	dy/dt
5	0.00	3.000	-1.800

	B	C	D
1	y₀	3	
2	dt	0.1	
3			
4	t	y	dy/dt
5	0	=y0	=-0.2*C5^2

FIGURE 7.33 Basic parameter values and initialization row for Euler method solution.

	B	C	D
4	t	y	dy/dt
5	0.00	3.000	-1.800
6	0.10	2.820	-1.590
7	0.20	2.661	-1.416
8	0.30	2.519	-1.269
9	0.40	2.392	-1.145
10	0.50	2.278	-1.038
51	4.60	0.781	-0.122
52	4.70	0.769	-0.118
53	4.80	0.757	-0.115
54	4.90	0.745	-0.111
55	5.00	0.734	-0.108

	B	C	D
4	t	y	dy/dt
5	0	=y0	=-0.2*C5^2
6	=B5+dt	=C5+D5*dt	=-0.2*C6^2
7	=B6+dt	=C6+D6*dt	=-0.2*C7^2
8	=B7+dt	=C7+D7*dt	=-0.2*C8^2
9	=B8+dt	=C8+D8*dt	=-0.2*C9^2
10	=B9+dt	=C9+D9*dt	=-0.2*C10^2
51	=B50+dt	=C50+D50*dt	=-0.2*C51^2
52	=B51+dt	=C51+D51*dt	=-0.2*C52^2
53	=B52+dt	=C52+D52*dt	=-0.2*C53^2
54	=B53+dt	=C53+D53*dt	=-0.2*C54^2
55	=B54+dt	=C54+D54*dt	=-0.2*C55^2

FIGURE 7.34 The second operational row is copied downward to complete the solution.

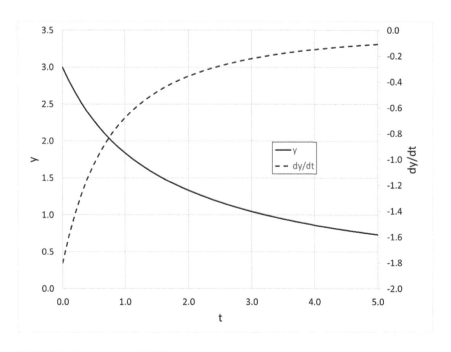

FIGURE 7.35   Solution of differential equation along with the derivative.

The solutions for y and its derivative are plotted in Figure 7.35. At each step of the Euler method, there is an error between the approximate and true solutions. These errors compound as the solution proceeds. If the step size is too large, this leads to a very inaccurate solution. A way to limit this is to reduce the step size. Numerically, one can make the step size so small that the results are impacted by *roundoff errors* due to the precision of Excel's number representation, and it is possible to run out of rows on the spreadsheet. These limitations are rarely encountered, however. More commonly, we want to have a step size that is small enough to ensure a representative solution. A practical approach to this is to start with a larger step size and reduce it until the solution no longer changes. Figure 7.36 shows results from such a procedure for our example differential equation.

We can observe that, for our original step size of 0.1, the solution is relatively close to that of a solution with step size of 0.01. However, for the larger step sizes, the error in the solution is too great. We can use our judgment to decide if the finest step size is required. In any case, the extent of the solution of the spreadsheet is manageable. A step size less than 0.01 would not be worthwhile.

As an aside, you should note that using smaller step sizes would necessitate manually extending the range of the operational rows depicted in Figure 7.34 in order to compute the desired range. Later in this book we will demonstrate how VBA macros can be designed to avoid such issues.

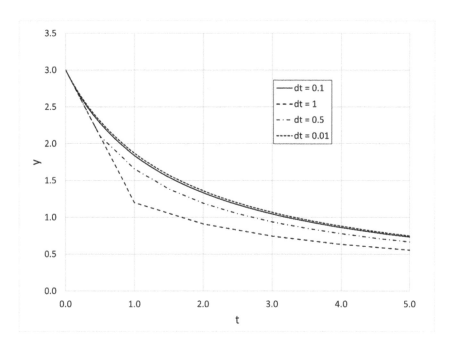

FIGURE 7.36 Euler method solution for different step sizes.

At this point, we will include a more complicated example. This is a second-order differential equation with split boundary conditions. This will illustrate how our Euler solution technique can be extended to multiple first-order differential equations and more complicated initial and final specifications.

Example 7.2 Solution of a Second-order Differential Equation with Split Boundary Conditions

Consider the following second-order differential equation with specifications on the dependent variable, y, at the beginning and end of the solution interval.

$$\frac{d^2y}{dt^2} = \frac{1}{4}\frac{dy}{dt} + y \qquad y(0) = 5 \qquad y(10) = 8 \qquad 0 \le t \le 10 \qquad (7.25)$$

The way we approach second-order equations is to decompose them into two first-order equations by introducing an additional dependent variable, y_1, as shown below.

$$\frac{dy}{dt} = y_1$$
$$\frac{dy_1}{dt} = \frac{1}{4}y_1 + y \qquad (7.26)$$

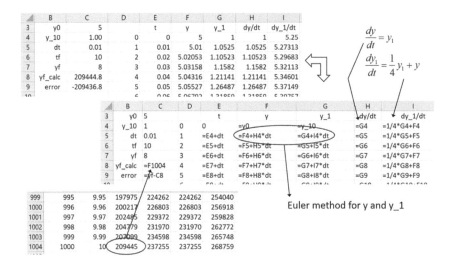

FIGURE 7.37 Initial attempt at solution for the split boundary value equations.

We are faced with the problem that we do not know the value of y_1 at $t = 0$. The first step of the solution scheme is to proceed with an estimate for that variable. Here, we will guess $y_1 = 1$. Figure 7.37 shows our initial attempt at the solution with a time step of 0.01. The two first-order differential equations are solved simultaneously with the Euler method.

The final value for y is about 209,000. It is supposed to be 8. So, on the spreadsheet, an error is computed between these two values. To meet our desired end condition, we want the error to be reduced to zero by adjusting our estimate of y_1. This can be done with Goal Seek or the Solver.

Figure 7.38 illustrates the use of Goal Seek to find the correct initial value of y_1. As the figure shows, this is approximately $y_1 = -4.41$. It should be mentioned that, with a similar application, the Solver fails. With some exploration, it is clear that the correct value of y_1 is between -4 and -5. If the initial estimate of y_1 is set at -4.5, and with constraints between -4 and -5, the Solver is successful. Goal Seek happens to solve this problem without any adjustments.

Figure 7.39 presents a plot of the solution of this differential equation. It includes the plot of the derivative of y, that being y_1. Also, you can see that the final condition of $y = 8$ is met. Note that, since we made use of Goal Seek to satisfy this condition, our solution is not completely live. If we change initial or final specifications, we will have to rerun Goal Seek to update the solution.

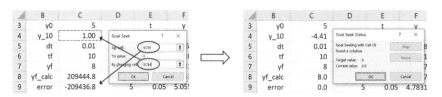

FIGURE 7.38 Application of Goal Seek to determine the value of y_1 that yields $y(10) = 8$.

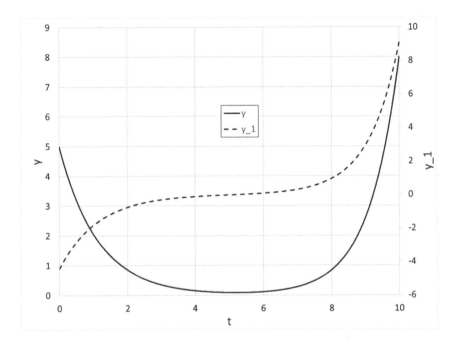

FIGURE 7.39 Solution of second-order differential equation with split boundary conditions.

Although Euler's method was adequate for the examples in this section, we have mentioned that it is lacking for some applications. If you are unfamiliar with numerical methods we urge you to learn as much as possible about the topic via courses, workshops, books, or the internet. From our own experience, this knowledge will serve you well as you develop your own Excel-based software for more complex problem solving.

7.6   IMPLICIT NONLINEAR RELATIONSHIPS YIELDING CIRCULAR CALCULATIONS

The situation we are addressing in this final section is illustrated schematically in Figure 7.40. We have a single formula or a sequence of formulas that require an input, x, but produce the same variable as an output. If we start by estimating

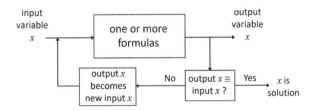

FIGURE 7.40 Implicit nonlinear relationship with substitution scheme for solution.

a value for the input and execute the calculations, usually the output does not agree with the input. A simple substitution technique, formally called *fixed-point iteration*, returns the output x as the new input x, and the procedure is repeated until the output approximately agrees with the input.

A simple example of an implicit relationship is

$$x = \frac{1}{\sin(x)} \tag{7.27}$$

It should be noted that, by moving the term on the right side of Equation 7.27 to the left side, we would have our general root-finding formulation $f(x) = 0$ scenario and could approach the solution using the methods of Section 7.1. But often, our calculations formulate naturally as $x = g(x)$, where for Equation 7.27, $g(x) = 1/\sin(x)$. In such cases, it is attractive to solve the implicit relationship using the technique illustrated in Figure 7.40. What makes this scheme even more attractive is that Excel provides us with a tool, the iterative solver, to implement this.

A problem with this approach is that it may not converge to the solution. Often it does, and when it doesn't it can sometimes be reformulated in a way that the scheme converges. We will not go further into the mathematical details of convergence at this point (see Chapra and Canale 2020 for the details), but instead focus on the implementation of the technique on the spreadsheet. The initial setup is shown in Figure 7.41. The initial estimate for the solution is 0.5, and the first evaluation of the right side of Equation 7.27 is 2.086. The next step would be to return this latter value to the x cell and repeat the calculation.

A way to automate this calculation is to enter a pointer formula, =B4, into cell B3, but Excel flags this as an error as soon as we do this. Figure 7.42 shows what happens. Excel recognizes that there is a *circular calculation*[5] and stalls with an error message in the Status Bar.

To remedy the circular reference error, we need to activate Excel's *iterative solver*. This is done via File ⇨ Options ⇨ Formulas as shown in Figure 7.43. Once this is completed, the first iteration of the circular substitution occurs on

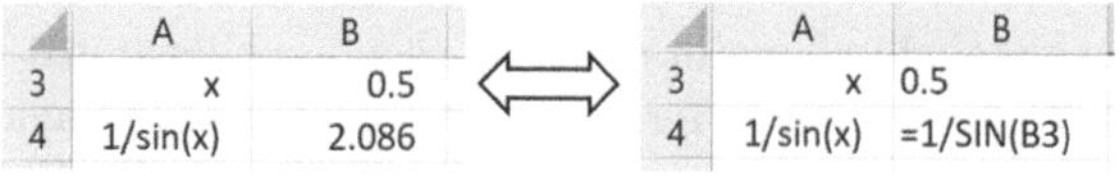

FIGURE 7.41   Initial spreadsheet for iterative solution of Equation 7.27.

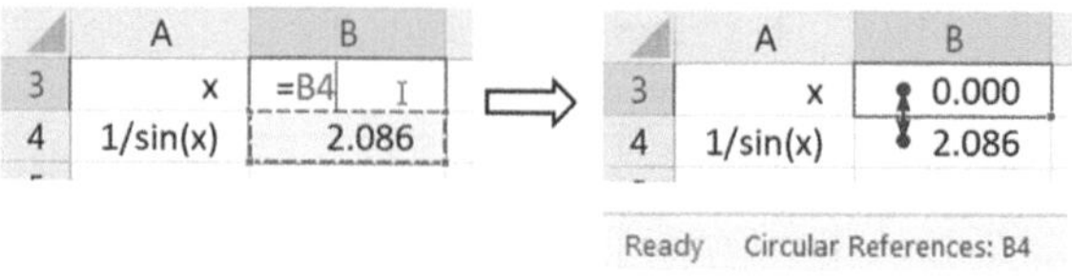

FIGURE 7.42   Circular reference error when pointer formula to cell B4 entered in cell B3.

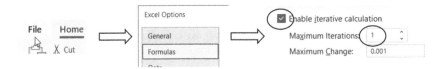

FIGURE 7.43 Activating the iterative solver with maximum iterations equal to one.

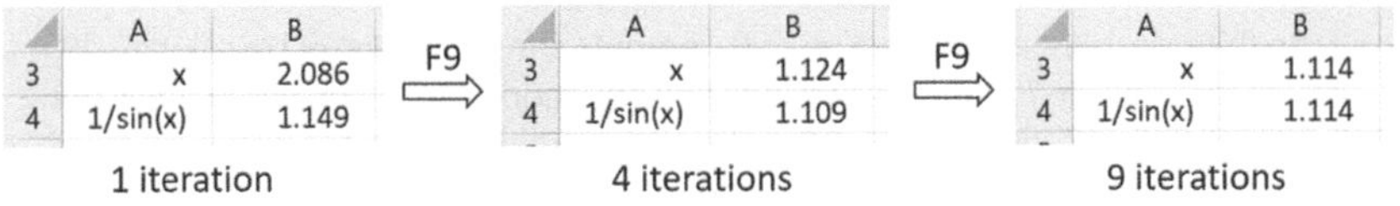

	A	B			A	B			A	B
3	x	2.086	F9	3	x	1.124	F9	3	x	1.114
4	1/sin(x)	1.149		4	1/sin(x)	1.109		4	1/sin(x)	1.114

1 iteration   4 iterations   9 iterations

FIGURE 7.44 Repeating the circular calculation manually to convergence.

the spreadsheet, and subsequent iterations can be executed, one by one, by pressing the calculate shortcut key, F9. Eventually, the process converges, as shown in Figure 7.44. This indicates that nine iterations are required for the input and output x's to agree to four significant figures.

Now that we understand that the process does converge and requires about 10 iterations, we can modify the Maximum Iterations shown on the right in Figure 7.43 to a value such as 25, and the convergence will take place in live fashion on the spreadsheet. By increasing the number of digits displayed, we can determine that convergence to six significant figures is achieved for this case. We now provide a practical example.

Example 7.3   Using the Iterative Solver to Determine the Liquid-Vapor Equilibrium of Water

For mechanical and chemical engineers, the equilibrium of liquid water and steam is important. Data on this equilibrium are traditionally available in "the steam tables." It is also possible, for moderate pressures and temperatures, to describe this equilibrium with two equations, the *ideal gas law* and the *Antoine equation* for vapor pressure.

$$PV = \frac{m_V}{MW}RT \quad \text{and} \quad \log_{10}(P) = A - \frac{B}{C+T} \qquad (7.28)$$

where with example units

P:	pressure in the vessel, kPa
V:	volume of the vapor space, m³
m_V:	mass of vapor, kg
MW:	molecular weight of water, 18.02 kg/kmol
R:	gas law constant, 8.314 m³ kPa/(kmol·K)
T:	temperature, K
A,B,C:	Antoine parameters for water: $A = 8.21$, $B = 2354.7$, $C = 7.56$

Given values for V and m_v, and the other constant parameter values, we want to be able to solve for P and T. We can approach this problem as the solution of two nonlinear algebraic equations; however, we can also formulate it as a circular calculation by solving the ideal gas law for T and the Antoine equation for P:

$$T = \frac{MW \cdot V}{R \cdot m_v} P$$

$$P = 10^{A - \frac{B}{C+T}}$$

(7.29)

In this way, we estimate a value of P, compute T from the first of Equations 7.29, and then compute P from the second. The equations are solved when the input value agrees with the output value. This is a circular calculation that involves two equations. Figure 7.45 presents the spreadsheet setup for this scheme and shows the result of the first calculation. Here, we have chosen 200 kPa as the initial pressure estimate.

We are concerned that the Pnew value is so much different than the initial P value. If we try 13.55 as the P value, the Pnew value is equal to zero. So, empirically, it seems that this formulation does not converge. If we run a case study on initial P values versus Pnew values using a data table, we can create a plot (Figure 7.46) that reveals the problem with this formulation. The initial estimate of 200 kPa evaluates to a low value for Pnew of about 13 kPa. The figure shows that this value is reflected off the 45°-line to become the next P value, and its evaluation is effectively zero. This is an example of a nonconvergent substitution scheme and reveals the cause of the instability. That is, the slope of the Pnew versus P line in the vicinity of the solution, where it crosses the 45°-line, is greater than one. A slope of less than one will converge, so we need to reformulate the problem.

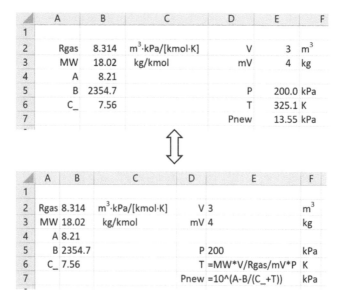

	A	B	C	D	E	F
1						
2	Rgas	8.314	m³·kPa/[kmol·K]	V	3	m³
3	MW	18.02	kg/kmol	mV	4	kg
4	A	8.21				
5	B	2354.7		P	200.0	kPa
6	C_	7.56		T	325.1	K
7				Pnew	13.55	kPa

	A	B	C	D	E	F
1						
2	Rgas	8.314	m³·kPa/[kmol·K]	V 3		m³
3	MW	18.02	kg/kmol	mV 4		kg
4	A	8.21				
5	B	2354.7		P 200		kPa
6	C_	7.56		T =MW*V/Rgas/mV*P		K
7				Pnew =10^(A-B/(C_+T))		kPa

FIGURE 7.45 Spreadsheet setup for water liquid-vapor equilibrium circular calculation.

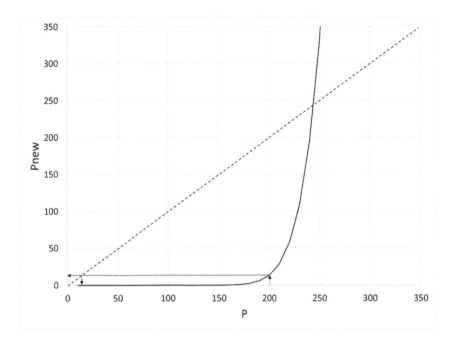

FIGURE 7.46 Graphical depiction of circular calculation.

For a general implicit relationship, $x = g(x)$, the requirement for a convergent solution is (Chapra and Canale 2020)

$$\left|\frac{dg}{dx}(x_{soln})\right| < 1 \tag{7.30}$$

To obtain another formulation, we solve the second of Equations 7.28 for T and the first for P. The result is

$$T = \frac{B}{A - \log_{10} P} - C$$

$$P = \frac{R \cdot m_v}{MW \cdot V} T \tag{7.31}$$

We can modify the spreadsheet shown in Figure 7.45 to implement this scheme, and the result is depicted in Figure 7.47.

	A	B	C	D	E	F
2	Rgas	8.314	m³·kPa/[kmol·K]	V	3	m³
3	MW	18.02	kg/kmol	mV	4	kg
4	A	8.21				
5	B	2354.7		P	200.0 kPa	
6	C_	7.56		T	390.9 K	
7				Pnew	240.49 kPa	

	D	E	F
2	V 3		m³
3	mV 4		kg
4			
5	P 200		kPa
6	T =B/(A-LOG10(P))-C_		K
7	Pnew =Rgas*mV/MW/V*T		kPa

FIGURE 7.47 Modified formulation for circular calculation.

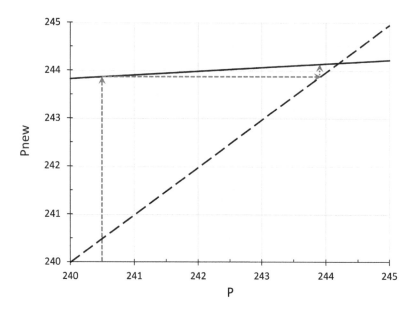

FIGURE 7.48 Graphical depiction of circular calculation for modified formulation indicating rapid convergence after only a few iterations.

Figure 7.48 repeats Figure 7.46 and shows why this formulation converges quickly. The slope of the curve, which for this case is a relatively flat straight line, is much less than one. Not only does this indicate convergence, but the mild slope also means convergence will be rapid.

If we replace the initial estimate with a pointer formula to the Pnew cell and enable the iterative solver, we can determine that the scheme converges quickly in just a few iterations to six significant figures. To do this, we adjust the iterative solver's Maximum Change value to 0.000001. This prevents the iterative solver from stalling prematurely. We can then set up the iterative solver for 10 iterations, and it provides automatic convergence.

As Figure 7.49 shows, the iterative calculation updates automatically if a parameter of the calculation is changed. Here, we use a two-variable data table with variations in V and m_V to illustrate this.

	D	E	F
2	V	3	m³
3	mV	4	kg
4			
5	P	244.138	kPa
6	T	396.9	K
7	Pnew	244.138	kPa

	N	O	P	Q	R	S	T
1					V		
2		244.14	2	2.5	3	3.5	4
3			3	277.34	217.84	178.87	151.45	131.13
4			3.5	327.77	257.38	211.30	178.87	154.85
5	mV		4	378.85	297.43	244.14	206.64	178.87
6			4.5	430.53	337.93	277.34	234.72	203.15
7			5	482.75	378.85	310.88	263.07	227.67

FIGURE 7.49 Completed iterative solution along with case study on m_V and V.

As we have introduced in this chapter, equations in engineering and science present themselves in distinct categories, and Excel spreadsheets can be developed to solve these using numerical methods. Often these are the simplest of the numerical methods available, yet they are adequate for the solutions required. We have seen advantages when live solution techniques are used, but there are situations where use of Goal Seek and the Solver would also be appropriate.

PROBLEMS

7.1 Employ the bisection method on a spreadsheet to solve for a root of

$$\sin(x+2)\cosh(x)+2=0$$

in the interval $1 \le x \le 3$.

7.2 Consider the modified "humps" function from Problem 5.5:

$$f(x)=\frac{1}{(x-0.3)^2+0.01}+\frac{1}{(x-0.9)^2+0..04}-26$$

   a. Create a plot of the function and identify approximate locations of roots of the function.

   b. Set up a live bisection solution of the equations and pick appropriate initial bracketing values of x to solve for each of the solutions identified in (**a**).

   c. Develop a live Newton-Raphson solution for a root of this function. Try initial solution estimates of $x = 0, 0.3, 0.6$, and 0.8, and interpret the results with help from the plot from part (**a**).

7.3 Figure P7.3 illustrates the mechanical linkage of four rigid bars with the two lower vertices being fixed. This was introduced in Problem 5.10. As this linkage is rotated back and forth, the two vertical members and the upper horizontal member move, but their movement is constrained by the linkage.

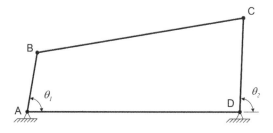

FIGURE P7.3 Four-bar linkage.

Following the diagram, as the angle θ_1 is increased from 0 to 180°, the angle θ_2 responds according to the linkage relationship. The equation that constrains the linkage movement is

$$\overline{AD} = \overline{AB}\cos\theta_1 - \overline{CD}\cos\theta_2 + \sqrt{\overline{BC}^2 - \left(\overline{CD}\sin\theta_2 - \overline{AB}\sin\theta_1\right)^2}$$

a. For the following values, set up a spreadsheet to solve for θ_2 given a value of θ_1 using the bisection method.

$$\overline{AB} = 1.5\text{m}, \ \overline{BC} = 6\text{m}, \ \overline{CD} = 3.5\text{m}, \text{ and } \overline{AD} = 5\text{m}$$

b. Create a case study for θ_1 from 0 to 180. Make a plot of θ_2 versus θ_1 for the solutions.

7.4 As mentioned at the end of Section 7.1.1, *false position* is a bracketing method that uses a straight line between the points $\{x_L, f(x_L)\}$ and $\{x_H, f(x_H)\}$ with the next solution estimate where the line intersects the x axis at $f(x) = 0$. The value of x at the intersection can be computed with the formula in Figure P7.4. Employing the same approach as for bisection, set up a spreadsheet using false position to compute the root of the spherical tank from Example 7.1 with the parameters and spreadsheet setup patterned after Figure 7.7.

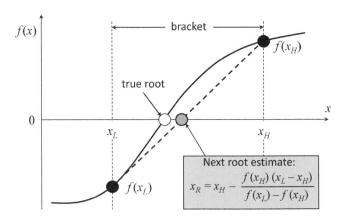

FIGURE P7.4   False position method.

7.5 *Van der Waals equation* of state for a gas is

$$P = \frac{RT}{\hat{V} - b} - \frac{a}{\hat{V}^2}$$

where, with typical units,

P: pressure, kPa
T: temperature, K
$\hat{V}$:   specific volume, L/mol
R: gas law constant, $\cong 8.3145$ L·kPa/(mol·K)
a, b: parameters depending on the gas under study

For carbon dioxide, $a = 365.4$ L²·kPa/mol² and $b = 0.04280$ L/mol.
  a. For $P = 950$ kPa and $T = 300$ K, set up a Newton-Raphson method on the spreadsheet to solve for $\hat{V}$.
  b. Rearrange the equation of state formula to solve for one of the $\hat{V}$ variables. Use the iterative solver to confirm your result from part (a) for the same P and T values.
7.6 A classic example of Equation 7.13 is the density function for the *standard normal (Gaussian) distribution* described by

$$\frac{dP}{dz} = f(z) \quad P(-\infty) = 0 \quad f(z) = \frac{1}{\sqrt{2\pi}} e^{-\frac{1}{2}z^2} \quad P(z) = \frac{1}{\sqrt{2\pi}} \int_{-\infty}^{z} e^{-\frac{1}{2}z^2} dz$$

where
z: a standard random variable with mean, $\mu = 0$, and standard deviation, $\sigma = 1$
P: *cumulative probability* of a random sample falling in the interval $-\infty$ to z

Practically,

$$f(-5) = \frac{1}{\sqrt{2\pi}} e^{-\frac{1}{2}5^2} \cong 1.5 \times 10^{-6}$$

so we can find an accurate, approximate value of P from

$$P(z) \cong \frac{1}{\sqrt{2\pi}} \int_{-5}^{z} e^{-\frac{1}{2}z^2} dz$$

For an interval, $\Delta z = 0.01$, create a spreadsheet that utilizes the trapezoidal rule to compute P versus z over the domain $-5 \leq z \leq 5$. Create a plot of the results.
7.7 Use Excel's built-in vector-matrix functions to solve the following set of linear algebraic equations.

$$-0.083x_1 - 0.875x_2 + 0.645x_3 + 0.675x_4 = 62.1$$
$$-0.258x_1 - 0.730x_2 + 0.898x_3 - 0.797x_4 = 91.8$$
$$0.262x_1 - 0.467x_2 + 0.251x_3 - 0.127x_4 = 74.9$$
$$0.582x_1 + 0.025x_2 - 0.907x_3 + 0.362x_4 = 130.9$$

7.8 A special matrix called the *Hilbert matrix* is famous for being *ill-conditioned*. Its form used in the following four equations in four unknowns is

$$
\begin{bmatrix}
1 & 1/2 & 1/3 & 1/4 \\
1/2 & 1/3 & 1/4 & 1/5 \\
1/3 & 1/4 & 1/5 & 1/6 \\
1/4 & 1/5 & 1/6 & 1/7
\end{bmatrix}
\mathbf{x} =
\begin{bmatrix}
2 \\
9/7 \\
1 \\
3/4
\end{bmatrix}
$$

a. On a spreadsheet, compute the determinant of the coefficient matrix. What do you observe?
b. Solve the system of equations using Excel's built-in vector-matrix functions.

7.9 For Equations 7.28 and the parameter values of Example 7.3, solve for pressure and temperature using the Solver and minimization of the sum of the squares of equation error, as explained in Section 7.4. Compare your solution to that of Example 7.3. If they are different, consider scaling one of the equations so the residual errors of the two equations are comparable in magnitude.

7.10 For the following set of nonlinear equations from Section 7.4,

$$x_1^2 - x_2 + 1 = 0$$
$$2\cos(x_1) - x_2 = 0$$

rearrange them for solution by circular calculation and implement that solution using Excel's iterative solver.

7.11 Create a spreadsheet to solve the differential equation

$$\frac{dy}{dx} = \ln(5 - 4\cos(x)) \qquad y(0) = 0$$

over the domain $0 \le x \le \pi$ using trapezoidal rule quadrature with $\Delta x = \pi/100$.

7.12 Solve this differential equation numerically:

$$\frac{dy}{dt} = -750y - 1100e^{-2t} \qquad y(0) = 5 \qquad 0 \le t \le 5$$

using Euler's method. Determine a step size, Δt, which provides an accurate solution. Create a plot of your results.

7.13 Differential equations are important in structural engineering. Figure P7.13 illustrates a cantilever beam (beam fixed at one end) subject to a load (force) at its end. The profile of deflection of the beam is described by a second-order differential equation,

$$\frac{d^2y}{dx^2} = -\frac{F}{EI}(L - x) \qquad y(0) = 0 \qquad \frac{dy}{dx}(0) = 0 \qquad 0 \le x \le L$$

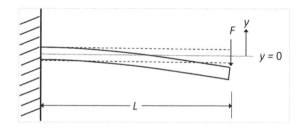

FIGURE P7.13 Deflection of cantilever beam under a load placed at the end.

where
y: vertical displacement of the beam at a given value of x, m,
x: location on the beam from 0 to L, m,
F: downward load placed at the beam end, N,
E: modulus of elasticity (stiffness) of the beam material, Pa,
I: moment of inertia of the beam cross-section, m⁴, and
L: length of the beam, m.

For the following parameter values and conditions,

$$L = 4 \text{ m}, \ F = 5{,}000 \text{ N}, \ E = 5 \times 10^{11} \text{ Pa}, \ I = 4 \times 10^{-4} \text{ m}^4,$$

solve the differential equation using the Euler method.

It turns out that there is an analytical solution to this differential equation,

$$y = -\frac{FL}{2EI} x^2 + \frac{F}{6EI} x^3$$

Plot and compare the analytical result along with your numerical solution to ensure that the latter is accurate.

7.14 A main reservoir of water, A, drains into two secondary reservoirs, B and C, as shown in Figure P7.14. Given the conditions and parameter values for this system, solve for the flow rates in each of the three pipes and the pressure at the node D.

Here is the information you will need. The flow rate in each pipe is governed by the engineering Bernoulli equation as

$$\text{Pipe 1:} \ \ g z_A = \frac{P_D}{\rho} + \frac{V_1^2}{2} + g z_D + \frac{f_1 L_1}{D} \frac{V_1^2}{2} \qquad V_1 = \frac{Q_1}{A}$$

$$\text{Pipe 2:} \ \ \frac{P_D}{\rho} + \frac{V_2^2}{2} = \frac{f_2 L_2}{D} \frac{V_2^2}{2} \qquad\qquad\ \ V_2 = \frac{Q_2}{A}$$

$$\text{Pipe 3:} \ \ \frac{P_D}{\rho} + \frac{V_3^2}{2} + g z_D = \frac{f_3 L_3}{D} \frac{V_3^2}{2} \qquad V_3 = \frac{Q_3}{A}$$

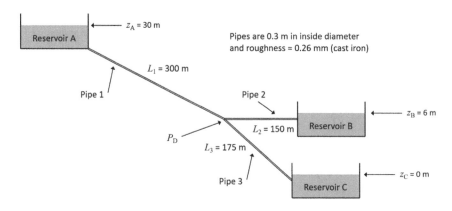

FIGURE P7.14   Reservoir drainage system.

where
V_i: average velocity in pipe i, m/s
Q_i: flow rate in pipe i, m³/s
D: pipe inside diameter, m
L_i: pipe length, m
A: pipe inside cross-section area, $= \pi D^2/4$, m²
g: gravitational acceleration, 9.807 m/s²
ρ: fluid density, 1,000 kg/m³
f_i: Moody friction factor in pipe i, as described below

Fluid friction in the pipes is described by the *Moody friction factor*, predicted by the *Haaland equation*

$$f = \frac{1}{\left(-1.8 \log_{10} \left(\left(\frac{\varepsilon/D}{3.7} \right)^{1.1} + \frac{6.9}{Re} \right) \right)^2} \qquad Re = \frac{\rho V D}{\mu}$$

where
ε: pipe roughness, for cast iron, 0.26 mm
Re: *Reynolds number*, dimensionless
μ: fluid viscosity, 0.00112 Pa·s

An additional relationship occurring at node D is that

$$Q_1 = Q_2 + Q_3$$

Suggested approach for the solution:
- Solve the three pipe equations analytically for P_D.
- Provide initial estimates for Q_1 and Q_2, for example, 1 m³/s and 0.4 m³/s, respectively.

- Compute Q_3 from the last formula above.
- Compute all velocities from the formulas on the right above.
- Compute all the Reynolds numbers and the three friction factors.
- Compute the three P_D values from the pipe equations.

It is likely that the three P_D values will not be equal. Compute the variance of these values using the VAR.S function. The variance will become zero when the three P_D values are equal.

Use the Solver to minimize the variance by adjusting the initial estimates of Q_1 and Q_2. With the variance close to zero, the problem is solved.

NOTES

1. An *algebraic equation* is an equality of two expressions formulated by applying algebraic operations (addition, subtraction, multiplication, division, raising to a power, and extraction of a root) to a set of variables. A *transcendental equation* also includes transcendental functions (such as exponential, logarithmic, trigonometric, or inverse trigonometric functions).
2. If the equation had been second order, we could easily find its two roots with the quadratic formula. Although there are schemes to determine the three roots of a cubic equation analytically, they are generally quite hard to implement.
3. Note: 2^{10} is 1,024. When counting bytes, it is a *kilobyte*. 2^{20} is 1,048,576, which is called a *megabyte*.
4. There are more sophisticated and preferable methods to evaluate whether a linear system of equations is either unsolvable (*singular*) or inaccurate (*ill-conditioned*). These are not available as built-in Excel functions but could be programmed with VBA. The theory and algorithms to do this are described elsewhere (for example, Chapra and Canale 2020).
5. In Excel, a circular reference occurs when a formula repeatedly refers to the same cell. As a result, this creates an endless loop or "circle" between multiple cells.

8 Applied Statistics

CHAPTER OBJECTIVES

- Be able to activate the Analysis Toolpak Data Analysis add-ins for statistical calculations
- Use built-in statistical functions and the Descriptive Statistics tool to characterize a data set
- Learn about statistical distributions and Excel's built-in functions along with random number generation using the Data Analysis tool
- Be able to implement typical statistical hypothesis tests and interpret their results
- Learn how to fit models to data with linear regression techniques using the Data Analysis regression tool
- Be able to employ the Solver to carry out nonlinear regression
- See how Excel's Trendline feature can be used for fitting empirical models

---

Although there are many capable software packages dedicated to applied statistics, most of them require purchase or annual license fees. An exception is the open-source R package, but that requires coding, which is not attractive to some users. Many of the day-to-day statistical calculations can be accomplished with Excel, which provides both built-in functions and the Analysis Toolpak add-in. Although the purpose of this chapter is not to teach applied statistics in depth, we will introduce common statistical calculations and how they can be implemented efficiently on the spreadsheet.

Most engineers and scientists have basic knowledge of probability and applied statistics but are not experts. Statistical calculations and their interpretation are particularly susceptible to misinterpretation. Many results are generated automatically, and care must be taken in reaching conclusions based on these as if the computer is an infallible black box. In this chapter, we will emphasize common applications of Excel to obtain statistical results and correct, cautious interpretation.

After reviewing how to install and activate the Analysis Toolpak add-ins, we introduce the use of Excel's built-in statistical functions for descriptive statistics, which generate live results, and compare these to the Descriptive Statistics tool, which provides fixed output on the spreadsheet. At this point, we introduce common statistical distributions, their built-in supporting functions, and the ability to generate random numbers according to the uniform and normal distributions using the Data Analysis tool.

DOI: 10.1201/9781003361053-8

Making claims based on data is an important activity for engineers and scientists. We introduce several, common hypothesis tests for this purpose, contrasting again built-in functions and Data Analysis tools. Following this, we move on to fitting models to data with regression analysis. Here, we include the use of the Trendline tool and point out its limitations. Again, there is a comparison with on-sheet live calculations using built-in functions, including the vector-matrix collection, and the Data Analysis Regression tool. Finally, we introduce how to build models with stepwise regression and assessing model adequacy.

To use the Data Analysis tools, you have to activate the Analysis Toolpak as an add-in. This is accomplished with the Add-Ins dialog box, which is available via File ⇨ Options ⇨ Add-ins ⇨ Go... or the shortcut Alt-T-I.[1] In the Add-ins dialog box, you should ensure that the Analysis Toolpak is checked, then click OK. Since we will use the Analysis Toolpak - VBA add-in later, you might as well activate that too.

To confirm the activation of the Analysis Toolpak, you can check the Analyze group on the Data tab of the Ribbon and see Data Analysis listed.

8.1 DESCRIPTIVE STATISTICS

At the heart of statistics are the concepts of the population and a sample, which are illustrated in Figure 8.1. The *population* may be finite or infinite. A finite population has a countable number of members, such as a census, whereas, an infinite population is typically a continuous property, such as temperature or pressure. Practical observations can convert an infinite population into a finite one. For example, if we are measuring temperature over a range from 0°C to 100°C with an instrument with a resolution of 0.1°C, there would only be 1,001 possible values. This is therefore finite. However, if the number of values is large, it will be a good approximation to an infinite population.

A statistical *sample* is a set of individual observations from within a population to estimate characteristics of the entire population. The number of items in

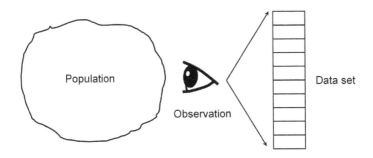

FIGURE 8.1   The population and a sample data set obtained through observation.

the data set, n, is called the *sample size*. Generally, the greater is n, the better we are able to characterize the population. However, making measurements can be costly and take time, so there is a tradeoff. Also, the nature of the population may change with time, called being *nonstationary*, making it difficult to obtain an adequate snapshot with the data set if it takes too long to obtain the observations.

If we are making measurements over a narrow range, the resolution of our instrument may be limiting. For example, if relative humidity varies from 30% to 35%, and our gauge has a resolution of 0.5%, our data set will have only 11 distinct values. If we require a finer resolution, we may need to purchase a humidistat that costs ten times more than our current instrument.

When it comes to measurement, apart from resolution, there is the issue of accuracy. *Accuracy* is the ability to measure truth and is judged by comparison to an established standard.[2] Instrument vendors will often certify accuracy by quantifying inaccuracy based on calibration to a standard. There are secondary standards available that have been calibrated against the primary standards. The difference between the expected value and the standard is sometimes called *bias*.

Data sets can be the measurement of a single quantity or simultaneous measurements of two or more different quantities. There are many characteristics of measured data, but two of the most important are central tendency and dispersion. We assume here that the items in a data set vary. There are different ways to characterize these concepts. One way is graphically with several types of plots: dot diagrams, stem-and-leaf charts, histogram charts, and scatter plots among them. We will describe the latter two later in this chapter. An example *dot diagram* is shown in Figure 8.2. You will notice that the dots are positioned on a horizontal axis according to their value and stacked where there are repeated values. For many scenarios, there will be a clustering of dots around a central value and fewer dispersed values away from the center.

For a set of n measurements, there are three common measures of *central value*: arithmetic average, median, and mode. We are familiar with the *arithmetic average*: the sum the n sample values and divide by n. The *median* is the middle value obtained by sorting the data set. If n is even, the median is the average of the two middle values. The *mode* is the value in the data set that occurs most frequently. The mode is more relevant for larger data sets.

An example data set shown in Figure 8.3 contains 25 measurements (in grams) of commercial sugar packets. The resolution of measurement for this example is 0.1 g. On the right, the measurements are sorted in ascending order and the median and mode are noted. The computation of the average is shown using

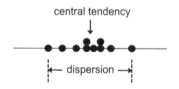

FIGURE 8.2   Example dot diagram.

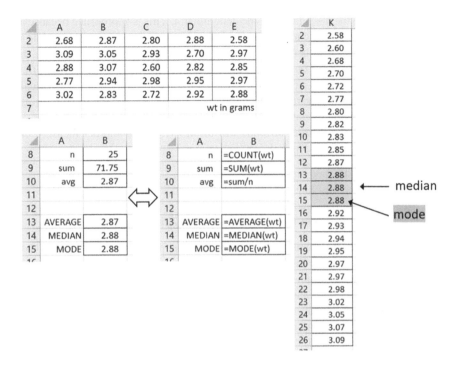

FIGURE 8.3 Sugar packet sample data set with central tendency calculations.

the COUNT and SUM functions. Confirming these, Excel's built-in AVERAGE, MEDIAN, and MODE functions are applied to the data set. For this data set, we note that the median and mode are the same values, and the average is slightly different, but close. The average can be significantly influenced by measurements at the boundaries of dispersion, especially *outliers*[3]. That is not the case with the median and mode.

There are additional measures of central tendency that are available. These include the geometric and harmonic means (built-in GEOMEAN and HARMEAN functions) and the *m*-estimator. The latter can be developed as a user-defined function with VBA.

Similar to central tendency, there are several common measures of *dispersion*. One often used for small data sets ($n \leq 5$) is the *range*, which is simply the difference between the largest and smallest items in the set. This can be divided by $\sqrt{n}$ for a rough estimate of the population standard deviation.

The most common measure is the *sample standard deviation*, computed by the formula

$$s = \sqrt{\frac{\sum_{i=1}^{n}(x_i - \bar{x})^2}{n-1}} \qquad (8.1)$$

You will note the $n - 1$ factor in the denominator. This is appropriate because a *degree of freedom* in the data has been removed by the calculation of the sample's arithmetic average, $\bar{x}$. Otherwise, the s value will be biased from the true value, σ, as $n \to \infty$. As an aside, if our sample represents an entire, discrete population, the denominator becomes n. Also, you see that the difference between the individual measurements and the average is squared. This means that the summation gives more weight to larger deviations.

We also note the *median absolute deviation* (MAD) as another measure of dispersion. It is computed as follows:

1. determine the median of the data set,
2. compute the absolute values of the differences between the data items and their median, and
3. divide the median of those deviations by 0.6745.

The 0.6745 factor is required so that the MAD approximates the σ value. The MAD is useful when the data set may contain extreme values. Outliers exaggerate the sample standard deviation, whereas the MAD is not influenced by them.

Excel provides two standard deviation functions, STDEV.S and STDEV.P. The latter is used when the data set represents the entire population. More commonly, STDEV.S is used. There are no built-in functions for the range or the MAD.

Figure 8.4 illustrates the calculation of the dispersion measures for the sugar packet data set. Note that an array formula is used here to compute the absolute differences from the median. As mentioned earlier, we can generate a rough

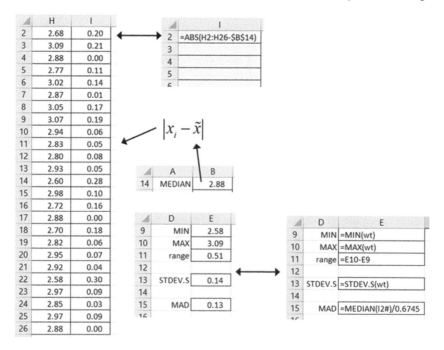

FIGURE 8.4 Calculation of range, sample standard deviation, and MAD.

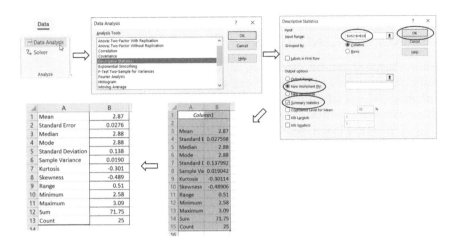

FIGURE 8.5 Application of the Data Analysis Descriptive Statistics tool to the sugar packet data.

estimate for s by dividing the range with $\sqrt{n}$, which for our example would be about $0.51/\sqrt{25} \cong 0.10$, somewhat smaller than the s and MAD values. For this case, the latter two are close as the data do not contain extreme outliers. Although we do calculate it in Figure 8.4 using several functions, computing the MAD would be more convenient if it were available as a built-in function. But, as we will see in Chapter 10, it can easily be generated with a user-defined VBA function.

There is a Data Analysis tool that provides several sample statistics. Figure 8.5 shows the use of the *Descriptive Statistics tool* for the sugar packet data. The Input Range in this case is a single column of the data set. We choose the option to have the results appear on a separate worksheet, and, importantly, *Summary Statistics* is checked.

The figure shows the initial output of the Descriptive Statistics tool. It is on a separate worksheet, selected, and not well formatted. With a little work, we can make the output more presentable, and this is shown in the lower left-hand corner of the figure. If we are going to use one of the Data Analysis tools frequently, it will be important to automate the formatting with a VBA macro.

The results presented by the Data Analysis tools are not live. If the data were to change, we would have to run the Descriptive Statistics tool again. We know most of the items presented. There are four that are worth comment:

1. **Standard Error:** The Standard Deviation (s) divided by the square root of the Count (n). It will come up in our discussion of distributions and hypothesis testing.
2. **Sample Variance:** The square of the Standard Deviation (s).
3. **Kurtosis and Skewness:** These relate to the deviation of the data from a "bell curve" distribution. Kurtosis < 0 indicates a flattened curve and Skewness < 0 a curve with a longer tail to the left.

8.2 STATISTICAL DISTRIBUTIONS AND RANDOM NUMBER GENERATION

Much of applied statistics depends on the ability to model the population and then estimate that model's parameters. Populations are modeled with *probabilistic distributions*. This is effective because distributions often provide adequate models. There are many distribution models, and Excel provides built-in functions and analysis tools to support these.

The random variables that distributions describe fall into two categories: discrete and continuous. *Discrete distributions*, such as binomial and Poisson, describe the probability of variables that only occur at countable values, such as counting integers. *Continuous distributions* describe variables which population is infinite, such as temperature, pressure, composition, etc. We will consider only continuous distributions here and concentrate on the uniform and normal.

8.2.1 STATISTICAL DISTRIBUTIONS

A continuous probability distribution for a continuous random variable X is defined by its *density function*, $f(x)$. The probability that a sampled value will fall in the interval $x_1 \leq x \leq x_2$ is given by

$$P = \int_{x_1}^{x_2} f(x)\,dx \tag{8.2}$$

Note: Theoretically, the probability that a sample value will be equal to a given value is zero. In practical settings, this is modified when a measurement system has a finite resolution.

The *uniform distribution* has a flat density function, as described in Figure 8.6. The area under the flat line is equal to one. This makes sense because that covers

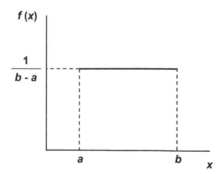

FIGURE 8.6 Density function for the uniform distribution.

all possible outcomes. As an example, if $a = 60$ and $b = 100$, the probability of a sample value being between 75 and 85 is given by

$$P = \int_{75}^{85} \frac{dx}{100-60} = 0.25 \text{ or } 25\% \qquad (8.3)$$

We do not encounter data sets that are well modeled by the uniform distribution very often; however, that is not the situation for the *normal* (or *Gaussian*) *distribution*. Its *probability density function* (*PDF*) is more complicated and is given by

$$f(x) = \frac{1}{\sigma\sqrt{2\pi}} e^{-\frac{1}{2}\left(\frac{x-\mu}{\sigma}\right)^2} \qquad -\infty \leq x \leq \infty \qquad (8.4)$$

where
    μ: mean
    σ: standard deviation

Note that the domain for the random variable is infinite, although probabilities are negligible outside the range $\mu - 5\sigma$ to $\mu + 5\sigma$.[4] A plot of the normal density is the well-known "bell curve," as shown in Figure 8.7. For $\mu = 0$ and $\sigma = 1$, or, using the standardized variable

$$z = \frac{x - \mu}{\sigma} \qquad (8.5)$$

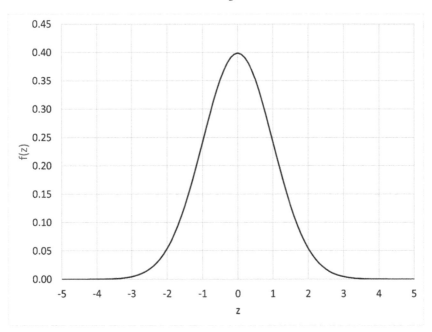

FIGURE 8.7 Density curve for the standard normal distribution.

The figure illustrates the practical limits of the normal density distribution. This gives rise to the approximate "normal probability rule" summarized as

$P(|z| \pm 1)$	68%
$P(|z| \pm 2)$	95%
$P(|z| \pm 3)$	99.7%

Determining probabilities for the normal distribution requires the integration of Equation 8.4. There is no analytical solution for this integral, so the probabilities must be estimated numerically. Excel provides a function for this purpose:

NORM.DIST(x, μ, σ, cumulative)

The cumulative argument is set TRUE to compute the cumulative probability from $-\infty$ to x. It is set FALSE to compute the value of the density function (Equation 8.4) at a specific value of x. Figure 8.8 shows the area or probability for $a \leq x \leq b$. This can be computed using two instances of the NORM.DIST function as

NORM.DIST(b, μ, σ,TRUE) − NORM.DIST(a, μ, σ,TRUE)

Figure 8.9 shows an example calculation for $\mu = 100$, $\sigma = 10$, $a = 110$, $b = 120$. The probabilities are formatted as percent.

It is worth mentioning two related built-in functions. The NORM.S.DIST function has the syntax

NORM.S.DIST(z, cumulative)

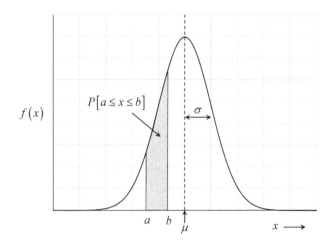

FIGURE 8.8 The probability of $a \leq x \leq b$ for the normal distribution with mean μ and standard deviation σ.

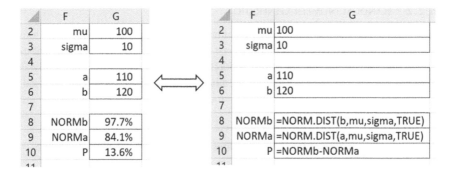

FIGURE 8.9   Using the NORM.DIST function to compute a probability.

and computes the probability $P[-\infty \rightarrow z]$ for the standard normal distribution with $\mu = 0$ and $\sigma = 1$. There are inverse functions for the normal and standard normal distributions. For the normal distribution,

NORM.INV(P, μ, σ)

returns the x value for the given probability, $P[-\infty \rightarrow x]$. This is illustrated in Figure 8.10. For $\mu = 100$ and $\sigma = 10$, the function determines an x value of 108.4 for a P-value[5] of 80%.

Excel provides similar functions for other continuous distributions:

* **Beta:** Useful for random variables with finite limits
* **Exponential:** Partner to the Poisson distribution
* **Gamma:** Allows for shaping asymmetric distributions
* **Log normal:** Useful for such applications as the distribution of particle sizes
* **Weibull:** Used for reliability, also an asymmetric distribution

and distributions for discrete random variables:

* **Binomial:** Used for pass/fail scenarios with infinite populations
* **Hypergeometric:** Similar to binomial but for finite populations
* **Negative binomial:** Models the number of failed samples before the first passed sample
* **Poisson:** Used for finite occurrences during given intervals

More information on these is available from Help via the Insert Function dialog window.

FIGURE 8.10   Use of the NORM.INV function to compute x from P.

8.2.2 RANDOM NUMBER GENERATION

A common activity in spreadsheet calculations is to generate a set of random numbers that appear to be sampled, at least approximately, from a given distribution. These sets can be used to test models or other calculations. The Analysis Toolpak add-in provides a Random Number Generation tool that can be used. Figure 8.11 illustrates the initial dialog window for this tool. The key settings are as follows:

- **Number of Variables:** This indicates the number of columns of random numbers, typically one.
- **Number of Random Numbers:** This is the number of rows for the random numbers – not required if a fixed Output Range is specified.
- **Distribution:** A drop-down list of the different distributions available – we have shown here the choice of the Uniform distribution.

Figure 8.12 shows how we use the Random Number Generation tool to produce 1,000 random numbers based on our normal distribution example above with $\mu = 100$ and $\sigma = 10$. We have also included the calculation of the average and the sample standard deviation for these numbers, and you can see that these agree quite closely with the distribution's parameter values.

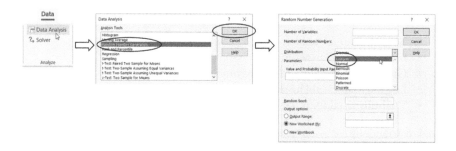

FIGURE 8.11 Initial view of the Random Number Generation dialog window.

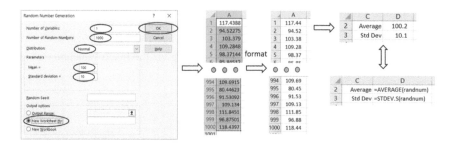

FIGURE 8.12 Generating 1,000 random numbers according to the normal distribution.

Apart from characterizing a data set, here a large set of random numbers with statistics such as the average and standard deviation, we can benefit from graphical depictions. A *histogram* is an analysis of the number of data items that fall within chosen intervals. These data are typically presented in the form of a bar chart. These charts are useful in assessing the distribution of the data and drawing conclusions about a theoretical distribution that might model the population.

Excel can calculate these numbers, also called *frequencies*, both with a built-in *FREQUENCY function* and a *Data Analysis Histogram tool*. Many histogram applications are for categorical, discrete data,[6] and Excel's tools are oriented to these. When we use the function of the Data Analysis tool for samples of continuous quantities, we must be careful with our interpretation. Histograms are best used when the number of data is large, such as with our random numbers.

The first consideration in constructing a histogram is specifying the intervals within which the frequencies are to be tabulated. For a data set with n items, there are two common formulas used to determine the number of intervals:

$$INT\left(\sqrt{n}\right) \quad \text{or} \quad INT\left(\log_2\left(n\right)-1\right) \tag{8.6}$$

The INT operation rounds a number down to the nearest integer. The first of these gives a greater number than the second. A strategy is to use these as brackets to guide us in picking a convenient number somewhere in between. For our random number data set, the two results are 31 and 9, respectively. If too many intervals are used, we won't be able to discern distribution patterns well because the details yield the so-called "broken comb look." Too few are also a problem because not enough pattern will be visible.

A second consideration is to establish the boundaries of these intervals. These can be guided by the range of the data. For our data set, the minimum value is about 64 and the maximum value is a little less than 132. The range is therefore 68. We might choose convenient interval boundaries (These intervals are commonly called bins.) starting at 60, with a width of 5, and ending at 135. These would account for 15 bins – within our bracketing guidelines.

A third issue is what happens with items that might fall exactly on a bin boundary and the question arises: within which interval will they be counted. For a large data set of random numbers (differing in 15 significant figures), this is not an issue. In the case where this is important, the bin boundaries can be set at half a resolution point to avoid this. For data resolution of 0.1, the boundaries would fall on values of 0.05.

Figure 8.13 shows the creation of the bin boundaries and the application of the built-in FREQUENCY function to compute the histogram data. The frequency results deserve interpretation. The initial zero in cell G2 indicates that there are no data less than or equal to the adjacent boundary 60. The one in cell G3 shows that there is one value > 60 and ≤ 65. The final zero in cell G18 shows there are no values greater than 135. We would expect the initial and final zero frequencies because of our interval design. The highest frequency occurs in the interval between 95 and 100. It is notable that the frequencies are live; that is, if any input data change, the frequency results will update immediately.

A	
1	117.44
2	94.52
3	103.38
4	109.28
5	98.37
c	oc oc
○	○ ○
994	109.69
995	80.45
996	91.53
997	109.13
998	111.45
999	96.88
1000	118.44

randnum

bins

	F	G
2	60	0
3	65	1
4	70	0
5	75	7
6	80	17
7	85	36
8	90	94
9	95	156
10	100	190
11	105	176
12	110	159
13	115	86
14	120	56
15	125	16
16	130	4
17	135	2
18		0

	F	G
2	60	=FREQUENCY(randnum,bins)
3	65	
4	70	
5	75	
6	80	
7	85	
8	90	
9	95	
10	100	
11	105	
12	110	
13	115	
14	120	
15	125	
16	130	
17	135	
18		
19		

FIGURE 8.13   Use of the FREQUENCY function to construct histogram data.

Using the *Data Analysis Histogram tool*, we can produce similar results. The Histogram dialog box provides more options, such as creating a histogram chart. This is illustrated in Figure 8.14. The Input Range and Bin Range are set in the same way as with the FREQUENCY function. As a variation, we choose for the results to be displayed on the current worksheet with origin cell M2, and we request Chart Output.

The table results displayed in the figure are the same as those shown previously. The chart output is convenient, but it is misleading. This is because the Histogram tool is ideally suited to categorical data and not continuous numbers. For example, the bar labeled 100 counts the data in the interval $90 < x \le 100$, so the label should actually be at the right side of the bar.

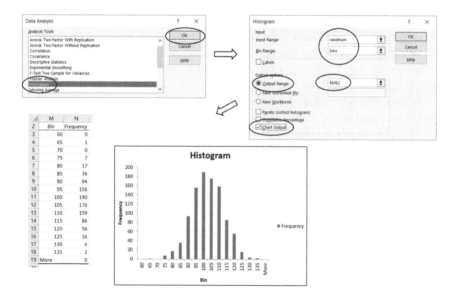

FIGURE 8.14   Use of the Data Analysis Histogram tool with the random numbers.

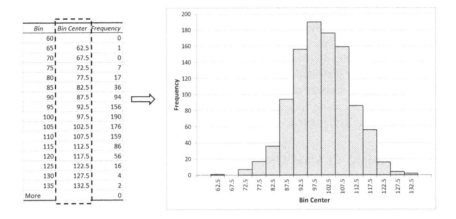

FIGURE 8.15 Use of bin centers for bar labels with bar chart with improved format.

There are several ways to improve the appearance and meaning of the histogram chart. One is to label the bars by the centers of the bins.[7] Figure 8.15 shows this and the resulting chart, which has been reformatted significantly to make it more presentable. The bin centers shown are calculated with formulas based on the bin values. These are used to replace the previous labels via the Chart Design ⇨ Select Data ⇨ Edit Horizontal Axis Labels. The bars have been widened to occupy the entire interval using Format Data Series ⇨ Series Options ⇨ Gap Width = 0, and a pattern fill has been employed for the bars.

An interesting aspect of the histogram chart is its comparison to a theoretical distribution. Of course, in our example, we expect close agreement. To carry this out, we want to compute the number of items that would occur in each bin interval given the normal distribution. To do this, we need estimates of the mean and standard deviation, and we take these as the average and sample standard deviation, respectively. Using the *NORM.DIST function*, we can calculate the probability for a bin and scale that up by the number of items in our data set. Figure 8.16 shows

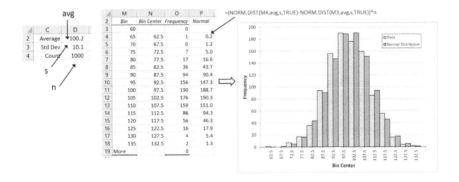

FIGURE 8.16 Adding normal distribution predictions to the histogram data and chart.

these calculations and the addition of the theoretical predictions to the chart. It is also possible to display the normal distribution series as a curve, utilizing the *Combo Clustered Column-Line chart type*.

For data that clearly aren't well modeled by the normal distribution, it is possible to fit another distribution to the histogram by varying its parameters to minimize the sum of squares of differences between the bin counts and theoretical predictions. This can be done using the Solver and is particularly applicable when the histogram is asymmetric.

8.3   MAKING CLAIMS USING STATISTICAL HYPOTHESIS TESTS

A common purpose of collecting data is to make claims based on statistics computed using the data set. Scenarios include claims on the quality (or lack thereof) of a product, judging the merit of improvements made to processes, meeting regulatory standards, and confirming environmental trends. As per the characteristics we discussed in Section 8.1, claims are made on the central value or dispersion of a quantity as observed via a set of data. If we can justify that the population is well modeled by a theoretical distribution, the tests we make to assess claims are called *parametric*. If that is not the case, there are *nonparametric* tests available that are not as discriminating in supporting a claim as the former.[8]

Since we are basing our assessments on data that likely include a random character and are an imperfect representation of the population, there is always uncertainty in making claims. This shows up in two types of errors that can be made:

- Type I: making an erroneous claim, one that isn't true (a false positive)
- Type II: failing to make a claim, one that is actually true (a false negative)

The tests we make allow us to adjust and interpret these risks.
A hypothesis test is comprised of two statements:

- H_0: null hypothesis, cannot be proven, can only be rejected or fail to be rejected
- H_1: alternate hypothesis, can be accepted, but with risk, or can be rejected, also with risk

The null hypothesis is always an equality statement. The alternate hypothesis is one of the following:

- $>$: a greater than claim
- $<$: a less than claim
- $\neq$: either $>$ or $<$, called a two-sided test

Most practical settings lead to either a $>$ or $<$ test.

8.3.1 HYPOTHESIS TESTS COMPARING TO A STANDARD

In this section, we will consider making claims associated with central value (or population mean) and dispersion (or population variance/standard deviation) when compared to a standard value. The latter is often associated with a product specification. The claim is either the product meets or exceeds a specification or the converse.

Let's illustrate this with our sugar packet data. The sample average for the 25 packets is 2.87 g and the sample standard deviation is 0.138 g. Let us say we want to make a claim that the mean packet weight is less than a standard value of 2.9 g. We would describe this in hypothesis test terms as

- H_0: $\mu = 2.9$g
- H_1: $\mu < 2.9$ g

If we are going to base our judgment on the normal distribution, we might look at a histogram of the data. This is presented in Figure 8.17. The distribution is asymmetric, skewed to the left. This might cause concern, but our ability to assess this observation is limited by the sample size. As it turns out, we are on safer ground for this hypothesis test because our statistic is the sample average. We can take advantage of the *central limit theorem* which posits that

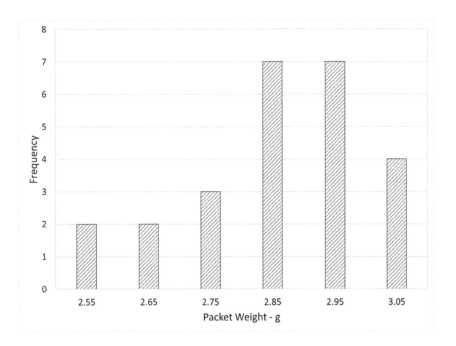

FIGURE 8.17 Histogram chart of sugar packet data.

the distribution of sample averages tends to be normal even if the distribution of the data is not. This is *not* the case when it comes to hypothesis tests on the variance, σ^2.

If we based our test directly on the normal distribution, we could evaluate how far away our sample average, 2.87 g, is from the standard, 2.90 g, in terms of the number of standard deviations, using our sample value of 0.138 g. The standard deviation of averages is less than that of the original data, given by $\sigma/\sqrt{n}$. Using our sample standard deviation, $s = 0.138$ g, our estimate of the standard deviation of averages would be 0.015 g. We would then estimate the number of standard deviations from 2.90 g to 2.87 g as $0.03/0.015 = 2$. So, how probable is it that, given H_0 is true, we would obtain a sample average of 2.87 or less? We can use the NORM.DIST function to compute this.

=NORM.DIST(2.87,2.90,0.015,TRUE) $\Rightarrow$ about 2.2%

Our "risk of being wrong" in claiming H_1 is then about 2% or one chance in 50. If this is acceptable, we claim H_1. In applied statistics, this probability is called a *P-value*. In this sense, the P-value is the bottom line of the test in that it summarizes the answer in a single number. That is, the lower the P-value, the more unlikely is the null hypothesis.

Now, to a few details, our analysis above is approximately correct, but it is imperfect because it is predicated on an estimate of the standard deviation that is dependent on the sample size, n. The distribution of averages with standard deviation computed from the data is different from the normal distribution. The estimate does approach the normal distribution as the sample size increases and, for $n = 25$, it is fairly close. But for smaller sample sizes, it can be significant. So, let's look at a way to mitigate the problem.

From Equations 8.4 and 8.5, if we define a standard random variable, Z, in terms of the random variable of our population, X,

$$Z = \frac{X - \mu}{\sigma} \tag{8.7}$$

the standard normal density function in terms of Z becomes

$$f(z) = \frac{1}{\sqrt{2\pi}} e^{-\frac{1}{2}z^2} \qquad -\infty \leq z \leq \infty \tag{8.8}$$

If instead we define another random variable, T, in terms of the random variables $\bar{X}$ and S, where S represents the general random variable associated with the sample standard deviation of a sample of n data,

$$T = \frac{\bar{X} - \mu}{S/\sqrt{n}} \tag{8.9}$$

The density function that describes T, which is called the *Student's t distribution*, is

$$f(x) = \frac{\Gamma\big((k+1)/2\big)}{\sqrt{\pi k}\,\Gamma(k/2)} \cdot \frac{1}{\big((x^2/2)+1\big)^{(k+1)/2}} \qquad (8.10)$$

where

   k: $n-1$
   $\Gamma(\bullet)$: gamma function

$$\Gamma(r) = \int_0^\infty x^{r-1} e^{-x}\, dx \text{ and, if } r \text{ is a positive integer, } \Gamma(r) = (r-1)!$$

Although Equation 8.10 might appear daunting to you, just like the normal and standard normal distributions, Excel provides built-in functions for the t distribution. One of these is

T.DIST(x, n-1, cumulative)

This provides the probability, or area under the density function, from $-\infty$ to x. There is another function T.DIST.RT that provides the probability from x to ∞. For our example,

T.DIST((2.87-2.90)/0.015,25-1,TRUE) $\Rightarrow$ P-value = about 2.7%

We see that our risk factor here is slightly greater than the 2.2% based on the normal distribution. For smaller sample sizes, this difference would be greater, and vice versa for larger samples. As a rule of thumb, for $n \geq 50$, we can safely use the normal distribution.

Next, we can consider comparing dispersion, as evidenced by our data and the sample variance or standard deviation, with a standard or specification. Recall that our sample standard deviation is $s = 0.138$ g, which can be squared to give the sample variance, $s^2 = 0.019$ g^2.

Imagine that there is a specification that the variability, σ, of sugar packet weight should be less than 0.12 g. Since our sample standard deviation is greater than this, we would like to claim that the population variability is greater than the standard; that is, it doesn't meet the specification. Of course, our value does exceed the standard, but there is uncertainty in using that value to make such a claim. The hypotheses here are as follows:

- H_0: $\sigma^2 = 0.12^2$
- H_1: $\sigma^2 > 0.12^2$

The random variable that is used to assess claims on the variance is chi-squared, X^2, and it is related to our sample variance random variable, S^2, by

$$X^2 = \frac{(n-1)S^2}{\sigma^2} \tag{8.11}$$

The ratio of S^2 to σ^2 makes sense in a way. If it is greater than one, it suggests that the actual variance may be greater than the standard value, and vice versa. But this is modified by $(n - 1)$ which creates a much larger statistic value. A reason for this is that this variable is modeled with a distribution. The density function for X^2 is

$$f(x) = \frac{1}{2^{k/2}\Gamma(k/2)} \cdot x^{(k/2)-1}e^{-x/2} \qquad x > 0 \tag{8.12}$$

where x: evaluation of X^2 for specific values of s^2.

Again, Excel provides handy functions for the chi-square distribution. One of these is

CHISQ.DIST(x, n-1, cumulative)

where x is the value at which Equation 8.11 provides the area under the density function; that is, the probability, from 0 to x. For our sugar packet data,

$$\frac{(n-1)s^2}{\sigma^2} = \frac{(25-1)0.138^2}{0.12^2} \cong 31.7 \tag{8.13}$$

and

CHISQ.DIST(31.7,25,TRUE) $\Rightarrow$ about 86.7%

This implies that, if the standard value of 0.12 were true, the probability of a sample of 25 having a value ≥ 0.138 would yield a P-value = 100% − 86.7% = 13.3%. This is the risk we will assume if we choose to make the claim that the packets are of adequate quality. For most situations, this risk is too high. Typically, we are looking for P-values less than 5%, or even less than 1% for greater certainty.[9] Therefore, we would fail to reject H_0 and not claim H_1.

There is some question about our application of the chi-square distribution here because it is based on our measurements being representative of a normal distribution. Recalling the skewed nature of Figure 8.17, there could be some reasonable doubt. Without expanding the content here, we will mention that there are statistical tests to support or reject a claim that data come from a normally distributed population. One of the more robust of these is called the *Anderson-Darling test*. Alternately, we can pursue a non-parametric test, one that doesn't depend on a distribution, to evaluate our claim. An option here is the *modified Levene test*. Recall, however, that non-parametric tests are not as discriminating as those that assume a population distribution. Neither of these named tests is directly supported by Excel functions, although they can be constructed on a spreadsheet. We refer you to texts on applied statistics text for more information (for example, the excellent text by Montgomery and Runger, 2018).

These two examples of comparing against a standard, for the mean and the variance, are the most commonly encountered. Tests involving other distributions are also possible and follow the same scheme.

8.3.2 Hypothesis Tests Comparing Two Samples

Tests involving two samples are highly important in engineering and science. The concept is making claims that the samples are different in a characteristic, such as mean for central tendency and variance for variability. In many cases, these can be called "before and after" tests. Changes are made to a process or procedure, and it is essential to evaluate whether characteristics of a product or result have been affected. An advantage here is that Excel provides tools to implement these tests.

A hypothesis test on means for two data sets, A and B, can be formulated as follows:

- H_0: $\mu_A = \mu_B$
- H_1: $\mu_A > \mu_B$ or $\mu_B > \mu_A$ depending on the data

The latter hypothesis can also be stated in a two-sided format:

- H_1: $\mu_A \neq \mu_B$

However, for most practical situations, our data indicate one or the other of the previous H_1 options.

An additional consideration for the test on means is whether the variabilities of the populations representing A and B are equal or distinct. When they appear to be equal, both samples can be combined to obtain a better estimate of σ^2 (or σ), and this leads to a more discriminative or tighter test. This motivates the need for a test on variances that is often applied to the two samples ahead of the test on means to determine the assumption of equal variances.

Excel's Data Analysis tools provide tests for all these situations. We summarize them here:

- F-test Two-Sample for Variances
- t-test Two-Sample Assuming Equal Variances
- t-test Two-Sample Assuming Unequal Variances
- z-test Two-Sample for Means

The z-test can be used when the sample sizes are large, and the t distribution is approximately equal to the standard normal distribution.

The F-test is based on a statistic involving the sample variances and their corresponding population variances:

$$F = \frac{S_A^2/\sigma_A^2}{S_B^2/\sigma_B^2} \qquad (8.14)$$

This statistic is governed by the F distribution, which has a density function we don't include here. The hypotheses for the test are as follows:

- $H_0: \sigma_A^2 = \sigma_B^2$
- $H_1: \sigma_A^2 > \sigma_B^2$ or $\sigma_A^2 < \sigma_B^2$ depending on the data set s^2 values

To test H_0, because the two variances are assumed equal, we evaluate

$$F_0 = \frac{S_A^2}{S_B^2} \qquad (8.15)$$

and to reject H_0, we expect that this value will be significantly different from one.

We will illustrate this test by considering two comparative data sets from a fluid flow scenario, water level, and flow rate. These data are shown in Table 8.1. The question we would like to address is whether the measurements in each category support a claim that the mean values are different from sample A to B.

TABLE 8.1

Measurements of Flow Rate and Liquid Level

Flow Rate – L/min		Level – cm	
A	B	A	B
3.891	3.795	12.502	13.338
3.878	3.817	12.900	13.277
3.886	3.785	12.339	12.767
3.889	3.833	11.830	12.533
3.902	3.797	12.472	12.309
3.854	3.783	11.687	12.727
3.873	3.830	13.032	12.808
3.938	3.826	14.714	12.676
3.907	3.807	12.217	12.400
3.892	3.802	12.186	12.176
3.915		12.563	

	Flow Rate - L/min		Level - cm	
	A	B	A	B
n	11	10	11	10
avg	3.893	3.808	12.586	12.701
s	0.022	0.018	0.811	0.380

FIGURE 8.18   Sample statistics for flow rate and level, samples A and B.

Before we apply the Data Analysis tests, we compute sample statistics for the four data sets. These are displayed in Figure 8.18.

Looking only at the average values, we observe that, for flow rate, sample A is greater than sample B, and, for level, sample B is greater than A. We also observe that, for flow rate, the sample standard deviations are similar, but for level, they differ by more than a factor of two.

For the first step of the analysis, we will use the Data Analysis F-test tool to evaluate a claim that the variances, for each measurement category, are different. Figure 8.19 shows the procedure in detail for flow rate. The flow rate data cells are named FlowA and FlowB, respectively. These are entered. There is a default Alpha value of 0.05 – to be explained presently. Clicking OK on the dialog window produces the poorly formatted output on a separate worksheet. This is reformatted manually with the result displayed on the right of the figure.

The first three lines of the output are the sample statistics for data sets A and B. The fourth line, dF, is called the degrees of freedom and is one less than the sample sizes. The F-value is simply the ratio of the A to the B variance. The P-value is 27.5% which represents the risk we would take if we were to claim that the variances are different. This is too high; therefore, we do not reject the H_0 hypothesis that the variances are equal, and we proceed to the means test on that basis. The final entry in the table is the F-value required for a risk of Alpha (5%). We can see that our F value from the data is well below that, again not justifying a claim of different variances.

If we apply the same test to the level data, we get a very different result. The F-value is about 4.5 and the P-value is 1.6%. That is an acceptable risk to claim that the variances are different, and we proceed to the means test for this situation.

It is worth mentioning that there is a built-in function *F.INV.RT* that will provide the F threshold value for a given Alpha risk. This will be the same as the F

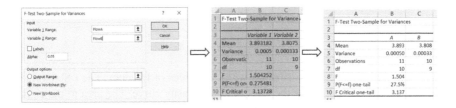

FIGURE 8.19   F-test comparison for flow rate data.

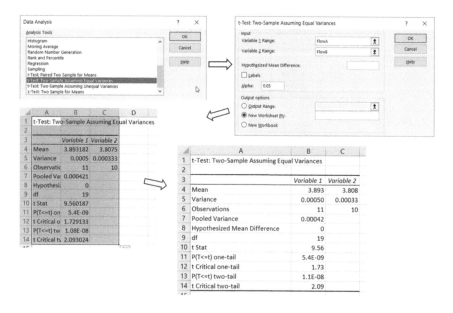

FIGURE 8.20 Means test for flow rate data.

critical from the F-test display. The RT addition to the function name is for "right tail," indicating the area to the right of the returned value containing, in this case, 5% of the total area under the distribution density curve.

The means test for the flow rate is illustrated in Figure 8.20. The test for equal variances is chosen because of the conclusion we drew from the F-test above. The results displayed on a new worksheet are reformatted and shown in the lower right of the figure. After the first three lines of sample statistics, the Pooled Variance is shown. This is a combination of the sample variances from A and B data sets and is computed using the formula

$$s_p^2 = \frac{(n_A - 1)s_A^2 + (n_B - 1)s_B^2}{n_A + n_B - 2}$$ (8.16)

The denominator is the degrees of freedom factor for the analysis, and this is shown on the df line. The t Stat value is computed with

$$t\ Stat = \frac{\bar{x}_A - \bar{x}_B}{s_p \sqrt{\dfrac{1}{n_A} + \dfrac{1}{n_B}}}$$ (8.17)

and is judged using the t distribution with df degrees of freedom. The t Stat value is 9.56 which places it far to the right on the t density curve. The area to the right of this is the P (one tail), 5.4×10⁻⁹. This tiny value leads us to conclude, with a great certainty, that the means of the two samples are different. The flow rate has changed.

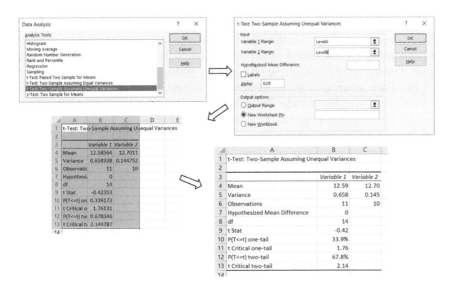

FIGURE 8.21 Means comparison test for level data sets.

You will note that the change in flow rate is relatively small, 0.086 L/min, but the variance is also small which leads to a greater certainty in making the distinction.

To conduct the means test for the level data sets, we choose another test from the Data Analysis list. This is for unequal variances. The procedure and results are shown in Figure 8.21. The formatted results are in the lower right of the figure. Again, the first three lines are sample statistics. The degrees of freedom are not as high as with the other test. It is computed using the compromise formula,

$$df = \frac{\left(\dfrac{s_A^2}{n_A} + \dfrac{s_B^2}{n_B} \right)^2}{\dfrac{\left(s_A^2/n_A \right)^2}{n_A - 1} + \dfrac{\left(s_B^2/n_B \right)^2}{n_B - 1}} \qquad df \text{ rounded down to an integer} \qquad (8.18)$$

The result here is 14. The t Stat value is computed using

$$t\,Stat = \frac{\bar{x}_A - \bar{x}_B}{\sqrt{\dfrac{s_A^2}{n_A} + \dfrac{s_B^2}{n_B}}} \qquad (8.19)$$

In this case, the result is negative because the average for sample A is less than that for sample B. This leads to an alternate hypothesis, H_1, that the mean of A is less than that of B. The value of -0.42 yields a P-value of about 34%, which indicates that we cannot support H_1. We cannot safely claim that the two averages are different. Notice here that the averages differ by only 0.11 compared to values around 12.

We observe in these examples the sequence of making a test on variances, F-test, first to determine whether they are evidently different, and then a *t*-test on means, which formulation depends on the conclusion from the F-test. This is a typical procedure for comparing two samples of data.

There are other tests worth mentioning. If data are collected in pairs, such as a measurement taken under the same conditions with two methods, there is a special t test that is available in the Data Analysis tools (t-test: Paired Two Sample for Means), a test on differences between the pairs of measurements. Recall that the F-test depends on the populations being described adequately by the normal distribution. The t tests can be applied to non-normal data to an extent because of the central limit theorem. There are also *non-parametric* tests, such as the *sign test* and the *Wilcoxon test.*

In summary, there are many hypothesis-testing Excel tools available to engineers and scientists. A caution is that these tools must be applied with care, and it is dangerous to accept and interpret results with little knowledge of the fundamentals. For this reason, we always recommend education in applied statistics, typically in the form of one or more academic courses, whether in-person or online.

An additional topic related to hypothesis testing has to do with the Type II error. Our focus above has been on P-values and Alpha (α) that relate to Type I errors, the risk of claiming H_1 when it is not true. The Type II error is missing that H_1 is true when the test doesn't lead to rejecting H_0. The Type II error probability is typically called Beta (β). This is usually interpreted by the design of data collection to be able to detect a given shift in a parameter, such as the mean. The control on this is sample size. So, a Type II analysis is to determine how many data to collect. Excel doesn't have built-in tools for this, but it can be computed using its basic statistical functions. Statistical software packages typically provide this capability. We do not take the time to elaborate on these calculations here.

8.4 USING REGRESSION TO FIT MODELS TO DATA

There is frequent and growing interest in mathematical models to describe processes and phenomena. Experimental data are collected to validate these models and tune them to be an accurate representation. Well-tuned models are valuable for understanding and prediction. The general scheme for fitting models is illustrated in Figure 8.22. Modeled processes generally have input factors that

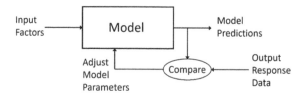

FIGURE 8.22 General scheme for model fitting.

are either controlled or measured, and they have one or more output responses. Commonly, we focus on one response. The mathematical model has adjustable parameters, and based on a set of values, we use the model to compute predictions to compare with the measured output response. By minimizing a fit criterion, typically the sum of squares of the differences between the model predictions and the response data, the model parameters are optimized. This is sometimes referred to as *model calibration*.

To describe the situation analytically, the model can be expressed as

$$y = f\left(x_1,\ldots,x_m,\beta_1,\ldots,\beta_k\right) \tag{8.20}$$

where

  y: response variable
  x_j: input variable j
  β_l: model parameter l

For the purposes of fitting, models can be categorized as nonlinear or linear in their parameters. The latter can be described further by

$$y = \beta_0 + \beta_1 f_1\left(x_j, j=1,\ldots,m\right) + \beta_2 f_2\left(x_j, j=1,\ldots,m\right) + \cdots + \beta_k f_k\left(x_j, j=1,\ldots,m\right) \tag{8.21}$$

The first parameter, β_0, is an intercept, and it is not always present. The linear form gives rise to the method of model fitting called *linear regression*, which is an efficient, one-time-through calculation. Methods for linear regression are built into Excel via the Data Analysis tools.

A very common linear model is one where the f_i's in Equation 8.21 are a single independent variable, x, where $f_i(x) = x^i$ for $i = 0, 1, \ldots, k$. That is, the model is a simple polynomial in x,

$$y = \beta_0 + \beta_1 x + \beta_2 x^2 + \cdots + \beta_k x^k \tag{8.22}$$

This is commonly called *polynomial regression*.

Another widely used linear model is one where the f_i's in Equation 8.21 are where there are multiple independent variables, x_i, where $f_i(x) = x_i$ for $i = 0, 1, \ldots, k$. That is, the model is,

$$y = \beta_0 + \beta_1 x_1 + \beta_2 x_2 + \cdots + \beta_k x_k \tag{8.23}$$

This is commonly called *multiple regression*.

Finally, methods for nonlinear regression are based on minimization via iterative techniques. We can accomplish this with Excel using the Solver, as we will illustrate later in this section.

8.4.1 LINEAR REGRESSION

The basis for model fitting or regression analysis is the collection of a data set. Often the data come from a designed experimental campaign, but they can also be collected "on the fly." What is important is that meaningful changes much occur in the x variables so that there is enough "information content" in the data set to make good estimates of the β parameters. A data set of n measurements can be represented as

$$\left\{ y_i, x_{1i}, \ldots, x_{mi}, i = 1, \ldots, n \right\} \tag{8.24}$$

It is convenient to portray the solution of the linear regression problem in terms of vectors and matrices. This is shown in Figure 8.23.

Given that the formulation of the model in Figure 8.23 is linear in the parameters, β, we can use calculus to solve for the parameter values that minimize the sum of squares criterion, V, using calculus:

$$\frac{\partial}{\partial \beta}(V) = \frac{\partial}{\partial \beta}\left(\mathbf{e}^T \mathbf{e}\right) = \frac{\partial}{\partial \beta}\left((\mathbf{y} - \mathbf{X}\beta)^T (\mathbf{y} - \mathbf{X}\beta)\right) = -2\mathbf{X}^T (\mathbf{y} - \mathbf{X}\beta) \Rightarrow = \mathbf{0} \tag{8.25}$$

We can determine the minimum sum of squares by setting the derivative equal to zero. The result is a set of linear algebraic equations, known as the *normal equations*.

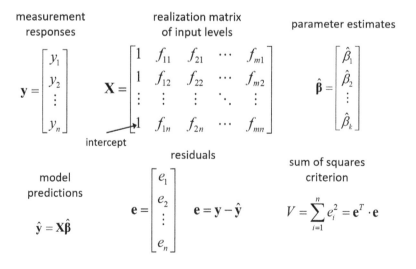

FIGURE 8.23   Vector-matrix formulation for linear regression.

TABLE 8.2
Viscosity of Water (mPa·s) versus
Temperature (°C)

Temperature (°C)	Viscosity (mPa·s)
0.0	1.794
4.4	1.546
10.0	1.310
15.6	1.129
21.1	0.982
26.7	0.862
32.2	0.764
37.8	0.682
48.9	0.559
60.0	0.470
71.1	0.401
82.2	0.347
93.3	0.305

Also, we can solve analytically for the optimal, least squares, parameter estimates, **b**,

$$\left(X^T X\right) b = X^T y \qquad \text{or} \qquad b = \left(X^T X\right)^{-1} X^T y \qquad (8.26)$$

We can solve for **b** on the spreadsheet using the built-in vector-matrix array functions. However, the Data Analysis Regression tool solves the set of linear equations in a more efficient manner. Further, there are other benefits to the Regression Tool that we will mention later.

We will illustrate linear regression with an example and introduce the concepts of goodness of fit and model adequacy. Table 8.2 displays measurements of the viscosity of water as a function of temperature. Figure 8.24 shows a plot of these data. Before choosing models to fit data, when feasible, it is always important to look at plots of data.

In Figure 8.24, we see that there is significant curvature in the data. We would not try to fit a straight line, but rather we might use a polynomial model, in its simplest form,

$$\mu = \beta_0 + \beta_1 T + \beta_2 T^2 \qquad (8.27)$$

where
  μ: viscosity, mPa·s
  T: temperature, °C

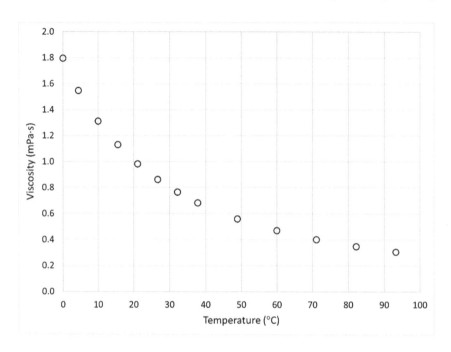

FIGURE 8.24 Plot of water viscosity data versus temperature.

The spreadsheet solution of the regression problem is shown in Figure 8.25. These include displays of the formulas used. The three values in the **b** vector are the estimated coefficients of the model.

We can use the **b** parameters and the temperatures to compute the predicted viscosities using the formula in the lower left of Figure 8.23 and add them as a curve on the plot. This is shown in Figure 8.26.

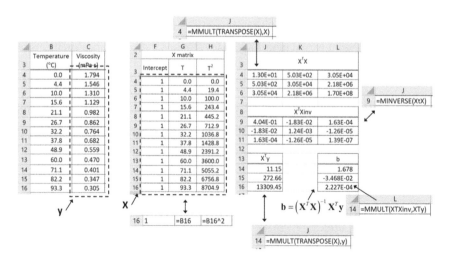

FIGURE 8.25 Linear regression solution for water viscosity data.

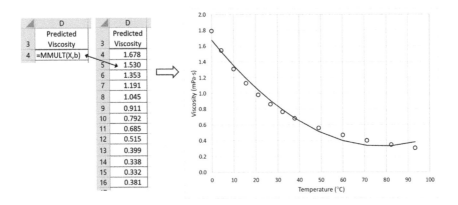

FIGURE 8.26 Predicted viscosities added to the plot as a curve.

We can observe that the curve captures curvature, but it isn't an adequate model. It is systematically low or high along the data. A higher-order polynomial is called for here. We will illustrate that using the Data Analysis Regression. To do that, the X matrix on the worksheet has to be modified. The column of ones is removed (Excel provides that), and columns are added out to the fifth order. For a fifth-order polynomial model, Figure 8.27 shows setting up the calculation.

You will notice that the numbers in the lower right corner of the X-Input range are large, $\sim 10^9$. As we use higher-order polynomials, this can cause numerical and statistical problems, especially since the powers of x are strongly correlated,

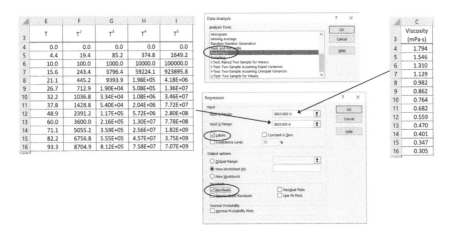

FIGURE 8.27 Setting up the Data Analysis Regression tool for a fifth-order polynomial model.

a phenomenon called *collinearity*. A remedy for this is to standardize the initial x values, T here, using the formula

$$z = \frac{T_i - \bar{T}}{s_T} \tag{8.28}$$

and formulate the linear regression in terms of z. The z values will be centered on zero and within a range of $-3 \leq z \leq 3$, so higher powers will not yield such large numbers.

The results are displayed on a new worksheet and require formatting. These are shown in Figure 8.28. The results are plentiful and deserve interpretation. First, the Coefficients column provides the parameter values for the fifth-order polynomial. The P-value column is worth attention to because these are probabilities that a given term in the model does not contribute significantly to the fit. You can see that these are small. The Lower 95% and Upper 95% columns provide confidence intervals for the parameters, and, without going into detail, if the interval defined by these boundaries includes zero, it is difficult to claim that the parameter value is different from zero, another criterion for eliminating a model term. The t Stat and Standard Error columns are related to these and can be interpreted similarly.

The ANOVA (analysis of variance) table addresses the question of whether the model contributes anything in interpreting the data. For most work in science and engineering, the answer to this is a clear yes, so this block is of little use. The Significance F number is a P-value related to the alternate hypothesis that the model "brings nothing to the party." You can see that this value is minuscule.

The Regression Statistics block deserves discussion. The total variability, also called the total corrected sum of squares, SS_T, of the output response, y, is defined as the sum of squares of differences with the sample mean,

$$SS_T = \sum_{i-1}^{n} (y_i - \bar{y})^2 \tag{8.29}$$

	A	B	C	D	E	F	G
1	SUMMARY OUTPUT						
2							
3	*Regression Statistics*						
4	Multiple R	0.99997					
5	R Square	0.99995					
6	Adjusted R Square	0.99991					
7	Standard Error	0.00455					
8	Observations	13					
9							
10	ANOVA						
11		*df*	*SS*	*MS*	*F*	*Significance F*	
12	Regression	5	2.698	0.540	26042.4	8.84E-15	
13	Residual	7	0.0001450	2.07E-05			
14	Total	12	2.698				
15							
16		*Coefficients*	*Standard Error*	*t Stat*	*P-value*	*Lower 95%*	*Upper 95%*
17	Intercept	1.7895	0.004171	429.043	9.9E-17	1.77965	1.79938
18	T	-0.059197	0.001082	-54.728	1.8E-10	-0.06175	-0.05664
19	T2	0.0013249	0.0000806	16.448	7.5E-07	0.00113	0.00152
20	T3	-1.9117E-05	2.32E-06	-8.227	7.6E-05	-2.46E-05	-1.36E-05
21	T4	1.5260E-07	2.83E-08	5.387	1.0E-03	8.56E-08	2.20E-07
22	T5	-4.9973E-10	1.22E-10	-4.081	4.7E-03	-7.89E-10	-2.10E-10

	A	B	C
26	RESIDUAL OUTPUT		
27			
28	*Observation*	*Predicted Viscosity (mPa·s)*	*Residuals*
29	1	1.790	0.00448
30	2	1.553	-0.00713
31	3	1.312	-0.00240
32	4	1.124	0.00451
33	5	0.979	0.00308
34	6	0.860	0.00160
35	7	0.766	-0.00165
36	8	0.685	-0.00349
37	9	0.561	-0.00150
38	10	0.467	0.00262
39	11	0.399	0.00186
40	12	0.350	-0.00279
41	13	0.304	0.00082

FIGURE 8.28 Formatted regression results.

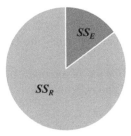

Total Variability
SS_T

FIGURE 8.29 Partition of total sum of squares.

The sum of squares of the residual errors is given as

$$SS_E = \sum_{i=1}^{n}\left(y_i - \hat{y}_i\right)^2 = \sum_{i=1}^{n} e_i^2 = \mathbf{e}^T\mathbf{e} \qquad (8.30)$$

The difference between SS_T and SS_E represents the portion of SS_T that was accounted for by the model, SS_R. This partition of SS_T can be conveniently represented in a pie chart, as shown in Figure 8.29. The ratio of SS_R to SS_T is called the *coefficient of determination* and represented as R^2 (typically called *R-squared*),

$$R^2 = \frac{SS_R}{SS_T} = 1 - \frac{SS_E}{SS_T} \qquad (8.31)$$

The coefficient of determination, R^2, is often termed *goodness of fit* and is used to quantify how well the model fits the data. This is flawed, and, in general, shouldn't be done. The SS_E captures both the inadequacy of the model to represent the data *and* the random error content of the data. If the data contain a considerable content of random noise that cannot be explained by the model, SS_E will be large and R^2 will be low, yet the model could nevertheless explain the systematic behavior in the data. Alternately, for data measured in the laboratory with little random variability, R^2 may be close to one, but the model may yet be inadequate.

Another reason that R^2 is not a good criterion for goodness of fit is that it depends on the number of model parameters when compared to the number of data. For example, with polynomial models, as the polynomial order approaches the number of data, R^2 will go to one, indicating a perfect fit. For n data, an $(n-1)$-order polynomial will have n parameters, and a least-squares fit will pass precisely through every data point, yielding $SS_E = 0$ and $R^2 = 1$. However, if one examines the model between data points, it will likely behave inappropriately, possibly with significant deviations from the trend of the data.

So, the question is, then, what else can we use to judge the performance of a model? There are several alternatives; in fact, we could dedicate a complete section to this, but we will focus on what the Data Analysis Regression tool provides us.

The *adjusted* R^2 penalizes the use of too many parameters in the model. It is a modification of Equation 8.29 and is defined as

$$R_{\text{adj}}^2 = 1 - \frac{SS_E/(n-k)}{SS_T/(n-1)} \tag{8.32}$$

where, as before, k is the number of model parameters. As we increase the number of model parameters, SS_E reduces, but its denominator reduces also. Adjusted R^2 is a good choice to assess model performance if n is not too large. For large n, the impact of increasing k is limited.

The standard error of the estimate, s_e, is given by

$$s_e = \sqrt{\frac{SS_E}{n-k}} \tag{8.33}$$

It also has a penalty as the number of model parameters increases. It can be effective in choosing between competing models, or model building where model terms are added one at a time.

In Figure 8.28, R^2 and Adjusted R^2 are shown in the Regression Statistics block. The Multiple R value is the square root of R^2, also known as the *Pearson correlation coefficient*. These three values are very close to one, indicating a tiny SS_E value. This is confirmed by the s_e value of 0.00455, which is very small compared to the y values.

If we return to our spreadsheet solution for a second-order polynomial, we can compute the following values:

$$R^2 = 0.984 \qquad R_{\text{adj}}^2 = 0.981 \qquad s_e = 0.0661 \tag{8.34}$$

Comparing these with our values from the fifth-order polynomial model, we see that both R^2 values here are lower and the s_e value is higher. Given these values, we conclude that the fifth-order model performs better. Similar to Figure 8.26, if we add the predicted viscosity values as a plotted curve to Figure 8.24, Figure 8.30 results. We observe a very good fit.

The remaining issue we would like to address here is *model adequacy*. This is assessed by the appearance of systematic behavior in the residual series produced by regression. If our model is not able to account for all the systematic behavior in the output response, y, it is inadequate. This can be assessed effectively by a "residuals versus fits" plot, a plot of the residual values, e, versus the model

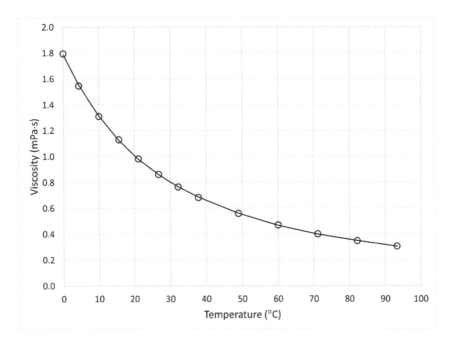

FIGURE 8.30   Viscosity data with fifth-order polynomial model.

predictions, $\hat{y}$. In Figure 8.31, we present these plots for the second-order and fifth-order models. The inadequacy of the second-order model is evident in the left plot of the figure. The right plot shows a random pattern with less systematic behavior. Also, the vertical scales of the two plots differ by an order of magnitude, indicating how much smaller the residuals are for the fifth-order model.

Data sets frequently exhibit much more random content than our example here of a physical property measured carefully in the laboratory. This is especially true

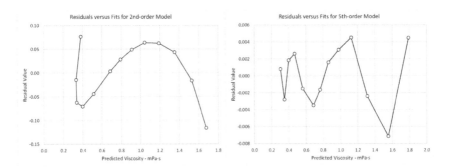

FIGURE 8.31   Residuals versus Fits plots for second-order and fifth-order polynomial models.

of environmental phenomena such as those related to climate. We would expect for such data that the fraction of SS_T represented by SS_E will be much larger; however, that would not change our approach to regression analysis.

A question to address now is, given the convenience of the Data Analysis Regression tool, why would we ever use the vector-matrix formulas on the spreadsheet? One answer would be hardly ever, but a distinction is that the on-spreadsheet formulas are live. That is, they would respond instantly to any changes in the data. There are applications where data are imported periodically to a spreadsheet, and a model fit needs to be updated. The on-spreadsheet formulas are preferred in this case. We should also comment that, once the **X** matrix is in place on the spreadsheet, the linear regression can be consolidated into one additional, lengthy, array formula.

8.4.2 NONLINEAR REGRESSION

Now that we have explored fitting of models that are linear in their parameters, we will complete the story by considering nonlinear regression. Nonlinear models take different forms. Some are relatively simple, as with algebraic equations, and others are more complicated, like the solution of differential equations. Also, a model like

$$y = a \cdot e^{b \cdot x} \tag{8.35}$$

appears to be nonlinear in the parameters a and b, but it can be transformed by the natural logarithm to

$$\ln(y) = \ln(a) + b x \tag{8.36}$$

and linear regression can be applied to $\ln(y)$ versus x. The parameter values determined are $\ln(a)$ and b. The estimate for a can then be determined from $\ln(a)$ using the exponential function as in $a = e^{\ln(a)}$.

For a nonlinear regression example, we will consider another physical property, the vapor pressure of water versus temperature at atmospheric pressure. Table 8.3 presents the data.[10] A model that is used for vapor pressure is the *Antoine equation*:

$$\log_{10}(P) = a - \frac{b}{c + T} \tag{8.37}$$

where
    P: vapor pressure, kPa
    T: temperature, °C
    a, b, c: model parameters

The model is obviously nonlinear in the model parameters.

TABLE 8.3
Vapor Pressure of Water (kPa) versus Temperature (°C) at Atmospheric Pressure

Temperature (°C)	Vapor Pressure (kPa)	Temperature (°C)	Vapor Pressure (kPa)
5	0.87	55	15.74
10	1.23	60	19.92
15	1.70	65	25.00
20	2.34	70	31.16
25	3.17	75	38.54
30	4.24	80	47.34
35	5.62	85	57.81
40	7.38	90	70.10
45	9.58	95	71.44
50	12.33	100	101.33

Figure 8.32 shows the setup on the spreadsheet for fitting the model to the data. The common logarithms of the vapor pressures are calculated in column P, and based on initial estimates of the model parameters, here named cells a, b, and c_, predicted $\log_{10}(P)$ values are computed in column Q. A plot was created to display the data and model predictions using the $\log_{10}$ values. This shows a significant offset between the model and the data.

By adjusting the parameter values manually, we can "tune" the model to the data, in effect, getting the model into the data's "ballpark." This will aid in the convergence of the Solver to determine the optimal parameter values. Figure 8.33 shows the modified parameters and resulting plot.

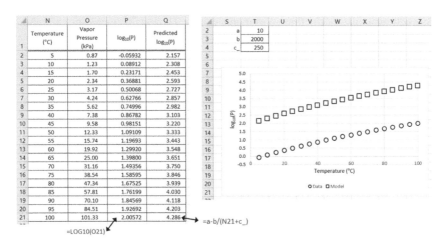

FIGURE 8.32   Spreadsheet for nonlinear regression of vapor pressure data.

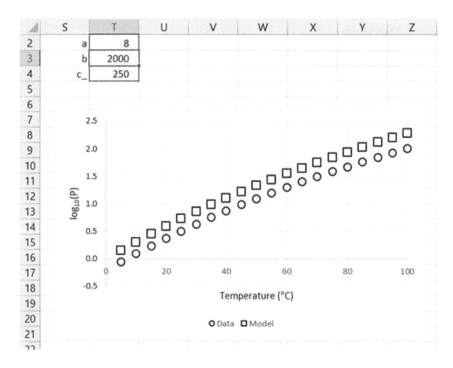

FIGURE 8.33 Tuned model, modifying the *a* parameter to 8.

We can now set up the scheme for using the Solver to determine the optimal parameter values. A column of errors is created, the difference between the $\log_{10}$ data and predictions, and the sum of squares of these errors, cell SSE, is computed using the SUMSQ function. This is shown in Figure 8.34.

With the arrangement in Figure 8.34, we invoke the Solver to minimize the SSE cell by adjusting the three parameter values. The results, including the modified plot (points changed to a line), are shown in Figure 8.35.

We observe that the SSE value has been reduced by a factor of over 5,000 from 1.21 to 0.00023. The model is an excellent fit, and, reliably, we would use the Antoine equation with the determined parameter values to predict vapor pressure over the range of temperature of the data, the full range of liquid water at atmospheric pressure.

We could compute the model performance measures and check adequacy, but that is hardly necessary here. Our results provide ample evidence of why the Antoine equation is used to model vapor pressure versus temperature.

8.4.3 FITTING MODELS TO DATA USING TRENDLINE

The final topic of this section and chapter is to review the use of Excel's *Trendline* feature on plots. Returning to Figure 8.24, we can right-click the data series and select the Add Trendline option, as shown in Figure 8.36.

	P	Q	R	S	T
1	log₁₀(P)	Predicted log₁₀(P)	Errors		
2	-0.05932	0.157	-0.216	a	8
3	0.08912	0.308	-0.219	b	2000
4	0.23171	0.453	-0.221	c_	250
5	0.36881	0.593	-0.224		
6	0.50068	0.727	-0.227	SSE	1.21
7	0.62766	0.857	-0.229		
8	0.74996	0.982	-0.232	=SUMSQ(e)	
9	0.86782	1.103	-0.236		
10	0.98151	1.220	-0.239		
11	1.09109	1.333	-0.242		
12	1.19693	1.443	-0.246		
13	1.29920	1.548	-0.249		e
14	1.39800	1.651	-0.253		
15	1.49356	1.750	-0.256		
16	1.58595	1.846	-0.260		
17	1.67525	1.939	-0.264		
18	1.76199	2.030	-0.268		
19	1.84569	2.118	-0.272		
20	1.92692	2.203	-0.276		
21	2.00572	2.286	-0.280		

FIGURE 8.34 Computation of residual errors and their sum of squares.

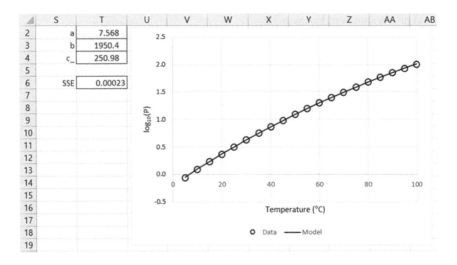

FIGURE 8.35 Nonlinear regression results from the application of the Solver.

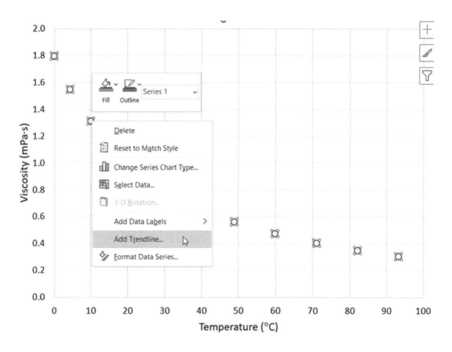

FIGURE 8.36   Selection of the Add Trendline command.

In Figure 8.37, the selection of a fifth-order polynomial as the model and an option to display the fitted equation on the plot are selected. The resulting plot is shown with the Trendline curve in place and the equation displayed (hard to discern).

The chart can be reformatted to show the Trendline curve better, enlarge the equation display, and improve the resolution of the model parameter values. This final plot is shown in Figure 8.38.

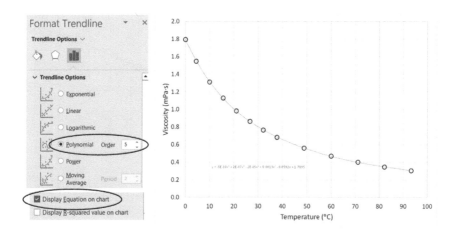

FIGURE 8.37   Selection of Trendline options and initial display on the plot.

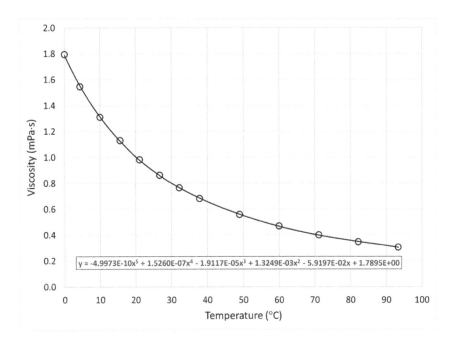

The equation shown in the figure:

$$y = -4.9973\text{E-}10x^5 + 1.5260\text{E-}07x^4 - 1.9117\text{E-}05x^3 + 1.3249\text{E-}03x^2 - 5.9197\text{E-}02x + 1.7895\text{E+}00$$

FIGURE 8.38 Final Trendline plot with improved equation display.

The use of the Trendline feature is certainly convenient. There are several empirical models that can be chosen, and the results are live and visible instantly. There are important limitations to Trendline, however, that lead us to prefer the earlier methods we have illustrated in this chapter:

- There are no accompanying statistics to help evaluate the fit except R^2, and we have already discussed its limitations.
- The displayed equation is text, and its coefficients must be copied manually onto the spreadsheet for use of the model.
- The initially displayed equation coefficients will not yield an accurate model. As shown here, we have increased the coefficient display to five significant figures. Even more may be required for manual copying to get a reliable model.
- The available models are not flexible. For example, we cannot generate a polynomial model with one or more terms missing.

On the other hand, a Trendline can be a starting point to explore different models before proceeding with a more detailed regression analysis.

The topic of regression analysis, as others in this chapter, merits a much more comprehensive treatment, but there is no space for that here. We recommend resources such as the texts by Montgomery and Runger (2018) and Chapra and Clough (2022). The latter also includes an in-depth coverage of the broader field of numerical methods.

PROBLEMS

8.1 Table P8.1 presents a set of biological oxygen demand (BOD) measurements made on water samples taken at two-hour intervals from a polluted river.

TABLE P8.1

BOD Measurements from a River

Sample Number	BOD (mg/L)
1	6.5
2	5.8
3	13.1
4	6.4
5	7.0
6	6.3
7	7.0
8	9.2
9	6.7
10	6.7

On a spreadsheet, compute the average, median, sample standard deviation, and MAD for the BOD data. Remove sample 3 and repeat the calculations. Comment on the differences between the two sets of calculations and the influence of sample 3 on the results.

8.2 Table P8.2 presents 40 measurements of the inside diameter (in mm) of engine piston rings.

TABLE P8.2

Measurements of Piston Ring Inside Diameter (mm)

74.030	74.010	74.009	73.987
73.994	74.001	74.000	74.035
73.982	74.002	74.012	74.008
73.994	74.006	74.001	73.984
73.995	74.015	73.995	74.008
74.004	74.015	73.994	74.017
74.004	73.992	73.995	73.998
74.008	73.984	74.015	74.000
73.988	73.982	73.985	74.003
73.983	74.030	74.006	74.010

---

a. Compute sample statistics for these data including average, median, sample standard deviation, and MAD.
b. Create a histogram chart of the data and interpret it.
c. Apply a hypothesis test to assess the claim that the mean diameter is less than 74.01 mm.

8.3 Obtain data on the number of Atlantic Ocean hurricanes per year from 1851 to the most recent year available from the Internet[11] and import these into a spreadsheet.

  a. Create a plot of the data without markers but with straight lines connecting the data points. Comment on the nature of the plot.

  b. Divide the data set into two equal halves for earlier and more recent years. Compute sample statistics for each half including average, median, sample standard deviation, and MAD.

  c. Use hypothesis tests to assess whether the variabilities and means of the two segments are significantly different.

8.4 Using the NORM.S.DIST and T.DIST functions, create a plot of the standard normal density curve and the t distribution density curve. For the latter, plot curves for $n = 5, 10, 25$, and 50. Display all curves on the same plot and include a legend.

8.5 Create a set of 1,000 random numbers based on the uniform distribution for the range $-1 \leq x \leq 1$. Compute frequency data and create a histogram chart for the data. Compare this chart with your expectation from the uniform density curve.

8.6 Create a set of 10,000 random numbers based on the uniform distribution for the range $0 \leq x \leq 100$, in the format of 10 columns wide and 1,000 rows deep. For each row of 10 random numbers, compute the average. Create a histogram of the averages and comment on whether it confirms the *central limit theorem*.

8.7 Measurements of equivalent fuel economy have been acquired for ten samples of two manufacturers' electric vehicles, designated A and B. These are shown in Table P8.7. The units are liters of gasoline saved per year. Test the claim that vehicle B provides better economy than A.

TABLE P8.7
Fuel Economy of Two
Electric Vehicles (L/yr)

A	B
1970	2009
1932	2018
1969	1972
2000	1991
1988	2019
1944	1995
1990	2018
1969	2008
1949	2023
1944	2016

8.8 Data indicate that the average temperature of the Earth's atmosphere has been increasing over the decades. Global temperature data can be obtained from https://www.ncdc.noaa.gov/cag/global/time-series for the years 1880–2021. These data are reported as a temperature anomaly, the difference from the 1951–1980 average temperature.

a. Download these data and import them into a spreadsheet. Create a plot of the data using straight lines, no markers. Comment on your observations of the plot.

b. Fit a straight-line model to the data. Report the R^2, Adjusted R^2, and s_e values for the regression. Plot the residual series and comment on the model's adequacy.

c. Fit a higher-order polynomial to the data and assess whether this model is preferred over the straight-line model.

8.9 Model building is an important approach to regression analysis. For the data in Table 8.2, implement the following strategies, but transform the temperatures first according to Equation 8.26.

a. Starting with a straight-line model, fit polynomials of increasing order with the final model being 7th-order. For each model, tabulate R^2, Adjusted R^2, and s_e. For the latter two criteria, which model(s) would be selected as those performing best?

b. Starting with a 7th-order polynomial model, examine the P-values from the regression output. Remove the polynomial term with the largest P-value above 1% and repeat the regression calculation. Conduct this procedure until all P-values are less than 1%. Compare the model you have determined with that from part (a). Create a plot of the data (markers), model (straight line segments), and residual series. Place the residual series on the secondary vertical axis.

8.10 Nitric oxide (NO) gas is absorbed into a reacting solution to produce a product. The data presented below are obtained from this process. A model has been proposed for this process, which is

$$y = b_0 e^{b_1 x} x^{b_2}$$

where
x: amount of NO absorbed, g/L
y: concentration of product, g/L

TABLE P8.10

Product Concentration versus Amount of NO Absorbed

x	0.09	0.32	0.69	1.51	2.29	3.06	3.39	3.63	3.77
y	15.1	57.3	103.3	174.6	191.5	193.2	178.7	172.3	167.5

---

a. Fit the model to the data using nonlinear regression. Create a plot of the data and the model line. Compute R^2, Adjusted R^2, and s_e. Create a plot of residuals versus fits, and comment on model adequacy.

b. Transform the model with the natural logarithm and fit it to the data using linear regression. Report the model coefficients, b_0, b_1, and b_2, and compare these to the results from part (a). Add the model curve obtained to the plot from part (a) and comment on any differences.

8.11 A designed experimental campaign specifies levels of input variables and collects output responses for combinations of those. One goal of the campaign is to develop a model relating the inputs to the output, and a common form of this is called a *response surface model*. Table P8.11 documents an experimental design and reports the results. In this case, the levels for the three input factors, x_1, x_2, and x_3, have been transformed into a range centered around zero. A second-order response surface model is proposed for this process,

$$y = a_0 + a_1x_1 + a_2x_2 + a_3x_3 + a_4x_1^2 + a_5x_2^2 + a_6x_3^2 + a_7x_1x_2 + a_8x_1x_3 + a_9x_2x_3$$

TABLE P8.11
Results of an Experimental Design

x_1	x_2	x_3	y
−1	−1	−1	1.77
1	−1	−1	2.1
−1	1	−1	3.37
1	1	−1	3.69
−1	−1	1	1.77
1	−1	1	2.17
−1	1	1	3.53
1	1	1	3.83
0	0	0	2.92
0	0	0	2.92
0	0	0	2.88
0	0	0	2.91
−1.633	0	0	2.35
1.633	0	0	3.05
0	−1.633	0	1.66
0	1.633	0	3.86
0	0	−1.633	2.61
0	0	1.633	3.04
0	0	0	2.92
0	0	0	2.87

   a. Using linear regression, fit this model to the data. Using the Data Analysis Regression tool, the X Input matrix will include nine columns according to the x groups in the terms of the model.

   b. Example the P-values for the model coefficients. If there is a P-value > 5%, remove the associated term from the model and repeat the regression calculation. Repeat this process until all P-values are < 5%.

   c. Create a markers-only plot of the predicted y values versus the measured y values. Ensure that the axis scales are equal. Add a 45° line to the plot. Comment on the performance of the model.

   d. Create a plot of the residuals versus fits and comment on the adequacy of the model.

NOTES

1. Hold down the Alt key while pressing T then I.
2. Primary standards are specified or maintained as an artifact by the *National Institute of Standards and Technology* (NIST) in the U.S. and internationally by the *Système International d'Unités* in France.
3. According to NIST, an outlier is an observation that lies an abnormal distance from other values in a random sample from a population. This definition leaves it up to the analyst (or a consensus process) to decide what will be considered abnormal. https://www.itl.nist.gov/div898/handbook/prc/section1/prc16.htm
4. The chance of data, events, athletic ability, etc. occurring outside this interval is about 1 in 3.5 million. For example, Typhoon Haiyan, perhaps the most powerful in recorded history, might be referred to casually as a "5-sigma" storm. Singular individuals such as da Vinci, Bach, Usain Bolt, Louis Armstrong, Maria Callas, Alan Turing, or Leonhard Euler might be referred to as "five-sigma" geniuses at their respective crafts. Of course, the term could also be applied to 5-sigma sociopaths such as Hitler, Stalin, Pol Pot, or Genghis Khan.
5. In this context, a P-value is the probability of how likely it is that a result of a statistical event, or one more extreme, occurring. In this example, it tells you that given the distribution and its parameters, there is an 80% probability that a value less than 108.4 will occur. We will return to this parameter and its interpretation when we explore hypothesis testing later in this chapter.
6. Categorical variables represent variables that may be divided into groups. Common examples of categorical variables are race, sex, age group, and educational level.
7. Statistical software packages often display tick marks and labels at the boundaries of the bins.
8. Although they are less powerful, non-parametric approaches have several advantages including that they may be the only alternative for small samples and when the population distribution is uncertain, they make fewer assumptions about the data, they are useful when the data are inherently categorical, and they often involve simpler computations and interpretations than parametric tests.
9. Assessing these risks may not seem that important to you, but in a case where your job depends on a correct conclusion, as they say, "the rubber hits the road."
10. These data are readily available from sources called "the steam tables."
11. http://tropical.atmos.colostate.edu/Realtime/index.php?arch&loc=northatlantic

9 Introduction to VBA and Macros

CHAPTER OBJECTIVES

- Learn how the VBA environment is separate from the Excel spreadsheet environment and how to navigate between the two
- Become familiar with the Visual Basic Editor and how to manage projects and VBA code windows
- Be able to record time-saving macros and debug them in the Visual Basic Editor
- Understand the rudiments of object-oriented syntax in VBA code
- Learn how to transfer information between the spreadsheet and VBA environments
- Be introduced to debugging methods in the Visual Basic Editor including the Locals window
- Learn how to use the Macro Recorder as a code "detective"

The origin of "macro" languages in spreadsheets was to record a series of keystrokes that could be replicated with a single keystroke combination. These languages were gradually extended by the inclusion of structured elements like decision and repetition. Eventually, they evolved into poorly designed programming languages.

In late 1993, Microsoft introduced Excel 5.0 as part of their Office software suite and accompanied that with a dramatic departure to a separate object-oriented programming language called Visual Basic for Applications (VBA). Separate "flavors" of VBA were created for the various components of Office, including Word, PowerPoint, Outlook, and Access. We focus here on the version of VBA for Excel.

After its introduction, VBA revolutionized Excel applications and, in part, was responsible for Excel's growing market share in the spreadsheet market, which eventually led to domination. Of course, the fact that Excel comes with the Office suite influenced that also. Yet, today most Excel users do not know much about VBA. The purpose of this and the following chapters is to change that.

The VBA programming language has tremendous potential to streamline your spreadsheet development and elevate the capabilities and efficiency of your spreadsheets. In this chapter, we will introduce the Visual Basic Editor (VBE), the separate programming environment for VBA. With the editor skills in hand, a major theme will be the recording of time-saving macros. Creating macros is

DOI: 10.1201/9781003361053-9

a good, first step into the use of VBA, and, generally, is greatly appreciated by users. Following the introduction of macros, we will experiment with some short VBA programs to learn more about VBA elements that obtain information from the spreadsheet, and send it back to Excel. The VBE has good tools to trace the execution of VBA code and detect errors. We will introduce those via examples. To finish this chapter, we will introduce the use of the Macro Recorder to discover VBA statements that perform certain actions on the spreadsheet.

9.1 VBA AND ITS ENVIRONMENT: THE VISUAL BASIC EDITOR

The first concept that is important in learning to use VBA is to understand that its environment is separate from that of Excel. This is pictured schematically in Figure 9.1. This separation motivates the need to communicate back and forth between "spreadsheet world" and "VBA world." Occasionally, we all will succumb to the assumption that the rules and practices in one "world" apply in the other. That is part of the process in learning to use VBA.

Prior to using VBA in the VBE application window, you need to enable the Developer tab on the Ribbon in Excel. You do this by selecting File ⇨ Options ⇨ Customize Ribbon, and checking Developer in the panel on the right. After clicking OK there, you should see the Developer tab on the Ribbon, which is illustrated in Figure 9.2. For our purposes in this chapter, we will use the Code group. Later, in Chapter 12, we will take advantage of the Controls group.

It is also useful when using VBA code, including macros of your own creation, to change the VBA security settings of Excel. You can do this in the following two steps from Excel:

- File ⇨ Options ⇨ Trust Center ⇨ Trust Center Settings... ⇨ Macro Settings ⇨ Enable VBA macros.[1]
- Also, check the box for Trust access to the VBA project object model.

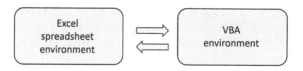

FIGURE 9.1   The Excel and VBA environments.

FIGURE 9.2   The Developer tab activated on the Ribbon in Excel.

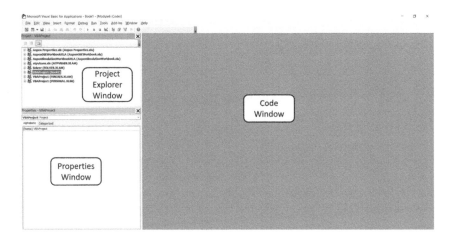

FIGURE 9.3   Visual Basic Editor window.

It may be, if your computer is managed by an IT organization, that you are not allowed to make these settings. But, at the least, you should be able to select *Disable VBA macros with notification.*

In Figure 9.2, the command icon on the far left is Visual Basic. Selecting this command activates the *VBE* application window. You will become accustomed to using the Alt-F11 shortcut to switch to the VBE and back to the Excel spreadsheet. Activate the VBE window, and you should see the display in Figure 9.3.

The VBE has a traditional menu at the top left with a toolbar just below it. There are three windows below the toolbar:

- The *Code Window* for typing VBA code
- The *Project Explorer Window* for managing Excel/VBA projects
- The *Properties Window* for object properties

We will not use the Properties window until Chapter 12. The Project Explorer window is important because it lists all the current projects. Your list will likely be different from the one displayed with fewer items. Since our workbook name back in Excel is Book1, there is a project item entitled VBAProject (Book1). It is important to have this project selected when we are developing code to be attached to this Excel workbook.

You will probably have these additional projects listed:

- atpvbaen.xls (ATPVBAEN.XLAM) – Analysis Toolpak – VBA add-in
- Solver (SOLVER.XLAM) – Solver add-in
- VBAProject (FUNCRES.XLAM) – Analysis Toolpak tools

These are related to the add-ins you installed earlier.

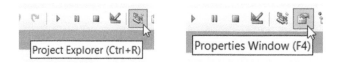

FIGURE 9.4   Project Explorer and Properties Window command icons and shortcut keys.

You will also note in the figure, an entry VBAProject (PERSONAL.XLSB). This is related to the *Personal Macro Workbook*. You may not have this yet, but it will be added later in this chapter. The Aspen items in the project list relate to a software package installed by one of the text authors.

Although not necessary, the Project Explorer and Properties windows can be removed from the display by clicking the ⊠ buttons in the upper right of each window. They can be returned to the display using command icons on the VBE toolbar, as shown in Figure 9.4. You will also note the shortcut keys for these commands.

Each of the projects listed has a + icon next to it. By clicking the +, it expands a branch showing a folder for Microsoft Excel Objects. This displays two objects: Sheet1 and ThisWorkbook. We will work with these objects in Chapter 11 when we introduce *event handlers*. For the moment, we do not enter VBA code related to these objects.

In order to add VBA code to a project, we need to insert a module. This is accomplished using the menu, Menu ⇨ Insert ⇨ Module. Be careful not to choose Userform or Class Module. We will employ the Userform choice in Chapter 12. After the module is inserted, the VBE window shows an empty Code Window and the Modules folder is added to the project tree with a Module1 object appended (Figure 9.5). The Code Window is open with an insertion cursor (|) blinking where code can be entered.

The name Module1 can be changed in the Properties window if it is selected in the Project Explorer. You can modify the Name field there. We usually don't do this though except when we get to userforms in Chapter 12.

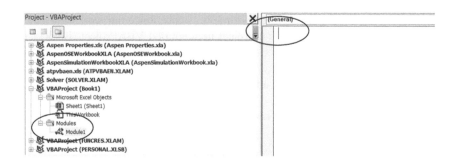

FIGURE 9.5   Visual Basic Editor window with module inserted.

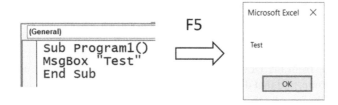

FIGURE 9.6 Simple VBA program executed with the F5 key.

We will start our VBA coding with a simple program, as shown in Figure 9.6. After we type in the first line of the program and press the Enter key, the VBE automatically adds an End Sub line. The Sub and End Sub are coded dark blue indicating they are *keywords* in VBA. Keywords are special words that are reserved. These cannot be used for variables, subroutines, or function names.

Figure 9.6 also shows how we run our program in the VBE using the F5 shortcut key. The output of the MsgBox command[2] appears on the spreadsheet, and we click the OK button to finish.

Sub is shorthand for *subroutine procedure*. The name of the Sub is Program1, and that is followed by a pair of parentheses. These may carry a list of arguments when the Sub is called from other VBA subprograms, but they are always empty when it is a program to be run directly from the VBE or Excel. VBA programs that can be run directly from Excel are called *macros*. MsgBox is short for message box, and you can see what it generates. We will explore more features of this command along the way and, in particular, in Chapter 12.

There are different ways to run a Sub. These are illustrated in Figure 9.7. In the VBE, there is a Run dropdown menu with the option Run Sub/Userform. On the

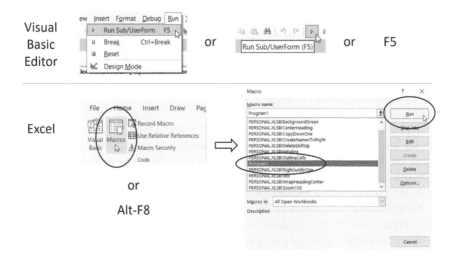

FIGURE 9.7 Different ways to run a VBA Sub.

toolbar, there is a Run Sub/Userform button, which is also called a play button. Both of these indicate the shortcut key, F5. We usually use the F5 key. If there are more than one Sub in the Code Window, the cursor should be somewhere in the code of that Sub before we attempt to run it from the VBE.

In Excel, from the Developer tab, we can click on the Macros command in the Code group, and that displays the Macro window. We will find our Program1 Sub listed there and can click the Run button. Alternately, we can display the Macro window quickly with the Alt-F8 key combination. You will note in the figure that there is a long list of macros with the preface PERSONAL. You will not likely have these until you create some of your "favorite" macros. That is coming.

As we mentioned before, you can "toggle" quickly back and forth between Excel and the VBE with the Alt-F11 key combination. That is most convenient. From the VBE, you can also click on the View Microsoft Excel button on the far left of the toolbar to switch back to Excel. Additionally, on the Windows taskbar, both environments will be visible if you hover the mouse cursor on the Excel command icon, and you can switch back and forth there.

9.2   RECORDING MACROS

Excel provides a feature whereby we can carry out a procedure on the spreadsheet, and a VBA Sub will be recorded to replicate it. In the "olden days" of spreadsheets, macro procedures were literal keystrokes and key letters on menus, so there was no ambiguity. Running the macro just replicated those keystrokes.[3] With Excel and VBA, it is different. As you execute a procedure on the spreadsheet, the Macro Recorder interprets what you are doing and writes a VBA Sub to replicate it. Frequently, it doesn't get it quite right, and you must examine the VBA code and modify it.

This brings us to a key point that will carry on through the rest of this book. VBA code is easy to read and understand, but it is detailed and difficult to write from scratch. VBA has thousands of keywords and commands. This leads us to develop a technique where we record as macros actions on the spreadsheet to see how they are coded in VBA. These may not be macros we are going to use but just as a code "detective" to see how to program a task in VBA. We will get back to this concept later in this chapter.

When should you consider creating a macro? The typical motivation is to recognize when you are repeating frequently a multi-step procedure on the spreadsheet. Being able to carry out that procedure with a single keystroke combination will save time and will actually be very pleasing! We will illustrate the procedure to create a macro with a useful example.

We often deal with tables of data that have headings. Headings often require more column width than the data. A common way to handle this cosmetically is to "wrap" the headings so they appear on more than one line. There is a multi-step procedure for this that will be converted into a macro. Figure 9.8 illustrates two adjacent headings in the upper left, Sample Number and Viscosity (mPa·s) that

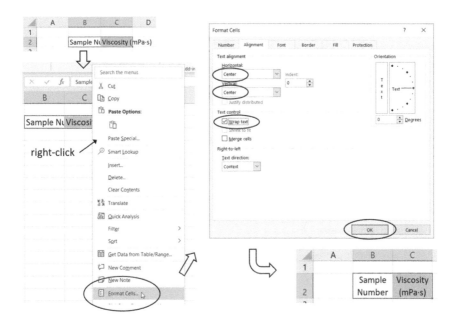

FIGURE 9.8   Procedure to format headings.

have been entered. They overlap so are not fully visible. The figure also illustrates a procedure to format the headings.

The first step is to have both cells selected and then right-click the selection to show the context-sensitive menu. There, the Format Cells option is selected, and in the Format Cells window, three settings are changed:

- Text Alignment
  - Horizontal ⇨ Center
  - Vertical ⇨ Center
- Text Control
  - Wrap text ⇨ checked

The OK button is then clicked, and the figure shows the formatted headings at the bottom right.

After rehearsing the procedure, you can reverse the formatting with the Undo button (or Ctrl-Z) and record a macro using the procedure.[4] Again, make sure you have the two cells selected. There are two ways to start the Macro Recorder, see Figure 9.9.

Upon executing the Record Macro command, the Record Macro window appears. Figure 9.10 shows how the entries are made in this window. A name is created for the macro, FormatHeading. This will be the name of the Sub created by the Recorder. A shortcut key is entered with an uppercase H. And, since we will want to use this macro in all our workbooks, the Store macro in: choice is

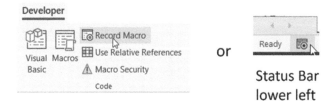

FIGURE 9.9   Ways to start the Macro Recorder.

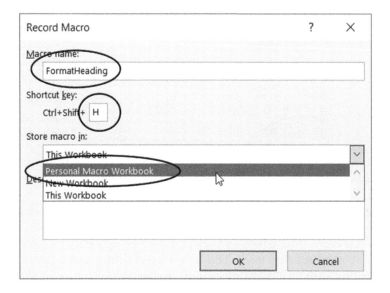

FIGURE 9.10   Record Macro window with settings made.

made for the *Personal Macro Workbook*. If you do not have any macros like this yet, you won't have a Personal Macro Workbook, but one will be created for you automatically when you finish the recording.

Some care must be taken in choosing the shortcut key for a macro. This is because its choice will override the same shortcut key built into Excel.[5] For example, if you choose Ctrl-c, that shortcut will no longer be available to initiate a copy. The lowercase letters e, j, l, m, q, and t do not have default assignments in Excel. It is better to use uppercase letters because the only ones reserved there are F, O, and P. Also, it is possible that other software applications may have a universal shortcut key that shouldn't be used, such as a *screen capture* application.

Once the Macro Recorder has started, you must be careful with the operations you perform in Excel because everything will be written into VBA code. A famous mistake (yes, made by the authors) is to forget to turn the Recorder off when you are finished with the procedure, and, after a few hours of work, having a VBA macro with a few thousand lines of code. With the Recorder active, we

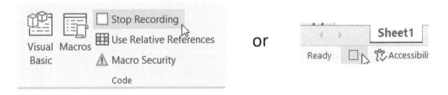

FIGURE 9.11   Options for stopping the Macro Recorder.

complete the heading format procedure. When complete we stop the Recorder. Two methods for doing that are shown in Figure 9.11. You will note that the original Record Macro command has been replaced with Stop Recording. Also, a similar change has occurred on the Status Bar below.

Before examining the VBA code, we can test the macro. To do this, we enter another heading, and with that cell selected, we employ the Ctrl-Shift-H shortcut to format the heading. The procedure is illustrated in Figure 9.12. It appears that the macro works.

Next, we would like to examine the VBA code generated by the Macro Recorder. Switching over to the VBE (Alt-F11), we should see a PERSONAL. XLSB project has been created, and we can find our macro code in a module under that project, see Figure 9.13.

As with our test program, the macro is a Sub. It has the title we gave the macro and a final End Sub statement. There are comments that begin with an ' symbol and are in green (although you can't see that color here). These non-executable comments are generated automatically and include the description of the keyboard shortcut. You can add your own comments on separate lines or at the end of a line of code.

FIGURE 9.12   Testing the macro.

FIGURE 9.13   Examining and modifying the VBA macro code.

We mention here that adding comments to code is important, both to aid others in understanding the code and to aid the author as they return to look at the code in the future. For the sake of compactness, we often do not include comments in the code in this text, but that is not because we believe comments are not important. We encourage you to add comments to your code.

The executable VBA code starts with a `With` statement referencing the object called the `Selection`. The latter refers to any selected cells. With this description, multiple headings can be formatted at once by the macro. The `With` statement is bracketed by an `End With` statement. Initially, there are nine properties of the `Selection` object assigned values, such as

```
.HorizontalAlignment = xlCenter
```

This block of code could be replaced with individual statements like

```
Selection.HorizontalAlignment = xlCenter
```

so the use of the `With ... End With` structure just saves repeating the object `Selection` nine times.

This is a typical syntax in VBA object-oriented references:

```
Object.Property
```

The nine properties listed reflect the options on the Format Cells window, Alignment tab. The first three deal with properties that we modified, and the last six merely show their default values. For this reason, we can remove the latter and not lose anything. This is shown with the VBA code on the right in Figure 9.13. On a grander scale, longer macros tend to have many excess, unneeded statements. For the sake of compactness and ease of understanding, it is good practice to trim these as was done here.

It does happen that macros have statements that are erroneous or don't quite capture what we intended. The VBE operates as a typical text editor like Notepad or Wordpad in Microsoft Windows. You do not need to save VBA projects and their modules separately. They are automatically saved when you save the Excel workbook. When you create the Personal Workbook project with your first macro, you will get a question whether to save it when you close Excel. Of course, you should select Yes. After that, it will happen automatically.

When developing your own VBA code, errors will inevitably occur, and it will be necessary to debug the code. The VBE provides some good tools for this, and that is the topic of the next section. You are given the opportunity to create a number of useful macros in the problems at the end of this chapter, and you will likely have your own preferences. Just remember that a good signal for creating a macro is when you recognize that you are repeating a sequence of operations in Excel over and over again. Also, if the macro is only associated with your current Excel workbook and will not be needed elsewhere, you can choose the This Workbook option on the Record Macro window instead of the Personal Macro Workbook. We will create macros often in the remaining chapters of this book.

9.3   DEBUGGING VBA CODE

Changes to VBA programs are more common when we write them ourselves and don't rely on the Macro Recorder to generate the code, although occasionally it is necessary to edit recorded macros. To illustrate debugging techniques, we will introduce an example that includes transmitting values back and forth between Excel and VBA.

From Chapter 5, we repeat Figure 5.25 as Figure 9.14 here. The figure describes a cylindrical bin with a conical base, typical for storage of granular materials and includes the pertinent formulas. We will create formulas on the spreadsheet to compute the total volume and surface area of the bin, not including the lid, and then create a VBA Sub to replicate those.

Figure 9.15 shows the spreadsheet calculation including the formulas for an example set of bin dimensions. The calculations are broken down following the formulas in Figure 9.14, and then sums are used to compute the total volume and

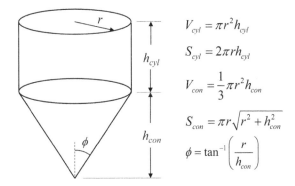

$$V_{cyl} = \pi r^2 h_{cyl}$$

$$S_{cyl} = 2\pi r h_{cyl}$$

$$V_{con} = \frac{1}{3}\pi r^2 h_{con}$$

$$S_{con} = \pi r \sqrt{r^2 + h_{con}^2}$$

$$\phi = \tan^{-1}\left(\frac{r}{h_{con}}\right)$$

FIGURE 9.14   Cylindrical bin with conical base including dimensions and formulas. In addition, V = volume, and S = surface area.

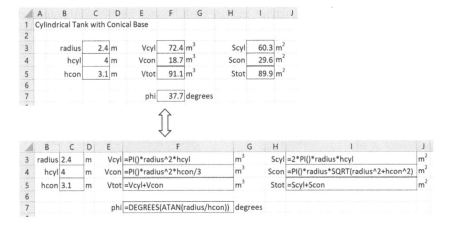

FIGURE 9.15   Spreadsheet calculations and formulas for bin volume and surface area.

```
Sub GrainBin()
r = 2.4
hcyl = 4
hcon = 3.1
Vcyl = Pi() * r ^ 2 * hcyl
Vcon = Pi() * r ^ 2 * hcon
Vtot = Vcyl + Vcon
Scyl = 2 * Pi() * r * hcyl
Scon = Pi() * r * SQRT(r ^ 2 + h ^ 2)
Stot = Scyl + Scon
Phi = Atn(r / hcon) * 180 / Pi()
End Sub
```

FIGURE 9.16   Initial VBA code for bin volume and surface area calculations.

surface area. The angle of the conical wall is also calculated as this may be compared to the angle of repose of the material stored to ensure it is steep enough to facilitate downward flow.

Initially, we will create a new module associated with the project for this workbook and type in an attempt to replicate the spreadsheet calculations. Figure 9.16 illustrates this. As you type in this code, you may encounter an error (shown in red) unless you put spaces around the ^ operator.[6]

The code in the GrainBin Sub has been entered naively. If we attempt to run the code (F5 key), we run into an error, as noted in Figure 9.17. Since, to obtain the value of π on the spreadsheet we use the PI() function, we have assumed that this will work in VBA. That function is not part of the VBA language and hence is not directly available in VBA.

There are a couple of ways to remedy the error in Figure 9.17. One is to compute π using VBA code,

```
Pi = 4*ATAN(1)
```

We will find that this generates another error because Excel's ATAN function is not available in VBA either! However, there is an Atn function. We will also find that there is not an SQRT function but rather Sqr.[7] Clicking OK on the error

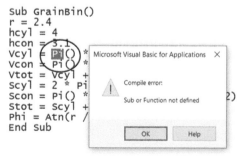

FIGURE 9.17   First execution error in VBA code.

```
Sub GrainBin()
Pi = 4 * Atn(1)
r = 2.4
hcyl = 4
hcon = 3.1
Vcyl = Pi * r ^ 2 * hcyl
Vcon = Pi * r ^ 2 * hcon
Vtot = Vcyl + Vcon
Scyl = 2 * Pi * r * hcyl
Scon = Pi * r * Sqr(r ^ 2 + h ^ 2)
Stot = Scyl + Scon
Phi = Atn(r / hcon) * 180 / Pi
End Sub
```

$\Longrightarrow$

```
Sub GrainBin()
Pi = 4 * Atn(1)
r = Range("radius").Value
hcyl = Range("hcyl").Value
hcon = Range("hcon").Value
Vcyl = Pi * r ^ 2 * hcyl
Vcon = Pi * r ^ 2 * hcon
Vtot = Vcyl + Vcon
Scyl = 2 * Pi * r * hcyl
Scon = Pi * r * Sqr(r ^ 2 + h ^ 2)
Stot = Scyl + Scon
Phi = Atn(r / hcon) * 180 / Pi
Range("vtot").Value = Vtot
Range("stot").Value = Stot
Range("phi").Value = Phi
End Sub
```

FIGURE 9.18 VBA code with errors removed and then spreadsheet interchange added.

message box will clear it, and the first line of the code will be highlighted in yellow. The usual procedure at this point is to reset VBA before correcting the code. That is accomplished with blue square *reset button* on the toolbar, ■.

If we make the required changes, the code will now appear as shown in Figure 9.18 on the left, and it will execute successfully with the F5 key. The only problem is that we haven't provided for any display of output. We could do that with one or more message boxes (MsgBox command as before); however, at this point, we will address how to obtain the dimensions (*r*, *hcyl*, and *hcon*) from the spreadsheet and return our results back to the spreadsheet.

In the VBA code on the right, we have replaced the statements for the dimensions with those using the Range type, making reference to the named cells on the spreadsheet. The syntax used here is similar to many statements in VBA:

```
object.property
Range(cell name or address).Value
```

Notice that radius is the cell name, but that is assigned to the VBA variable r. We couldn't use a cell name r on the spreadsheet, but we can use that in VBA. Also, the hcyl and hcon references in the Range type statements are for the cell names, and the hcyl and hcon variables to which they are assigned are in VBA and separate from the cell names on the spreadsheet.

The first three statements boxed in the figure transfer the spreadsheet values to variables in VBA, and the last three boxed statements transfer the result variable values back to the named cells on the spreadsheet. Figure 9.19 shows worksheet cells before and after the VBA Sub's execution.

Given the spreadsheet in Figure 9.19, if we were to change one or more values in the dimensions, the results would not update until we would run the Sub again. That is a distinction from the live formulas in Figure 9.15. We can also run the Sub from the Excel environment, as shown in Figure 9.20. This is done via the Macro window. Although these calculations are of a scale that they can be conveniently implemented on the spreadsheet, we will encounter others that are difficult-to-impossible, and VBA will be a natural choice.

	B	C	D	E	F	G
3	radius	2.4	m	Vtot		m³
4	hcyl	4	m	Stot		m²
5	hcon	3.1	m	phi		degrees

⇓ F5

	B	C	D	E	F	G
3	radius	2.4	m	Vtot	128.5	m³
4	hcyl	4	m	Stot	89.9	m²
5	hcon	3.1	m	phi	37.7	degrees

FIGURE 9.19 Execution of the VBA Sub with results appearing on the spreadsheet.

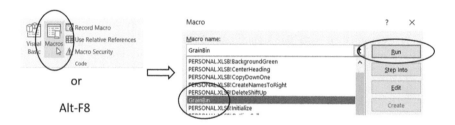

FIGURE 9.20 Executing the GrainBin Sub from the Excel environment.

Next, we will delve into debugging tools that the VBE provides. The first of these is *stepwise execution* (invoked with the F8 key). If we start execution with the F8 key, instead of F5, the first executable line of the Sub is highlighted yellow. By pressing the F8 key again and again, the code will be executed one line at a time. Figure 9.21 shows this in before-and-after screenshots. As you move past

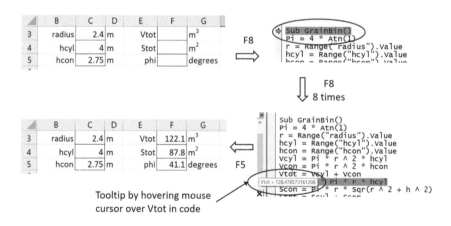

FIGURE 9.21 Stepwise execution of VBA code using the F8 key.

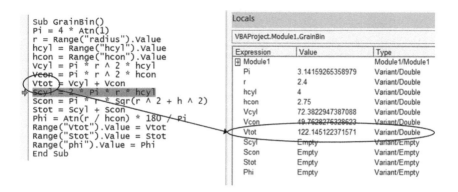

```
Sub GrainBin()
Pi = 4 * Atn(1)
r = Range("radius").Value
hcyl = Range("hcyl").Value
hcon = Range("hcon").Value
Vcyl = Pi * r ^ 2 * hcyl
Vcon = Pi * r ^ 2 * hcon
Vtot = Vcyl + Vcon
Scyl = 2 * Pi * r * hcyl
Scon = Pi * r * Sqr(r ^ 2 + h ^ 2)
Stot = Scyl + Scon
Phi = Atn(r / hcon) * 180 / Pi
Range("vtot").Value = Vtot
Range("Stot").Value = Stot
Range("phi").Value = Phi
End Sub
```

Locals		
VBAProject.Module1.GrainBin		
Expression	Value	Type
⊞ Module1		Module1/Module1
Pi	3.14159265358979	Variant/Double
r	2.4	Variant/Double
hcyl	4	Variant/Double
hcon	2.75	Variant/Double
Vcyl	72.3822947387088	Variant/Double
Vcon	49.7628276328623	Variant/Double
Vtot	122.145122371571	Variant/Double
Scyl	Empty	Variant/Empty
Scon	Empty	Variant/Empty
Stot	Empty	Variant/Empty
Phi	Empty	Variant/Empty

FIGURE 9.22   The Locals window displaying variable values during single-stepping code.

statements in the code, you can hover the mouse cursor over a variable, and it will display the current value in a *tooltip*. As the figure shows, we can finish with a continuous execution at any point with the F5 key. If an error is found, we can reset the execution with the blue square reset button on the toolbar.

There is another useful way to observe the variable values as the VBA code is executed step-by-step. This is to use the *Locals window*. You can enable this from the VBE menu ⇨ View ⇨ Locals Window, and it will appear below the Code window. Figure 9.22 illustrates the Locals window along with the Code window with the single-step progression shown in Figure 9.22 on the left. In the figure, the Locals window has been increased in height and the column widths narrowed to improve visibility. The variable values are displayed in full precision (15 significant figures), and the variable type is shown. We will discuss variable types in Chapter 11.

Single-stepping code can become tedious with longer programs, especially those that have loops that may repeat many times. To remedy this, we introduce two additional debugging techniques. First, instead of using just the F8 key, we can use the *Ctrl-F8* combination to *run to cursor*. As Figure 9.23 shows, we can

```
Sub GrainBin()
Pi = 4 * Atn(1)
r = Range("radius").Value
hcyl = Range("hcyl").Value
hcon = Range("hcon").Value
Vcyl = Pi * r ^ 2 * hcyl
Vcon = Pi * r ^ 2 * hcon
Vtot = Vcyl + Vcon
```

Ctrl
F8

⇨

```
Sub GrainBin()
Pi = 4 * Atn(1)
r = Range("radius").Value
hcyl = Range("hcyl").Value
hcon = Range("hcon").Value
Vcyl = Pi * r ^ 2 * hcyl
Vcon = Pi * r ^ 2 * hcon
Vtot = Vcyl + Vcon
```

cursor

FIGURE 9.23   Execute to cursor with the Ctrl-F8 key combination.

FIGURE 9.24   Placing a breakpoint and executing to it.

place the cursor at a line of the code, press Ctrl-F8, and execution will stop at that line. From there, we can proceed with single-stepping. This process can be repeated as needed in longer codes.

A second technique is to use *breakpoints*. These can be placed on one or more lines of the code, and execution will stop there. From the breakpoint, one can single-step (F8) or run to the next breakpoint or end of the code with the F5 key. There are two ways to create a breakpoint. With the cursor on a line, press F9, and a red dot appears in the margin to the left of the line. The same key will clear the breakpoint. Alternately, we can just click in the margin next to a line, and the breakpoint will be placed. Clicking again will remove it. Figure 9.24 illustrates breakpoints.

During debugging, if you have placed several breakpoints, you can clear all of them with a single command, *Ctrl-Shift-F9*. The various debugging options are also available under the Debug menu in the VBE.

Many VBA syntax errors will be noted as soon as a statement is entered. Figure 9.25 illustrates such an error. VBA turns the line with the error red and

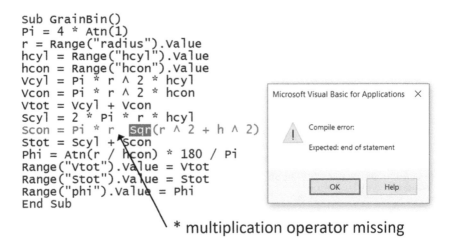

FIGURE 9.25   Syntax error noted when VBA statement is entered.

presents a diagnostic message as shown. These are sometimes, but not always, clear and helpful. In this example, we have neglected to type the * between r and Sqr. The error message complains that there is no end of statement, which does not really describe what went wrong. The best we can do in cases like this is to review the statement carefully to discover the error and correct it.

A typographical error in entering a variable name may not be directly discovered by the VBE. This is shown in Figure 9.26. Here we have entered hcn instead of hcon in the formula for Vcon. As a result, since hcn is created as an "empty" variable, zero is used in the formula, and zero results for the value of Vcon. This makes the value for Vtot incorrect.

Such errors are hard to recognize. We all make typographical errors on occasion. However, VBA has an important feature to catch and protect against these errors. It is called *variable declaration* and is entered as *Option Explicit* at the top of the module before the Sub statement. Figure 9.27 illustrates this and the consequences when we attempt to run the Sub. A requirement of the Option Explicit statement is that all variables must be declared in *Dim statements*. These have been added to the beginning of the Sub. This is customary. The Dim statements could be anywhere as long as they occur before the variable is used. You note in the figure that the hcn variable has been flagged as an error because it hasn't been declared in a Dim statement. This is because it is a typographical error, and this has been pointed out.

It is good practice, in fact, *very* good practice to declare all variables you are using in your code. This is also a requirement of other programming languages. Later in Chapter 11, we will show how to declare variables of different types. For the present, we think this is such a good practice that VBA allows you to enforce it by making a setting in the VBE. This is shown in Figure 9.28. When you set this

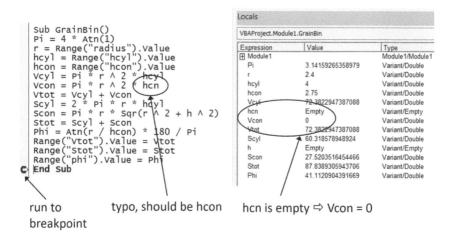

run to
breakpoint   typo, should be hcon   hcn is empty ⇨ Vcon = 0

FIGURE 9.26   Typographical error and its consequences.

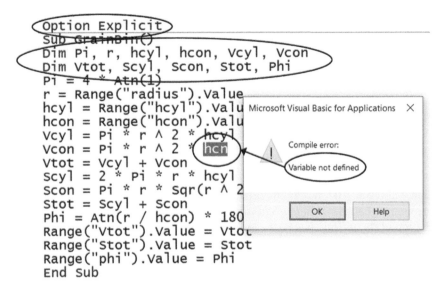

```
Option Explicit
Sub GrainBin()
Dim Pi, r, hcyl, hcon, Vcyl, Vcon
Dim Vtot, scyl, scon, Stot, Phi
Pi = 4 * Atn(1)
r = Range("radius").Value
hcyl = Range("hcyl").Value
hcon = Range("hcon").Value
Vcyl = Pi * r ^ 2 * hcyl
Vcon = Pi * r ^ 2 * hcon
Vtot = Vcyl + Vcon
scyl = 2 * Pi * r * hcyl
Scon = Pi * r * Sqr(r ^ 2
Stot = Scyl + Scon
Phi = Atn(r / hcon) * 180
Range("Vtot").Value = Vtot
Range("Stot").Value = Stot
Range("phi").Value = Phi
End Sub
```

Microsoft Visual Basic for Applications ✕

⚠ Compile error:

Variable not defined

[OK]   [Help]

FIGURE 9.27 Option Explicit statement and Dim statements prevent typos in variable names.

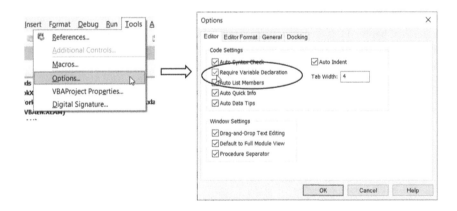

FIGURE 9.28 Setting the option in the VBE to require variable declaration.

option, every time you insert a module, the VBE will automatically provide the Option Explicit statement at the top of the Code Window. We encourage you to make this setting, and you will see the use of this statement and Dim statements from now on in the text.

In Chapter 11, we will introduce the use of the Locals Window in the VBE, but, for now, you have most of the useful tools for debugging VBA code. Happy debugging!

When you save an Excel workbook that has associated VBA code, you cannot use the .xlsx file extension, rather the *.xlsm extension*. Using the Save As command (F12) from the Excel environment, select the Save as type: field as Excel Macro-Enabled Workbook (*.xlsm).

9.4 USING THE MACRO RECORDER AS A CODE DETECTIVE

The VBA language has many commands, keywords, and features. It is not possible to study and learn all these. A common need we have is to carry out operations in the Excel environment using commands from the VBA environment. These commands are often obscure. The help facility in the VBE may not be that "helpful." A detailed reference is always useful to have nearby, such as one of the books by Alexander and Kusleika (2019) and former editions by Walkenbach. A good habit to learn is to use the Macro Recorder to discover how to program a particular operation. We will illustrate this with several examples.

Figure 9.29 illustrates a column of data and the spreadsheet operations to select the entire column. These steps work with a column of any number of cells, as long as there aren't any blank cells within the column. After selecting the first cell in the column, the Ctrl-Shift-↓ keystroke command[8] is issued, and the entire column of numbers is selected.

Having rehearsed this operation, we now create a brief macro to replicate it. This is shown in Figure 9.30 including the VBA code created. You will note that we do not change the default name Macro1, and we allow the macro to be stored with This Workbook. Once the recorder is activated, with the top cell of the column selected, the keystroke combination is executed, and then the Stop Recording command is clicked. The result is a brief macro containing the key VBA statement:

```
Range(Selection, Selection.End(xlDown)).Select
```

FIGURE 9.29 Spreadsheet command to select an entire column of filled cells.

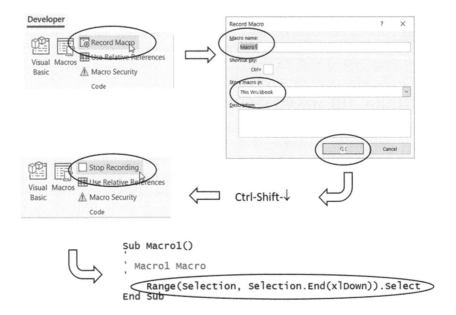

```
Sub Macro1()
' Macro1 Macro
'
    Range(Selection, Selection.End(xlDown)).Select
End Sub
```

FIGURE 9.30 Recording a macro to select a column of filled cells.

This statement can be saved for future use or copied into another VBA program. It is not a statement we would have thought of, but now we know. The first part of the statement identifies an object that is the entire filled column, and the second part uses the Select method to accomplish the selection. In object-oriented code, an alternate format is

object.method

Where a method is an action taken on the object. As we have seen, the other format is where a property of the object is referenced.

As we know from Chapter 7, Goal Seek and the Solver produce fixed (not "live") results. If any values on the spreadsheet change that would affect those results, the application needs to be run again. That brings up the question how we would run, for example, Goal Seek from VBA. We can investigate this with the Macro Recorder.

If we consider the example equation from Problem 7.1,

$$\sin(x+2)\cosh(x)+2=0 \tag{9.1}$$

and finding a root in the interval $1 \leq x \leq 3$. We can formulate the solution of this on a spreadsheet using Goal Seek. This is illustrated in Figure 9.31. The solution is found to be approximately 1.82289 with the equation error, the difference to zero, being about -2×10^{-5}.

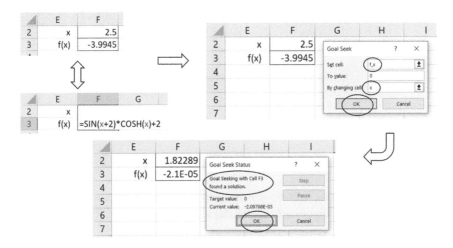

FIGURE 9.31 Application of Goal Seek to find the solution of an equation.

We now wish to know how to execute Goal Seek in this way from VBA. To do this, we record a macro that duplicates the procedure shown in Figure 9.31. The result is shown in Figure 9.32. The key VBA statement is seen there. Seeing it, we now know how to invoke Goal Seek from VBA.

As a final example, we will investigate how to add a new worksheet to the workbook in Excel. Again, we rehearse this on the spreadsheet, as shown in Figure 9.33. In addition to adding the worksheet with the default name, here Sheet2, we change the name to Results.

```
Option Explicit
Sub Macro2()
'
' Macro2 Macro
'

'
    Application.CutCopyMode = False
    Application.CutCopyMode = False
    Application.CutCopyMode = False
    Range("F3").GoalSeek Goal:=0, ChangingCell:=Range("F2")
End Sub
```

FIGURE 9.32 Recorded macro to execute Goal Seek.

FIGURE 9.33 Procedure to add a worksheet to the workbook in Excel and name the new worksheet Results.

```
Sub Macro3()
'
' Macro3 Macro
'
    Sheets.Add After:=ActiveSheet
    Application.Run "getUnits"
    Sheets("Sheet3").Select
    Sheets("Sheet3").Name = "Results"
End Sub
```

FIGURE 9.34 Recorded macro to add and name a worksheet to the Excel workbook.

We now use the Macro Recorder to repeat the operation of adding the worksheet and naming it. Figure 9.34 shows the resulting VBA code. The main statement to add a worksheet is

```
Sheets.Add
```

The `After` assignment is not required but specifies the position of the added sheet. If it is not there, the sheet added will be at the far left. The `Application.Run` statement is not required either, nor is the `Sheets.Select` statement. To name the new sheet, the VBA code is

```
Sheets("Sheet3").Name = "Results"
```

If we try to use this code generally, there will be a problem because the new worksheet will not always be named `Sheet3`. When a new worksheet is added, it becomes the `ActiveSheet`. Consequently, the code can be simplified to

```
Sheets.Add
ActiveSheet.Name = "Results"
```

This illustrates that the Macro Recorder doesn't always give you exactly what you need, rather you need to adapt the code produced using a bit of VBA knowledge.

Most macros created to determine VBA code are "throw-aways" once they are created. The discovered code can be copied to other modules in other projects. Additionally, an entire module can be dragged and dropped into another project.

In this chapter, we have introduced some VBA code without getting into too much detail. Also, you have learned some techniques for finding and fixing errors. It is not a good strategy to study VBA in depth all at once. It works much better to build your knowledge as you go along. Repetition will cause you to learn aspects of VBA that you need, and there are ways to expand your knowledge with your programming activities.

A good strategy for learning and working with VBA is to find examples of code elsewhere and ask questions of those who know more. You can find a lot of information on the Internet, and you should tap the knowledge of colleagues and fellow students. Finally, this process of expanding your knowledge requires some patience. Now that we have discussed the creation of macros and general Subs, in Chapter 10, we will move on to user-defined functions in VBA.

PROBLEMS

9.1 Create macros in the Personal Macro Workbook for the following procedures with the shortcut keys suggested. Document the macros by testing.

  a. Format one or more selected cells with horizontal text alignment right indent of one space. Ctrl-J

  b. Transfer the labels in one or more selected cells as names on the cells to their right. Ctrl-N

  c. Create a green background color for one or more cells selected. Ctrl-G

  d. Remove any background color for one or more cells selection. Ctrl-B

  e. Create all borders for one or more selected cells. Ctrl-L

  f. Copy a cell to the next cell below it. Use Ctrl-C, ↓, and Ctrl-V. Do not use a drag copy. Ctrl-D

  g. Change the worksheet zoom setting to 150%. Ctrl-Z

9.2 Based on your own experience, create a macro that would be useful to you. Assign a shortcut key combination to it. Review and, if feasible, simplify the VBA code. Document the code and its application.

9.3 Describe in ordinary language what the following Sub accomplishes.

```
Sub Problem9_3()
Temp = Range("B2").Value
Range("B2").Value = Range("C2").Value
Range("C2").Value = Temp
End Sub
```

9.4 The radius, r, of the largest circle that can be inscribed in a triangle with sides a, b, and c, is given by

$$r = \sqrt{\frac{(s-a)(s-b)(s-c)}{s}} \quad \text{where} \quad s = \frac{a+b+c}{2}$$

Develop and test a VBA Sub that obtains a, b, and c, from the spreadsheet and returns r to the spreadsheet. Investigate what happens when a, b, and c cannot form a triangle.

9.5 A bin with square cross-section and pyramidal base is shown in Figure P9.5. Excluding the top of the bin, create a VBA Sub that computes the area of material required and the volume of the bin.

  a. Using the variables noted, and for values of $s = 3$ m, $h = 4.5$ m, and $d = 2.5$ m, established on the spreadsheet, return the area and volume to cells on the spreadsheet.

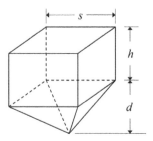

FIGURE P9.5   Bin with square cross-section and pyramidal base.

    b. Using debugging techniques, single-step the Sub and note the values of the components of the area and volume.

    c. Capture and reproduce the Locals window at the end of the Sub.

9.6 The distance between two points on the surface of the Earth can be expressed based on spherical trigonometry as[9]

$$d = 2\sin^{-1}\left(\sqrt{\left(\sin\left(\frac{lt_1 - lt_2}{2}\right)\right)^2 + \cos(lt_1)\cos(lt_2)\left(\sin\left(\frac{ln_1 - ln_2}{2}\right)\right)^2}\right) r$$

where

r: Earth's radius, approximately 6371 km

lt_1, lt_2:   latitude of positions 1 and 2 in radians

ln_1, ln_2: longitude of positions 1 and 2 in radians

d: distance between positions 1 and 2, km

Although the Earth isn't quite a perfect sphere, this formula is widely used and fairly accurate. By convention, longitude is measured with respect to the Prime Meridian in Greenwich, England, with angles to the west taken as positive and those to the east as negative. Latitude is measured with respect to the Equator and is positive to the north and negative to the south.

    Create a spreadsheet where the two positions are specified by latitude and longitude in degrees in named cells. Develop a VBA Sub that computes the distance between the points in km and returns this to a cell on the spreadsheet. Test your Sub for the following locations:

• San Francisco International Airport (SFO)
• Auckland, New Zealand Airport (AKL)

Compare your results to the published distance. Fix any discrepancies.

9.7 Use the Macro Recorder to discover how to code the following spreadsheet operations in VBA.

   a. Move the ActiveCell down one row and one column to the right. Repeat the recording, but this time select Use Relative References in the Code group of the Developer tab of the Ribbon. Compare the two macros and comment.

   b. Select the cell range B2:D4. Format the selected cell with the following specifications:
- All Borders
- Background color light blue
- Font Courier, 10 pt, bold, italic
- Center entry in the cell

   c. Start with the ActiveCell as the home cell, A1. Select the cell range B5:E15. Repeat the recording, but this time select Use Relative References in the Code group of the Developer tab of the Ribbon. Compare the two macros and comment.

   d. First, before the recording, create a column of ten random numbers using Data Analysis Random Number Generation, uniform distribution, range $0 \to 1$. Then, record a macro to sort these numbers in ascending order.

9.8 Repeat the example in Section 9.3 that recorded a macro using Goal Seek to solve for the root of Equation 9.1 but do so with the Solver instead.

9.9 The following VBA Sub is intended to compute the volume of liquid in a horizontal cylindrical vessel with hemispherical ends, as illustrated in Figure P9.9.

```
Sub Cylvessel()
Radius = 2
Length = 5
h = 2.7
Vcyl = (Radius ^ 2 * Cos((Radius - h) / h) _
    - (R - h) * sqrt(2 * R * h - h ^ 2)) * Length
Vsph = Pi * h ^ 2 * (3 * Radius - h) / 3
V = Vcyl + Vsph
MsgBox "Liquid Volume = " & V
End Sub
```

   a. Debug this Sub and document the value of V displayed. Document the corrections made.

   b. Set up separate calculations on the spreadsheet to confirm your result from part (a).

   c. Modify the Sub to obtain the Radius, Length, and h values from named cells on the spreadsheet, and, instead of the message box output, return the values of Vcyl, Vsph, and V to cells on the spreadsheet.

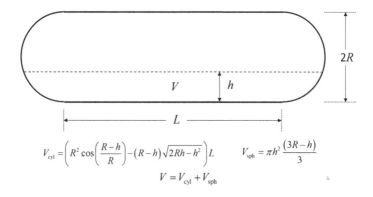

$$V_{cyl} = \left(R^2 \cos\left(\frac{R-h}{R}\right) - (R-h)\sqrt{2Rh - h^2} \right) L \qquad V_{sph} = \pi h^2 \frac{(3R-h)}{3}$$

$$V = V_{cyl} + V_{sph}$$

FIGURE P9.9   Horizontal cylindrical vessel with hemispherical ends storing a liquid.

9.10 Based on the first example in Section 9.3, where we discovered the code to select a column of data, write a short VBA Sub to select a row of data instead. Test this Sub. If it doesn't work correctly, use the Macro Recorder to discover the correct statement. Explain the difference between your first attempt and the result from the Macro Recorder.

NOTES

1. Note: The warning about dangerous code that might pop up on your screen is primarily directed at macros you might import from other sources. It is not that relevant for the VBA code you develop yourself.
2. A MsgBox is a dialog box that is used to display a custom message (in our example, the word "Test") or to elicit some basic input, such as Yes/No or OK/Cancel. While the MsgBox dialog box is displayed, your VBA code is paused. You need to click any of the buttons in the MsgBox to run the remaining code.
3. You can still use key letters to operate the Excel Ribbon menus and commands. If you press the Alt key, the letters appear on the Ribbon, and you can select them to execute a command. For example, Alt ⇨ A ⇨ W ⇨ G executes the Goal Seek command. Try it. Some users still commit letter sequences to memory to execute commands more quickly.
4. Note: You can't undo the execution of a macro.
5. Appendix B lists the Excel and VBA shortcut keys that relate to the material covered in this book.
6. This is an idiosyncrasy of the 64-bit versions of Windows and Office. The earlier 32-bit versions did not require the spaces.
7. These three-letter functions are inherited from earlier versions of the BASIC language, all the way back to the 1960s. That language was developed at Dartmouth by Kemeny and Kurtz to be a simpler tool than Fortran. It also operated as an interpreter, not requiring a compilation step as did (and does) Fortran.
8. Note that we use the arrow symbols to represent the arrow keys. Hence, ↓ signifies the arrow down key.
9. The derivation of this formula is available from many sources. One good one is Aviation Formulary V1.46 (www.edwilliams.org).

10 User-defined Functions

CHAPTER OBJECTIVES

- Learn how to automate engineering formulas by creating *user-defined functions*
- Understand how to debug user-defined functions in the VBA Editor
- Be able to include cell-range arguments in user-defined functions
- Recognize how to access Excel functions in VBA code
- Learn how to create *user-defined array functions*
- Be able to package families of user-defined functions in an add-in

Scientists and engineers find great benefit in the hundreds of built-in functions provided by Excel. On repeated occasions, they have a need and search those functions to find what will resolve it. These searches are necessary because it is challenging to command knowledge of all the functions available. And then, there are times when the search comes to a dead end because no built-in function is available to satisfy the particular need. But there is still hope! The purpose of this chapter is to teach you how to create your own functions for this dilemma.

The first type of user-defined function we introduce allows you to program an engineering or scientific formula so that you don't have to enter it as a cell formula each time you want to use it. This is useful for convenience and improved reliability, as there is now a single source for the calculation. Occasionally, we need to add a little bit of programming logic to these functions, and we will illustrate that. Second, there are times when we need a user-defined function to act on one or more arguments that are blocks or ranges of cells. We address that need by illustrating how a calculation developed on the spreadsheet can be "elevated" into a VBA user-defined function to make it more flexible and efficient. Along the way, we show how you can tap into Excel's built-in functions within VBA code. This lets you take advantage of Excel functions that aren't normally available in the VBA environment.

In Chapter 3, we introduced array formulas, and in Chapter 7, you saw how built-in array functions could be used for matrix calculations. That leads to the next step: learning how to build user-defined array functions. Finally, you will learn how to combine a number of related user-defined functions into a packaged add-in. This is similar to using the Personal Macro Workbook for common macros. An add-in can be installed so the functions are always available and can be moved to multiple computers and shared with colleagues.

DOI: 10.1201/9781003361053-10

10.1 AUTOMATING ENGINEERING FORMULAS AS USER-DEFINED FUNCTIONS

Engineering formulas can get lengthy and having to recreate them for use in different cells, worksheets, or even other workbooks can be tedious and raise the likelihood of making errors. Take for example a formula we have encountered earlier in the text for computing the volume of liquid in a partially filled, *horizontal, cylindrical tank*, as illustrated in Figure 10.1,

$$V = \left(R^2 \cos\left(\frac{R-h}{R}\right) - (R-h)\sqrt{2Rh-h^2} \right) L \qquad (10.1)$$

where
  R: inside radius of tank
  h: depth of liquid at tank centerline
  L: length of tank
  V: volume of liquid

If we enter this formula in a spreadsheet cell, using named cells for R, h, and L, it would look like

`=(R_^2*COS((R_-h)/R_)-(R_-h)*SQRT(2*R_*h-h^2))*L`

This is a good case for simplifying this with a user-defined function,

`=CylTankVol(R_,h,L)`

You will note that we have used a cell name R_, since, as you know, R is not allowed.

The user-defined function can be created by opening the Visual Basic Editor, inserting a new module in the current project, and entering the code

```
Function CylTankVol(R, h, L)
CylTankVol = (R ^ 2*Cos((R - h)/R) _ - (R - h)*Sqr(2*R*h _
- (R - h)*Sqr(2*R*h - h ^ 2)) * L
End Function
```

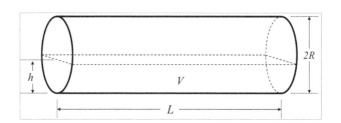

FIGURE 10.1   Horizontal cylindrical tank partially filled with liquid.

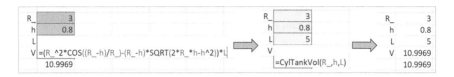

FIGURE 10.2   Validating the CylTankVol user-defined function with the on-sheet formula.

Based on our discussion in Chapter 9, we have explicitly put spaces around the ∧ operator. When we enter the code, the VBE will automatically add additional spaces. Also, you see that we can use the variable name R because the spreadsheet-based naming conventions do not apply to VBA code.

To validate our function, we can compare it with the cell-based formula for a given set of specifications. This is shown in Figure 10.2. The two results are the same.

You see from the code for the function that, instead of starting the procedure off with the Sub statement and terminating it with End Sub, we use Function and End Function. In the Function statement, we sequence the arguments in the desired order, which must be followed when the function is invoked from the spreadsheet. We do not have to use the same variable names for the arguments that we used for the cell names on the spreadsheet. In fact, we can use cell addresses on the spreadsheet or even literal numbers (that is, constants) to call the function. For the Function's VBA code to return a result, we must assign a value, here the result of the formula, to the name of the function; otherwise, zero will appear in the cell.

It is also good practice for longer formulas to break them up into several statements. This provides for more readable code and ease of debugging. A modification of the CylTankVol function to accomplish this is shown below.

```
Function CylTankVol(R, h, L)
Term1 = R ∧ 2 * Cos((R - h) / R)
Term2 = (R - h) * Sqr(2 * R * h - h ∧ 2)
CylTankVol = (Term1 - Term2) * L
End Function
```

When any changes are made to a function, there will be no change on the spreadsheet until the function is made to recalculate. This can be done in two ways. First, if the value of any input argument is changed, the function will automatically recalculate. Second, by editing the cell where the function is located and pressing the Enter key, a recalculation is caused. A quick way to accomplish the latter is to select the cell, press the Edit key (F2), and press Enter.

At this juncture, it is useful to describe the technique used to *debug a function* in the VBA Editor. With a Sub, we could initiate execution in the Editor with the Run (F5) or Single Step Code (F8) commands. A way that works is to place a breakpoint on the Function statement and initiate execution as described in the previous paragraph. In that case, execution will stall on the Function statement, and the various debugging steps, such as single step (F8), can be carried out. When debugging is complete, the breakpoint can be removed by clicking on it or pressing F9.

Note that if the same function is invoked in multiple cells on the spreadsheet, the results will refer to the specific function that you invoked as per the previous paragraph. However, if you recalculate all the worksheets within all of the workbooks that a user has open (F9 key), the order in which functions are executed is based on advanced algorithms to order the calculations in the most efficient sequence. If you build an application with the same function invoked in multiple cells that are interdependent, a subtle error could occur without it being obvious which function contains the error. In such cases, you might have to do some detective work based on examining each of the function calls to pinpoint and fix the mistake.

In creating a user-defined function, different formulas may apply depending on one or more of the input arguments. A good example is the *Earth atmosphere model*[1] used to predict temperature, pressure, and air density depending on the altitude, h. There are three regions modeled:

troposphere	$h \le 11{,}000$ m
lower stratosphere	$11{,}000 < h \le 25{,}000$ m
upper stratosphere	$h > 25{,}000$ m

and the predicted temperature (°C) is

troposphere	$T = 15.04 - 0.00649\,h$
lower stratosphere	$T = -56.46$
upper stratosphere	$T = -131.21 + 0.00299\,h$

To accommodate the three different models in a function, our user-defined function must contain program logic to determine, based on an altitude argument, which model to use. Here, we provide a preview of the *multi-alternative If structure* to accomplish this. We will be introducing all the program structure elements in detail in Chapter 11.

```
Function AtmTemp(h)
If h > 25000 Then
    AtmTemp = -131.21 + 0.00299 * h
ElseIf h > 11000 Then
    AtmTemp = -56.46
Else
    AtmTemp = 15.04 - 0.00649 * h
End If
End Function
```

The program flow here is, first, to check if the altitude is greater than 25,000 m, and, if so, use the appropriate formula. Once that is done, the program exits the

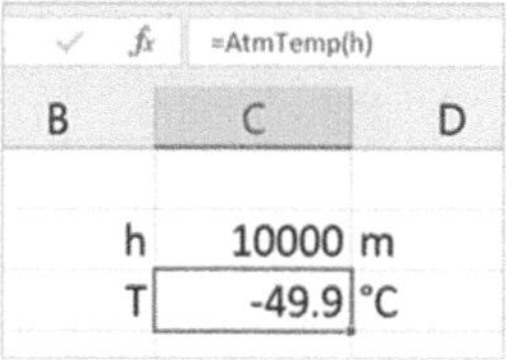

	f_x	=AtmTemp(h)
B	**C**	**D**
h	10000 m	
T	-49.9 °C	

FIGURE 10.3   Test of the AtmTemp function.

If structure to the executable statement immediately following the End If state-ment. If this is not true, we know the altitude is less than or equal to 25,000 m, and the ElseIf statement executes. If the ElseIf test is true, the second formula is used because we know that the altitude is between 11,000 and 25,000 m, so that formula is executed, and the program exits the If structure below the End If statement. If the ElseIf test is false, execution skips to the Else clause. Since we know that the altitude is less than or equal to 11,000 m, the third formula is evalu-ated, and then the If structure is exited via the End If statement.

Figure 10.3 illustrates a single test of the AtmTemp function. To exercise all the options of the function and illustrate how it can be invoked multiple times on the spreadsheet, Figure 10.4 shows the application of the function for several altitude values. Here, cell addresses are used as the argument to facilitate copy-ing the formula down. By testing with independent calculations, either on the spreadsheet or separately, e.g., with a calculator, we confirm that the function is performing correctly.

Before leaving this section, we review here a couple of fine points. First, the AtmTemp function has no protection against erroneous inputs. For instance, if we entered an altitude of −10,000 m, the function would return a temperature of about 80°C – obviously incorrect as this is 10 km into the Earth's crust. We will see in Chapter 11 how to include protection against incorrect entries.

Second, for convenience, as we enter a formula with our AtmTemp function and start to type the function name, a tooltip appears showing the name of the func-tion (and perhaps others). By pressing the Tab key, the name entry is completed. This is illustrated in Figure 10.5.

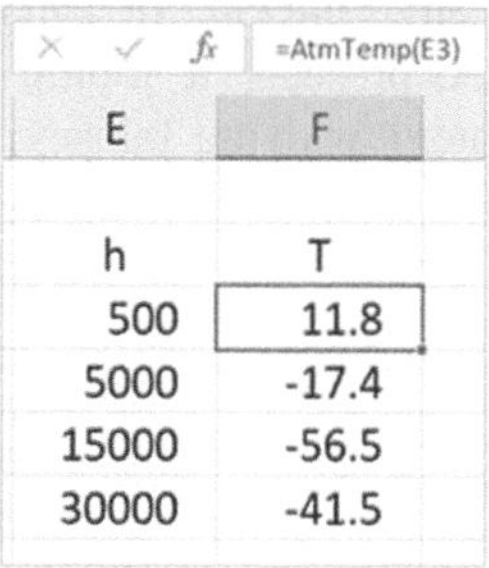

×	✓	f_x	=AtmTemp(E3)
	E	**F**	
	h	T	
	500	11.8	
	5000	-17.4	
	15000	-56.5	
	30000	-41.5	

FIGURE 10.4   Test of the AtmTemp function for various altitudes.

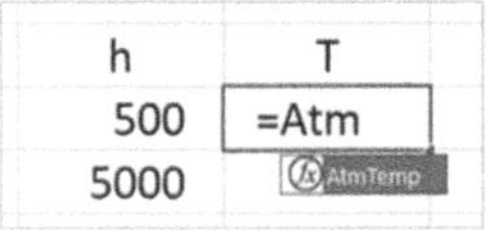

h	T
500	=Atm
5000	AtmTemp

FIGURE 10.5 Tooltip completion of function name entry.

10.2 USER-DEFINED FUNCTIONS WITH CELL RANGE ARGUMENTS

We use many built-in functions in Excel that operate on arguments that are blocks or ranges of cells. Figure 10.6 illustrates several sample statistical formulas applied to a data set. In these cases, cell addresses are used as the arguments, although names representing these might have been employed.

A dilemma arises when we would like to use a function that operates on a range of cells, and Excel does not provide it. A good resolution to that predicament is to develop a user-defined function to accomplish the calculation. We will illustrate this with two examples. The first example is simple, but it illustrates an important feature of VBA.

A common statistic used to characterize small sets of data is the *range*, that is, the algebraic difference between the maximum and minimum values. This, of course, can be obtained by entering a cell formula using the MAX and MIN built-in functions. However, we will consider here implementing a user-defined function, RNG that accomplishes the calculation. Note: We do not use the name RANGE for the function, as this is a reserved word in VBA.

Inserting a module in the project associated with the spreadsheet shown in Figure 10.6, we enter the simple code for the user-defined function, RNG.

```
Function RNG(data_array)
maxval = Application.WorksheetFunction.Max(data_array)
minval = Application.WorksheetFunction.Min(data_array)
RNG = maxval - minval
End Function
```

What is new here are the statements that "borrow" the MAX and MIN functions from the spreadsheet environment. In entering those statements, after typing Application., VBA presents a tooltip that allows you to select WorksheetFunction and enter it with the Tab key. It is also possible to enter the

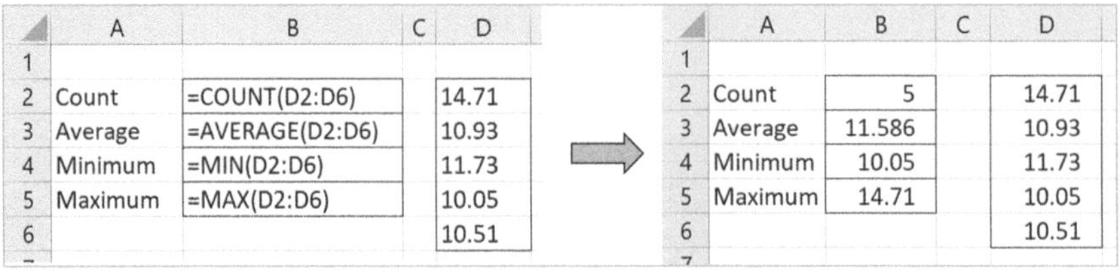

	A	B	C	D		A	B	C	D
1					1				
2	Count	=COUNT(D2:D6)		14.71	2	Count	5		14.71
3	Average	=AVERAGE(D2:D6)		10.93	3	Average	11.586		10.93
4	Minimum	=MIN(D2:D6)		11.73	4	Minimum	10.05		11.73
5	Maximum	=MAX(D2:D6)		10.05	5	Maximum	14.71		10.05
6				10.51	6				10.51

FIGURE 10.6 Built-in functions with cell range arguments.

	A	B	C	D			A	B	C	D
1						1				
2	Count	=COUNT(D2:D6)		14.71		2	Count	5		14.71
3	Average	=AVERAGE(D2:D6)		10.93		3	Average	11.586		10.93
4	Minimum	=MIN(D2:D6)		11.73		4	Minimum	10.05		11.73
5	Maximum	=MAX(D2:D6)		10.05		5	Maximum	14.71		10.05
6	Range	=RNG(D2:D6)		10.51		6	Range	4.66		10.51

FIGURE 10.7 Adding the RNG function to the example cells from Figure 10.9.

statements without the Application. part, but, in that case, no tooltip is made available. It would be possible to enter code to find the maximum and minimum values in data_array, but this is simplified by borrowing the functions from "spreadsheet world."

Figure 10.6 is then expanded to include a formula using the new function as shown in Figure 10.7. Of course, the RNG function can be used multiple times throughout the workbook.

Next, we will look at an application for a user-defined function that is more involved. We would like to create a MAD function (as explained previously in Chapter 8) that computes the *median absolute deviation* as an estimate of the spread or standard deviation of a data set. These are the steps in calculating the MAD:

- compute the median value of the data set
- compute the absolute values of the deviations of the data from the median
- compute the median of those absolute values
- divide the result by 0.6745

The *median*, or "middle value," is obtained by sorting the data in ascending (or descending) order and selecting, for an odd number of data, the middle value. If there is an even number of data, the two middle values are averaged to provide the median. The median is a measure of central tendency that is an alternative to the average and is influenced less by outliers in the data. The division by 0.6745 is based on theory to achieve a valid estimate of the standard deviation.

A good strategy to implement this statistic is to set up a prototype calculation on the spreadsheet and then "elevate" that into a user-defined function. Figure 10.8 shows the prototype on the spreadsheet for the small data set used previously in Figures 10.6 and 10.7.

	A	B	C	D	E	F			A	B	C	D	E	F
1				data		absolute deviations		1				data		absolute deviations
2	median	=MEDIAN(D2:D6)		14.71		=ABS(D2-$B$2)		2	median	10.93		14.71		3.78
3				10.93		=ABS(D3-$B$2)		3				10.93		0
4	MAD	=MEDIAN(F2:F6)/0.6745		11.73		=ABS(D4-$B$2)		4	MAD	1.19		11.73		0.8
5				10.05		=ABS(D5-$B$2)		5				10.05		0.88
6				10.51		=ABS(D6-$B$2)		6				10.51		0.42

FIGURE 10.8 Spreadsheet prototype of MAD calculation.

Based on the prototype, we create a user-defined function in a VBA module associated with the workbook.

```
Function MAD(data_array)
Dim med_array()
n = data_array.Count
ReDim med_array(n)
data_median = _
Application.WorksheetFunction.Median(data_array)
For i = 1 To n
    med_array(i) = Abs(data_array(i) - data_median)
Next i
MAD = _
Application.WorksheetFunction.Median(med_array) / 0.6745
End Function
```

As our prototype shows, VBA needs to create an array for the absolute deviations from the median. An issue in creating the code is that we do not know the size of the data_array argument. The technique used to accomplish this is to create an empty array, med_array, using the Dim statement, determine the number of data in data_array using its Count property, and then expand the empty med_array to the appropriate size using the ReDim statement.

The median of the input argument data_array is computed by borrowing the Median function from the spreadsheet environment. Next, we introduce a For loop that fills the med_array with the absolute values of the differences between the data_array elements and their median. The index of the For loop is i, and that is used as the subscript (or index) of the respective arrays. We will see more about For loops in Chapter 11. Finally, the function name, MAD, is assigned the median of med_array, again borrowing the Median function divided by 0.6745.

The implementation of the MAD function is illustrated in Figure 10.9. We can use the on-sheet prototype calculation to validate the user-defined function.

Our MAD function will also compute a result for a rectangular block of cells. This is shown in Figure 10.10. The spreadsheet in the figure contrasts the built-in sample standard deviation function, STDEV.S, with our user-defined MAD function.

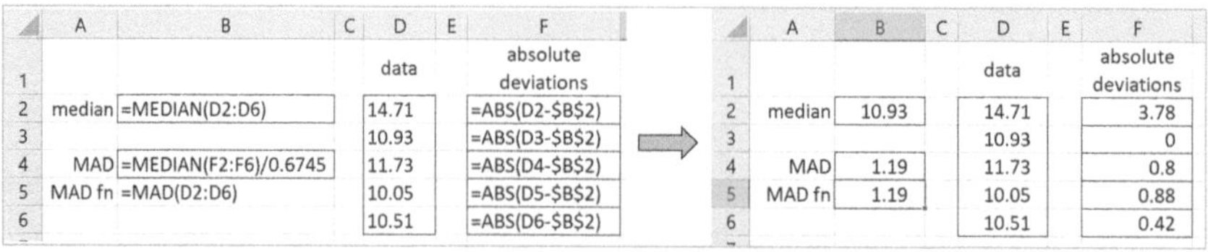

FIGURE 10.9   Implementation of the user-defined MAD function.

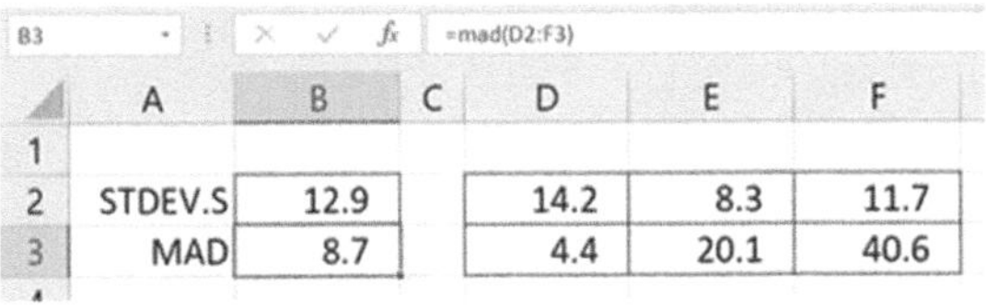

FIGURE 10.10   MAD and STDEV.S functions with rectangular cell range argument.

Intentionally, there is an outlier in the data set (40.6). This inflates the STDEV.S result more than the MAD result and illustrates that the MAD is less sensitive to outliers in the data.

There are occasions where a user-defined function is desired to implement a specialized calculation and perhaps do so for a large collection of data. We complete this section with such a "real world" example illustrated in Figure 10.11. Here, we have side-by-side arrangements of five cells. The objective is to find the maximum value in the set to the left and use its location to select the value in the set to the right, placing that value in the location of the function. We will call our function MAXLOC, and it will have two arguments, the two cell ranges. We can create the function so that it will deal with side-by-side ranges containing more or fewer than five cells.

The VBA code for the function is as follows:

```
Option Base 1
Function MAXLOC(data1, data2)
n = data1.Count
n2 = data2.Count
If n <> n2 Then
    MAXLOC = "error, arrays don't match"
Else
    data1max = Application.WorksheetFunction.Max(data1)
    For i = 1 To n
        If data1(i) = data1max Then
            maxi = i
            Exit For
        End If
    Next i
    MAXLOC = data2(maxi)
End If
End Function
```

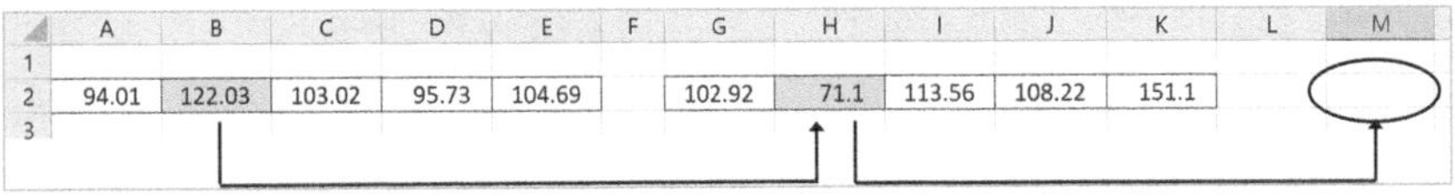

FIGURE 10.11   Example application of the MAXLOC user-defined function.

M2		▾	×	✓	fx	=MAXLOC(A2:E2,G2:K2)						

▲	A	B	C	D	E	F	G	H	I	J	K	L	M
1													
2	94.01	122.03	103.02	95.73	104.69		102.92	71.1	113.56	108.22	151.1		71.1

FIGURE 10.12 Using the MAXLOC function on the spreadsheet.

Prior to the Function statement, you see Option Base 1. This is called a *declaration* statement. Although not mandatory, it is typically placed at the beginning of a macro and, as was the case for comments, it is non-executable. It indicates that the index or subscript of all arrays starts at one, not zero, which is the default. Most mathematical descriptions use 1-based subscripting, whereas computer languages such as C/C++ and Python use 0-based. Other packages/languages, such as MATLAB® and Fortran, use 1-based indexing. With the Option Base 1 declaration, VBA gives you the choice.

The function first checks the number of items in each array argument and returns an error message if they are not equal. This is accomplished using the Count property and a two-way If structure. The Else part of the If structure handles the normal case. First, the maximum element in the data1 array is found by borrowing the Max function from the spreadsheet environment. Then, via a For loop, the item matching the maximum and its location, maxi, are determined. Finally, using the maxi location, that element from the data2 array is assigned to the function name, MAXLOC. One note: If there is more than one maximum (equal values) in data1, the function will proceed with the location of the first one.

Figure 10.12 illustrates the implementation of the MAXLOC function. Notice the M2 formula in the Formula Bar. If instead one of the argument ranges is reduced by one cell, the error message shows up in cell M2. Also, with multiple rows of data, the formula can be copied down to provide the correct result for each row. This is illustrated in Figure 10.13.

The point of Figure 10.13 is that, if we were only going to carry out the calculation once, we could do so by inspection and a simple pointer formula. The value of the function comes through multiple applications. Imagine, for example, if we had a block of several hundred or thousand rows of data like those presented in the figure. Using the MAXLOC function, with one double-click of the fill handle in cell M2, all the 100 or 1,000 values would be calculated and displayed immediately. That is a big payoff!

▲	A	B	C	D	E	F	G	H	I	J	K	L	M
1													
2	94.01	122.03	103.02	95.73	104.69		102.92	71.1	113.56	108.22	151.1		71.1
3	104.55	91.13	110.13	86.43	133.25		97.05	73.89	105.65	116.47	68.6		68.6
4	88.65	118.17	101.88	81.17	78.62		90.34	96.84	116.99	115.8	87.84		96.84
5	91.34	115.71	127.77	125.31	115.93		155.04	89.02	96.05	84.15	76.03		96.05
6	84.54	92.11	153.04	102.27	78.74		104.76	68.71	76.05	106.68	95.62		76.05

FIGURE 10.13 Multiple use of the MAXLOC function by copying down.

10.3   CREATING USER-DEFINED ARRAY FUNCTIONS

We have seen the value of array functions in previous chapters. The TABLE function, which is generated automatically using the Data Table command on the spreadsheet, provides a useful tool for case studies. The "M-functions," e.g., MMULT, MINVERSE, MDETERM, and TRANSPOSE, facilitate vector/matrix calculations on the spreadsheet. And we have seen how multiple formulas created by copying down can be replaced by a single array formula that produces results in multiple cells. Consequently, the prospect of creating user-defined functions that can return results to multiple cells is attractive and intriguing.

We will start with a simple example and then return to our MAXLOC function from the previous section. Consider that we have a column of numbers on the spreadsheet, as shown in Figure 10.14. In an adjacent column, we would like to enter an *array function*, called ARRAYFRAC, that displays the fraction each number is of the total.

We can add the code below to a module created in the project associated with the spreadsheet in Figure 10.14. As before, the Option Base 1 declaration is used to specify that the index for the first element of an array is one. Also, we use the Dim, Count, ReDim combination to create an empty frax array the same size as the argument array, array_data. The sum of the elements in array_data is obtained by borrowing the Sum function from the spreadsheet environment and assigned to the array_sum variable. Then, there is a For loop that computes the elements of the frax array. The final executable statement borrows the Transpose function from the spreadsheet environment and applies it to the frax array, assigning the result to the name of the function. The transpose operation is required because, by default, the array function will return its result to a row on the spreadsheet. In this case, we want it displayed in a column.

```
Option Base 1
Function ARRAYFRAC(array_data)
Dim frax()
n = array_data.Count
ReDim frax(n)
array_sum = Application.WorksheetFunction.Sum(array_data)
For i = 1 To n
    frax(i) = array_data(i) / array_sum
Next i
ARRAYFRAC = Application.WorksheetFunction.Transpose(frax)
End Function
```

	A	B
1		
2		850.3
3		902.2
4		881.6
5		920.4
6		174.3

FIGURE 10.14   Column of numbers for ARRAYFRAC array function.

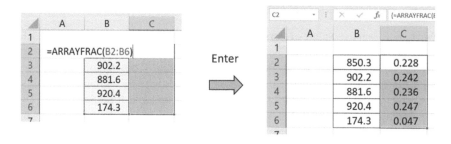

FIGURE 10.15 Using the ARRAYFRAC function on the spreadsheet.

Implementation of the ARRAYFRAC function is illustrated in Figure 10.15. As we have demonstrated before with array functions, we first select the range of cells as the destination for the function's results. Then, the formula is entered (retaining the selection), and the process is completed with the Enter key (or Ctrl-Shift-Enter key combination for a legacy function implementation). We could change the format of the displayed results to percentages (%) if we wished.

The next array function example takes our MAXLOC function from the previous section and extends it so that, instead of the multiple formulas illustrated in Figure 10.13, a single array function MAXLOCA can be used to complete the calculation. The modified function is shown below.

```
Option Base 1
Function MAXLOCA(data1, data2)
Dim datamax(), maxout()
nc = data1.Columns.Count
nc2 = data2.Columns.Count
nr = data1.Rows.Count
nr2 = data2.Rows.Count
If nc <> nc2 Or nr <> nr2 Then
    MAXLOCA = "error, arrays don't match"
Else
    ReDim datamax(nr), maxout(nr)
    For j = 1 To nr
        datamax(j) = data1(j, 1)
        For i = 2 To nc
            If data1(j, i) > datamax(j) Then
                datamax(j) = data1(j, i)
            End If
        Next i
        For i = 1 To nc
            If data1(j, i) = datamax(j) Then
                maxi = i
                Exit For
            End If
```

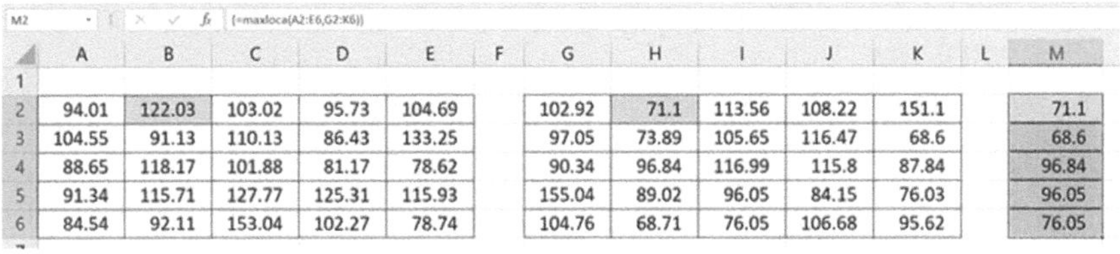

M2 | fx [=maxloca(A2:E6,G2:K6)]

	A	B	C	D	E	F	G	H	I	J	K	L	M
1													
2	94.01	122.03	103.02	95.73	104.69		102.92	71.1	113.56	108.22	151.1		71.1
3	104.55	91.13	110.13	86.43	133.25		97.05	73.89	105.65	116.47	68.6		68.6
4	88.65	118.17	101.88	81.17	78.62		90.34	96.84	116.99	115.8	87.84		96.84
5	91.34	115.71	127.77	125.31	115.93		155.04	89.02	96.05	84.15	76.03		96.05
6	84.54	92.11	153.04	102.27	78.74		104.76	68.71	76.05	106.68	95.62		76.05

FIGURE 10.16 Using MAXLOCA in an array formula on the spreadsheet.

```
      Next i
      maxout(j) = data2(j, maxi)
   Next j
End If
MAXLOCA = Application.WorksheetFunction.Transpose(maxout)
End Function
```

First, the function declares (Dim statement) two empty arrays that will store the maximum values for each row of data1 and the corresponding values at the same position on data2. This is followed by checks on the number of columns and rows of the two input arrays to make certain they have the same shape; otherwise, an error message is returned.

If the data1 and data2 array arguments are compatible, the Else clause of the If statement executes. The datamax and maxout arrays are ReDim'd to the number of rows, and then a For loop computes the appropriate value for each row. This is where the MAXLOC function is extended to multiple rows. Within the main For loop, there are two other loops. The first one finds the maximum value in the current row and stores it in the datamax array. (We cannot use the borrowed Max function for this.) The second For loop locates the datamax item for the current row and extracts the corresponding value from the data2 array, assigning it to the element of the maxout array for the current row. Finally, the maxout array is transposed and returned to the spreadsheet.

The MAXLOCA function is entered on the spreadsheet as an array formula, as shown in Figure 10.16. A takeaway message is that it is important to look for opportunities where array functions would be appropriate and advantageous.

10.4   PACKAGING FUNCTIONS IN AN ADD-IN

As we saw in Chapter 9, when we create macros as subprogram procedures (Subs), we had the choice of storing them with the current Excel workbook project or in the Personal Macro Workbook. The latter allowed for portability; that is, the macros could be accessed by any Excel workbook opened on the computer and the Personal Macro Workbook could even be transferred to other computers. The Personal Macro Workbook is not a location where user-defined functions can be stored. Rather, we can package a family of functions (or even just one function) in a separate workbook called an *add-in*. We will illustrate this here.

TABLE 10.1
VBA's Built-in Mathematical Functions

Function	Description	Function	Description
Abs	absolute value	Log	natural logarithm
Atn	arctangent	Rnd	random number
Cos	cosine	Sgn	sign function
Exp	exponential	Sin	sine
Fix	converts to integer*	Sqr	square root
Int	converts to integer*	Tan	tangent

* Both remove the fractional part of a number and return the resulting integer. They differ for negative numbers as Int returns the first negative integer less than or equal to the number, whereas Fix returns the first negative integer greater than or equal to the number.

We know that VBA has a very limited set of built-in mathematical functions, as listed in Table 10.1. A couple of notes on these:

1. The Atn function is a two-quadrant function. It returns angles from quadrants I and IV, that is, from $-\pi/2$ ($-90°$) to $+\pi/2$ ($+90°$). See below for a user-defined four-quadrant arctangent function.
2. The Fix and Int functions are similar but not the same for negative numbers. The Fix function returns the integer to the left of the argument number on the number line, e.g., Fix(-2.5) $\rightarrow$ -3. The Int function truncates (i.e., chops off) the fractional part, e.g., Int(-2.5) $\rightarrow$ -2.
3. The Rnd function has no argument, i.e., Rnd(), and returns a pseudorandom, uniformly distributed, random number in the range 0 to 1.

From our spreadsheet experience, we know that there are far more *built-in mathematical functions* available in "spreadsheet world," and we can always borrow those in VBA by using the Application.WorksheetFunction. prefix. But that is somewhat unwieldy, and, if we have applications that require math functions, it would be convenient to have these available in VBA. Creating an add-in is the answer.

We will create a math functions add-in based on a list of *derived functions*.[2] Trigonometric functions are listed in Table 10.2. Hyperbolic functions are listed in Table 10.3. Three additional functions are shown in Table 10.4.

Starting with a new, blank Excel workbook, we open a new module in the associated VBA project and enter a lengthy series of function definitions.[3] These are shown in Figure 10.17.

Next, we return to the spreadsheet environment. With the spreadsheet empty, but the function collection part of the associated project, we initiate a Save As (F12), change the Save as type: field by selecting Excel Add-in (*.xlam), and enter a filename, such as MathFunctions.xlam. The SaveAs window is displayed in Figure 10.18. You will note that Excel has automatically shifted to the default

TABLE 10.2
Derived Trigonometric Functions

Function	VBA Name	How Related to Built-in VBA Functions
secant	Sec(x)	1/Cos(x)
cosecant	CoSec(x)	1/Sin(x)
cotangent	CoTan(x)	1/Tan(x)
inverse sine	ArcSin(x)	Atn(x/Sqr(-x*x+1))
inverse cosine	ArcCos(x)	2*Atn(1)-Atn(x/Sqr(-x*x+1))
inverse secant	ArcSec(x)	x >= 0, Atn(Sqr(x*x-1))
		x < 0, Atn(Sqr(x*x+1))-4*Atn(1)
inverse cosecant	ArcCoSec(x)	x >= 0, Atn(1/Sqr(x*x-1))
		x < 0, Atn(1/Sqr(x*x+1))-4*Atn(1)
inverse cotangent	ArcCot(x)	2*Atn(1)-Atn(x)

TABLE 10.3
Derived Hyperbolic Functions

Function	VBA Name	How Related to Built-in VBA Functions
hyperbolic sine	HSin(x)	(Exp(x)-Exp(-x))/2
hyperbolic cosine	HCos(x)	(Exp(x)+Exp(-x))/2
hyperbolic tangent	HTan(x)	(Exp(x)-Exp(-x))/(Exp(x)+Exp(-x))
hyperbolic secant	HSec(x)	2/(Exp(x)+Exp(-x))
hyperbolic cosecant	HCoSec(x)	2/(Exp(x)-Exp(-x))
inverse hyperbolic sine	HArcSin(x)	Log(x+Sqr(x*x+1))
inverse hyperbolic cosine	HArcCos(x)	Log(x+Sqr(x*x-1))
inverse hyperbolic tangent	HArcTan(x)	Log((1+x)/(1-x))/2
inverse hyperbolic secant	HArcSec(x)	Log((1+Sqr(1-x*x))/x)
inverse hyperbolic cosecant	HArcCsc(x)	x > 0, Log((1+Sqr(1+x*x))/x)
		x < 0, Log((1-Sqr(1+x*x))/x)
inverse hyperbolic cotangent	HArcCot(x)	Log((x+1)/(x-1))/2

TABLE 10.4
Additional Functions. Note that the Pi () Function Is Based on the Fact that the Tangent of 1 Is Equal to $\pi/4$

Function	VBA Name	How Related to Built-in VBA Functions
logarithm to base n	LogN(x,n)	Log(x)/Log(n)
4-quadrant arctangent	Atn2(x,y)	x > 0, Atn(y/x)
		x < 0, Atn(-y/x) + 4*Atn(1)
value of π	Pi ()	4*Atn(1)

```
Function Sec(x)  'Secant
Sec = 1 / Cos(x)
End Function
Function CoSec(x)  'Cosecant
CoSec = 1 / Sin(x)
End Function
Function ArcSin(x)  'Inverse sine
ArcSin = Atn(x / Sqr(-x * x + 1))
End Function
Function ArcCos(x)  'Inverse cosine
ArcCos = 2 * Atn(1) - Atn(x / Sqr(-x * x + 1))
End Function
Function ArcSec(x)  'Inverse secant
If x >= 0 Then
    ArcSec = Atn(Sqr(x * x - 1))
Else
    ArcSec = Atn(Sqr(x * x - 1)) - 4 * Atn(1)
End If
End Function
Function ArcCoSec(x)  'Inverse cosecant
If x >= 0 Then
    ArcCoSec = Atn(1 / Sqr(x * x - 1))
Else
    ArcCoSec = Atn(1 / Sqr(x * x - 1)) - 4 * Atn(1)
End If
End Function
Function ArcCot(x)  'Inverse cotangent
ArcCot = 2 * Atn(1) - Atn(x)
End Function
Function HSin(x)  'Hyperbolic sine
HSin = (Exp(x) - Exp(-x)) / 2
End Function
Function HCos(x)  'Hyperbolic cosine
HCos = (Exp(x) + Exp(-x)) / 2
End Function
Function HTan(x)  'Hyperbolic tangent
HTan = (Exp(x) - Exp(-x)) / (Exp(x) + Exp(-x))
End Function

Function HSec(x)  'Hyperbolic secant
HSec = 2 / (Exp(x) + Exp(-x))
End Function
Function HCoSec(x)  'Hyperbolic cosecant
HCoSec = 2 / (Exp(x) - Exp(-x))
End Function
Function HArcSin(x)  'Inverse hyperbolic sine
HArcSin = Log(x + Sqr(x * x + 1))
End Function
Function HArcCos(x)  'Inverse hyperbolic cosine
HArcCos = Log(x + Sqr(x * x - 1))
End Function
Function HArcTan(x)  'Inverse hyperbolic tangent
HArcTan = Log((1 + x) / (1 - x)) / 2
End Function
Function HArcSec(x)  'Inverse hyperbolic secant
HArcSec = Log((1 + Sqr(1 - x * x)) / x)
End Function
Function HArcCsc(x)  'Inverse hyperbolic cosecant
If x > 0 Then
    HArcCsc = Log((1 + Sqr(1 + x * x)) / x)
Else
    HArcCsc = Log((1 - Sqr(1 + x * x)) / x)
End If
End Function
Function HArcCot(x)  'Inverse hyperbolic cotangent
HArcCot = Log((x + 1) / (x - 1)) / 2
End Function
Function LogN(x, n)  'Logarithm to base n
LogN = Log(x) / Log(n)
End Function
Function Atn2(x, y) '4-quadrant arctangent
If x > 0 Then
    Atn2 = Atn(y / x)
Else
    Atn2 = Atn(-y / x) + 4 * Atn(1)
End Function
Function Pi()  'value of pi
Pi = 4 * Atn(1)
End Function
```

FIGURE 10.17   VBA code for collection of math functions.

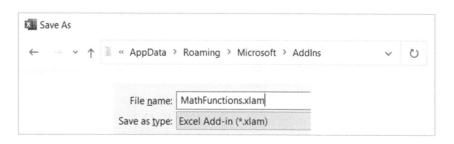

FIGURE 10.18   Save As window for saving an add-in.

folder, AddIns, and the path to the folder is denoted. You can change to another folder if you wish, but that will modify how you install the add-in in the next step. In any case, you can always locate the add-in by a Windows search for the filename. For our purposes here, it will be important to note the path, e.g., something like C:\Users\yourname\AppData\Roaming\Microsoft\AddIns. Click Save to store the add-in.

You can close the add-in workbook now and open a blank workbook. Next, to activate the add-in, follow the steps, File ⇨ Options ⇨ Add-ins ⇨ Go... and the Add-ins window will appear. There is a shortcut, Alt-T-I, which will make the Add-ins window appear immediately. If you see your add-in listed (it should be if you stored it in the default location), you can select it by checking its box and click Ok. This is shown in Figure 10.19. Note: There are likely other add-ins in the list, such as the Analysis Toolpak (Data Analysis tools) and the Solver. In the figure, there are others that are blanked out. Having done that, any of the functions in the add-in are available on the spreadsheet for any workbook you open, just as macros in the Personal Macro Workbook similarly available.

The functions in our add-in are most often used within VBA code rather than on the spreadsheet because there are equivalent functions for most available already on the spreadsheet. To do this, the VBA References must be modified to include the add-in. This is similar to using Solver commands in VBA. To do that, in the VBA Editor, select Tools ⇨ References. Use the Browse button to locate the MathFunctions add-in file. You must change the file type to Microsoft Excel Files to do this. Select the file and click Open. Then, the AddIn files are accessible from VBA code. Note: The References process must be carried out for every new workbook project.

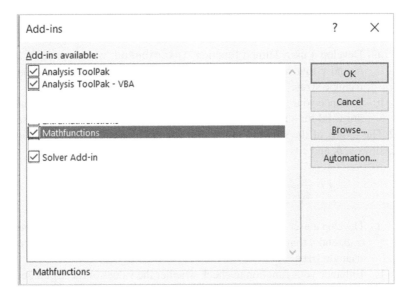

FIGURE 10.19   Add-ins window with Math functions add-in selected.

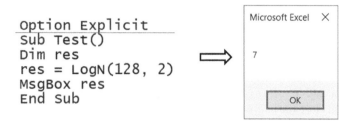

```
Option Explicit
Sub Test()
Dim res
res = LogN(128, 2)
MsgBox res
End Sub
```

FIGURE 10.20   Employing the LogN function from the MathFunctions add-in.

The following VBA Sub, located in a new workbook project, illustrates the use of one of these functions. The Sub displays a result in a message box on the spreadsheet, as shown in Figure 10.20.

Frequently, add-ins are created as a collection of user-defined functions with a common theme. These are then used primarily on the spreadsheet. Examples would be a family of engineering calculations having to do with types of equipment or process phenomena, such as fluid flow and heat transfer. At a higher level of development, it is possible to add groups and commands to the Ribbon in Excel that incorporate the use of add-ins.

PROBLEMS

10.1 The radius of the largest circle that can be inscribed in a triangle with sides a, b, and c, is given by

$$r = \sqrt{\frac{(s-a)(s-b)(s-c)}{s}} \quad \text{where} \quad s = \frac{a+b+c}{2}$$

    a. Develop a user-defined function, InscribeRad, with arguments a, b, and c that returns the value of r. Test your function for various triangle sizes.

    b. Enhance your function to check whether the values of the arguments are valid for a triangle. Return an error message for invalid inputs.

10.2 The radius of the smallest circle that can circumscribe a triangle with sides a, b, and c is given by

$$r = \frac{abc}{4\sqrt{s(s-a)(s-b)(s-c)}} \quad \text{where} \quad s = \frac{a+b+c}{2}$$

    a. Develop a user-defined function, CircumscribeRad, with arguments a, b, and c that returns the value of r. Test your function for various triangle sizes.

    b. Enhance your function to check whether the values of the arguments are valid for a triangle. Return an error message for invalid inputs.

10.3 Write a VBA function, called Rgas, that returns the value of the *gas law constant* for several different systems of units based on an input argument shown in Table P10.3.

TABLE P10.3

Values of the Gas Law Constant for Different Units

Input Argument	Gas Law Constant	Units
1	8.314472	J/(kg·mol)
2	0.0820574587	L·atm/(mol·K)
3	1.987	cal/(mol·K)
4	10.73159	ft³·psi/(°R·lbmol)
5	62.36367	L·torr/(K·mol)

10.4 Create a user-defined function named MyFunc, which has one argument, x. The function should return a result according to the graph shown in Figure P10.4.

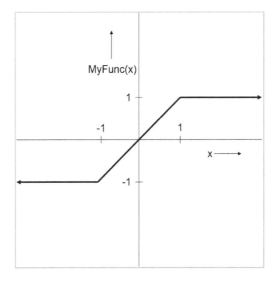

FIGURE P10.4 Graphical description of MyFunc function.

Test your MyFunc function on the spreadsheet with a set of values of x in a column from −2 to 2 in steps of 0.5. Enter a formula with your function in the cell adjacent to $x = −2$, and copy the formula down to evaluate the rest of the x values.

10.5 Develop a user-defined function, called IQR, that computes the *interquartile range* of a set of values provided in its argument. The interquartile range is the difference between the 75th and 25th percentile. Your function can borrow the PERCENTILE.EXC function from the spreadsheet world to complete the calculation.

10.6 For a set of student grades, it is desired to compute an average grade forgiving a number of the lowest grades in the set. Develop a VBA function, called NetGrade, with arguments for the range of cells containing the grades and the number of low scores to be forgiven. Your function should check the number to be forgiven to make certain it is less than the number of grades. Test your function using the grade set below and the forgiveness numbers 2, 3, and 4.

TABLE P10.6
Set of Student Grades

89.1	86.1	65.7	90.3	78.9
73.0	77.3	90.7	93.0	77.9
80.4	58.3	79.9	81.2	68.8

10.7 Create two user-defined, enhanced rounding functions, RndToEvens and RndTo5, described as follows:

a. The ROUND function on the spreadsheet always rounds a number like 6.5 to 7 and 7.5 to 8. Thus, it always rounds up to the higher number. A more common strategy is "round to evens," e.g., 6.5 to 6 and 7.5 to 8. Create a user-defined function, RndToEvens, that accomplishes this.

b. Create a user-defined function RndTo5 that rounds to the nearest 0.5, i.e., 6.7 to 6.5 and 6.8 to 7. Test your function for (a) 5.649, (b) 8.8, and (c) 3.75.

These functions should handle the digits argument as does the ROUND function.

10.8 There are alternate methods for estimating the central tendency of a data set other than the average or the median. One time-tested, short-cut method is summarized as follows for a data set with n members sorted in ascending order:

$$n \leq 5 \qquad \frac{\text{largest} + \text{smallest}}{2} = \frac{x_1 + x_n}{2}$$

$$6 \leq n \leq 20 \qquad \frac{x_3 + x_{n-2}}{2}$$

$$n > 20 \qquad \frac{x_{25\%} + x_{75\%}}{2}$$

Create a user-defined function named MeanEst that has an argument of an array of numbers and implements the strategy above. Note: You can borrow the sort routine from the spreadsheet environment to sort the argument array into another array of ascending order, and you can borrow the PERCENTILE.EXC function for the 25% and 75% determinations. You may need to record a macro with the sort routine to see how it is coded in VBA.

10.9 If air resistance is neglected, the following formula predicts the range (R) of a projectile fired at an angle θ with an initial velocity u.

$$R = \frac{2u^2}{g} \sin(\theta)\cos(\theta)$$

Develop and test a user-defined function, ProjRange, that computes the range, R, given the initial angle, θ, and velocity, u. If the muzzle velocity of a projectile from a rifle is 990 m/s, use your function on the spreadsheet to compute the range for a given angle. Then, use a targeting method (Goal Seek, Solver) to find the angle that yields a range of 5 km. You may conclude from your result that it is important to include air resistance in these calculations but don't attempt to do that here.

10.10 In some engineering and business applications, the situation arises where one needs to calculate the area of two overlapping circles. This seems like it should be a simple calculation.

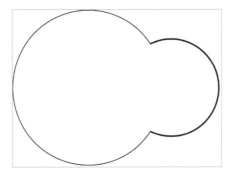

FIGURE P10.10a Enclosed area of two overlapping circles.

The problem is to find the enclosed area, as shown in Figure P10.10a. This is shown in more detail in the diagram in Figure P10.10b:

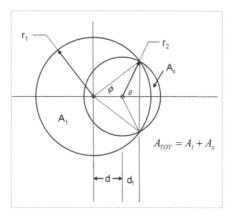

FIGURE P10.10b   Overlapping circles with detail.

For this diagram, the following equations hold:

$$r_1 \cdot \sin(\phi) = r_2 \cdot \sin(\theta) \qquad r_1 \cdot \cos(\phi) = d + d_1 \qquad r_2 \cdot \cos(\theta) = d_1$$

These three are solved for d_1, ϕ, and θ to yield:

$$d_1 = \frac{r_1^2 - d^2 - r_2^2}{2d}$$

$$\phi = \tan^{-1}\left(\frac{\sqrt{2d^2 r_1^2 - r_2^4 + 2d^2 r_2^2 + 2r_1^2 r_2^2 - d^4 - r_1^4}}{r_1^2 + d^2 - r_2^2} \right)$$

$$\theta = \tan^{-1}\left(\frac{\sqrt{2d^2 r_1^2 - r_2^4 + 2d^2 r_2^2 + 2r_1^2 r_2^2 - d^4 - r_1^4}}{r_1^2 - d^2 - r_2^2} \right)$$

which can be used to determine

$$A_x = \frac{1}{2}\left[r_2^2 \left(2\theta - \sin(2\theta) \right) - r_1^2 \left(2\phi - \sin(2\phi) \right) \right]$$

a. First, develop a spreadsheet that calculates A_{TOT} given cells with values of the two radii and the center distance of separation.

b. Then, add to this workbook a VBA function called Overlap with input arguments of the two radii and the center distance of separation. The function should return the net area. Your function should protect against a d value that is large enough that the circles are separated and in such a case it should return the sum of the areas of the two circles. It should also protect against a d value that has the smaller circle completely inside the larger circle and in such a case it should return the area of the larger circle. Hint: The arctangent function used must be 4-quadrant and not 2-quadrant.

c. Finally, carry out a case study, using your `Overlap` function, for $r_1 = 10$, $r_2 = 5$, and a range of d values from 0 to 15. Create a plot of A_{TOT} versus d.

10.11 A viscous oil is allowed to drain out of a cylindrical tank through a vertical pipe. This is pictured in Figure P10.11.

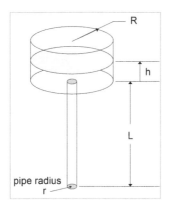

FIGURE P10.11 Cylindrical tank with pipe drain.

The time it takes the tank to empty can be calculated by the formula:

$$t_{eff} = \frac{8\mu L R^2}{\rho g r^4} \ln\left[\frac{L+h}{L}\right]$$

where

t_{eff}: time of efflux, s
μ: fluid viscosity, Pa·s
L: length of exit pipe, m
R: inside radius of tank, m
ρ: fluid density, kg/m³
g: gravitational acceleration, m/s²
r: inside radius of exit pipe, m
h: fluid depth in tank, m

The example dimensions for the tank and pipe are

$$L = 0.6 \text{ m} \qquad R = 0.25 \text{ m} \qquad r = 0.012 \text{ m}$$

Then, for a given value of h, for example, 0.4 m, it is possible to calculate the *efflux time*, t_{eff}.

Develop a VBA function, called EffluxTime, that computes and returns t_{eff} given arguments, μ, L, R, ρ, r, and h. Gravitational acceleration is fixed at 9.81 m/s^2. Set up a case study on the spreadsheet for a range of L values and present a graph of t_{eff} versus L. Use a silicone oil with density of 936 kg/m^3 and viscosity of 0.35 Pa·s.

10.12 When water is flowing down a sloped, open channel, a phenomenon called a *hydraulic jump* can occur where the water with a depth, h_1, and a flow velocity, v_1, suddenly jumps up to a depth, h_2, with a slower velocity, v_2. This is shown in Figure P10.12.

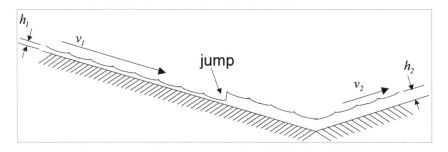

FIGURE P10.12   Open channel with hydraulic jump.

An equation that describes a hydraulic jump is

$$h_2 = \frac{h_1}{2}\left[\sqrt{1 + \frac{8v_1^2}{gh_1}} - 1\right]$$

and such a jump is possible only if

$$v_1 > \sqrt{gh_1}$$

Develop a user-defined function, called HydrJump, that computes h_2 given arguments h_1 and v_1. If the inequality is not satisfied, the function returns an error message, "no jump."

Given $h_1 = 0.1$ m, test your function for typical range of velocities: e.g., $v_1 = 0.5$ to 1.5 m/s in steps of 0.25 m/s.

10.13 The coordinates for points on the plane are given in either Cartesian (x, y) or polar (r, θ) terms. The relationships between the two systems are

$$r = \sqrt{x^2 + y^2} \qquad \theta = \tan^{-1}\left(\frac{y}{x}\right)$$

$$x = r\cos(\theta) \qquad y = r\sin(\theta)$$

Develop two user-defined array functions, called `CartToPolar` and `PolarToCart`, that have the two corresponding arguments and produce the appropriate coordinates. To test the functions, the input arguments on the spreadsheet should be two adjacent cells on the same row, and the function should produce the two corresponding coordinates on two horizontal cells. Note: The inverse tangent for the calculation of θ should be the *four-quadrant arctangent function*.

10.14 In this chapter, we developed the `AtmTemp` function based on the Earth atmosphere model. That model also presents relationships between air pressure and density based on altitude. Develop an array function, `EarthAtm`, that returns in a column on the spreadsheet the temperature, pressure, and density of air, based on an input argument of altitude. Test your function for various altitudes.

10.15 Create an add-in, called `CentroidCalcs`, that includes array functions for the following calculations of centroids for five different planar shapes:

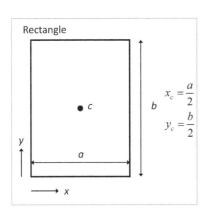

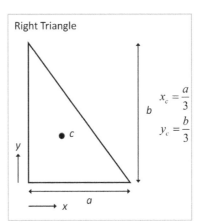

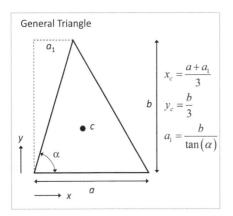

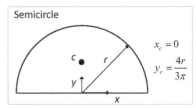

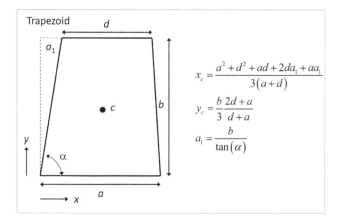

On a new blank workbook spreadsheet, show tests of the functions and validations with on-sheet formulas.

Note: For planar shapes with uniform mass density, it is the shape's *barycenter* or *center of mass*. In simple terms, it can be understood as the point at which a cutout of the shape could be perfectly balanced on the tip of a pin. Mathematically, the coordinates can be determined by the following two area integrals:

$$x_c = \frac{1}{A} \int_A x \cdot dA \qquad y_c = \frac{1}{A} \int_A y \cdot dA$$

10.16 A practical industrial calculation is that of determining the volume of fluid in a tank of given geometry where the depth of the fluid is known. Create an add-in, called Tankvolumes, that contains a collection of functions that implement these calculations for the tank types described below. On a new blank workbook spreadsheet, show tests of the functions and validations with on-sheet formulas.

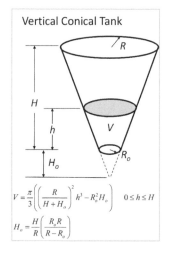

Vertical Conical Tank

$$V = \frac{\pi}{3}\left(\left(\frac{R}{H+H_o}\right)^2 h^3 - R_o^2 H_o\right) \qquad 0 \le h \le H$$

$$H_o = \frac{H}{R}\left(\frac{R_o R}{R - R_o}\right)$$

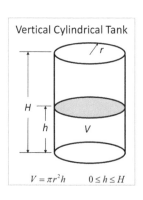

Vertical Cylindrical Tank

$$V = \pi r^2 h \qquad 0 \le h \le H$$

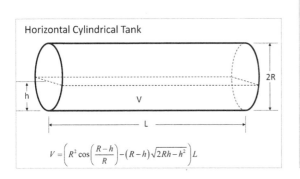

Horizontal Cylindrical Tank

$$V = \left(R^2 \cos\left(\frac{R-h}{R}\right) - (R-h)\sqrt{2Rh - h^2} \right) L$$

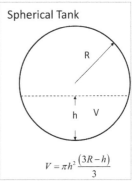

Spherical Tank

$$V = \pi h^2 \frac{(3R - h)}{3}$$

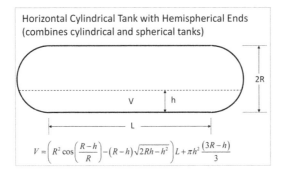

Horizontal Cylindrical Tank with Hemispherical Ends (combines cylindrical and spherical tanks)

$$V = \left(R^2 \cos\left(\frac{R-h}{R}\right) - (R-h)\sqrt{2Rh - h^2} \right) L + \pi h^2 \frac{(3R - h)}{3}$$

NOTES

1. https://www.grc.nasa.gov/www/k-12/airplane/atmosmet.html
2. Obtained from Zwillinger, D., 2018. *CRC Standard Mathematical Tables and Formulas*, 33rd Ed., Chapman Hall, London, UK.
3. Note: If you are not interested in the entire set of functions, you can certainly just enter a few and then learn the steps to create the add-in.

11 VBA Programming

CHAPTER OBJECTIVES

- Learn VBA's various data types and structures and where these are used appropriately in programs
- Gain appreciation for the structure of VBA objects including the use of the Object Explorer in the Visual Basic Editor
- Understand the concept of variable scope and how to control it in VBA projects
- Learn the features of structured programming in VBA including selection and repetition
- See how to use VBA Subs and Functions to implement calculations in a modular fashion
- Experience prototyping tasks on the spreadsheet and then elevating them into VBA programs

Chapters 9 and 10 introduced VBA Subs and Functions without getting too involved with the details of the VBA programming language. In this chapter, we will delve into those details. This will expand your ability to create VBA programs to tackle more complicated tasks. This chapter should also serve as a useful reference as you develop VBA applications in the future.

In contrast to other programming languages, VBA has a large family of data types, different modes for storing numerical and other information. Data structures include single variables and arrays, including one- and two-dimensional. For engineering and scientific applications, we do not commonly use all the data types, but they are part of the overall VBA environment, and we will introduce them here. We have shown examples of VBA objects, properties, and methods in previous chapters, and we will expand that knowledge here. The concept of variable scope is essential to creating larger VBA codes with multiple modules, Subs, and Functions. You will learn about scope here and how to take advantage of it.

The ability to implement decisions and iteration in VBA program code is essential. The topics of selection and repetition introduce the fundamental VBA structures that facilitate this. We will provide both examples and guidance in the use of these. Later in this chapter, you will see how these structures are put to work. To this point, we have used Subs as macros and user-defined functions to be invoked from the Excel environment. In this chapter, we will introduce modular programming where these are called from other routines within VBA.

An important skill in Excel and VBA application development is to start by prototyping calculations on the spreadsheet. These are useful but are often clumsy

DOI: 10.1201/9781003361053-11

and inflexible. The next step then is to transform these prototypes into VBA code to provide more efficient, portable, and flexible means for implementing these calculations. This will be the final topic of this chapter.

11.1   VBA DATA TYPES

Data are stored in computer memory, which is based on binary digits (bits), 8-bit bytes, and words, which are collections of bytes. The typical word lengths are 16, 32, and 64 bits (2, 4, and 8 bytes). Data types define how information is stored in relation to computer memory. We can specify the type in a Dim statement. General categories include numbers, text, and true/false (Boolean) data. For numbers, we have real numbers with fractional parts and exponents (also called *floating point numbers*), and integers.[1] Table 11.1 lists VBA's data types and provides information about them.

TABLE 11.1
VBA's Data Types

Data Type	Description
Variant	• VBA's default data type • adjusts automatically to the data type required • a "chameleon" data type that can change during program execution • memory storage requirements vary with data type
Double	• used for numbers with decimal fractions and possibly exponents • used where high precision or extended range are required • precision: about 15-to-16 significant figures • range: from about 10^{308} down to 10^{-309}, both positive and negative • uses 8 bytes of memory • Double is the most common "real number" data type used in VBA as it matches the precision and range used on the spreadsheet
Single	• used for numbers with decimal fractions and possibly exponents • precision: about 6-to-7 significant decimal figures • range: from about 10^{38} down to 10^{-39}, both + and − • uses 4 bytes of memory
Integer	• used for most counting numbers, like subscripts or loop counters • range: from −32,768 to 32,767 • uses 2 bytes of memory
Long	• used for integers where extended range is required • range is about −2 billion (10^9) to +2 billion • uses 4 bytes of memory
String	• used to store text • 1 byte is used for each character in the string • characters are stored according to the standard ASCII code
Boolean	• used for True/False information • constants are True and False • uses 2 bytes for each Boolean quantity

---

There are several additional data types that we encounter occasionally. These include Object. We can declare a variable name to represent an object in VBA. When we assign an object to that variable, we must use the syntax with the initial word `Set`,

```
Set VariableName = ObjectName
```

There is also a data type `Range`. This can be used to store a reference to a range of one or more cells on the spreadsheet. There are also data types for currency and date.

Variables that are not typed specifically with `Dim` statements are automatically give the Variant type. As the variables store values, their type is adapted to the information stored. Figure 11.1 illustrates this by highlighting the Locals window during the stepwise execution of the VBA Sub shown. You can see that the type of the variable x remains Variant, but it adapts to the required sub-type as different assignments take place.

When we declare a variable to be of a specific type, like Integer, an error may occur if we try to assign a value of another type to the variable. See Figure 11.2. In this case, x is declared as type Double, and a real number value is assigned to it. Then, assigning a string is attempted, and the error occurs. Note that, if we attempt to assign an integer to x, that is successful as VBA automatically converts the integer value to the Double type.

One might question whether there is any need to declare variables to be of a specific type as the Variant type allows variables to adapt to the type required. One good reason to declare types is to prevent any errors from occurring if the

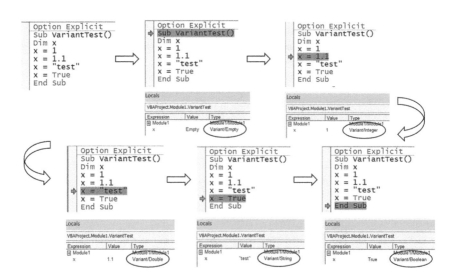

FIGURE 11.1 A Variant variable (x) changing type during stepwise execution (F8) of a Sub.

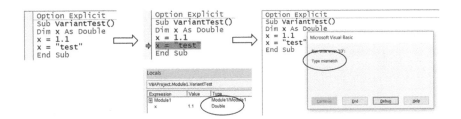

FIGURE 11.2 Error occurring when a string value is assigned to a Double variable, x.

variable is erroneously assigned a value different from the type in the Dim statement. A typical practice here is to declare counting variables, such as those that occur in loops (we'll see these later in this chapter), as Integer. Also, it is more explicit to declare variables of type String or Boolean because that makes the VBA code clearer.

Generally, we do not use the *Single type*. Although it saves memory, it actually does not speed up calculations. One would expect that, but tests will show that calculations with Double variables occur more quickly than Single. This has to do with the internal code in Excel and VBA. Also, there is little need to declare a variable as Long unless the counting numbers it stores exceed the range of the Integer type, about ±32,700. For these reasons, you will see us declare variables as Double, Integer, String, and Boolean as required. One can also declare a variable as Variant, but that will occur anyway if the declaration is left out of the Dim statement.

Next, we come to the matter of data structure. In the case of VBA, we distinguish variables and storage of single values with ranges of values in arrays.

11.2 ARRAYS IN VBA

We use array variables frequently in VBA, just like in other programming languages. A very simple example of an array is illustrated in Figure 11.3. The array A is declared in a Dim statement with the extent 10. This actually allocates 11 elements to A which are indexed 0 through 10. As the figure shows, single-stepping to the End Sub statement reveals in the Locals window that 1 has been assigned

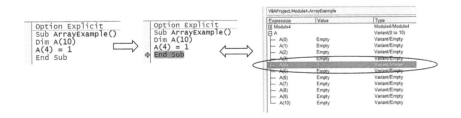

FIGURE 11.3 Assigning a value to an element of an array.

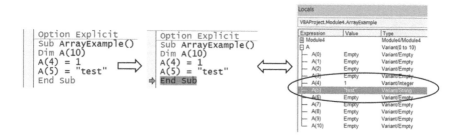

FIGURE 11.4   Assigning values of different types to elements of an array.

to the fifth element of A, that being A(4). Also, the type of that element alone has been changed to Variant/Integer. All other unassigned elements are Variant/ Empty.

This raises an interesting question, which is answered in Figure 11.4. Can you store information of different types in a Variant array? The answer is yes, as you can see with 1 stored in element 4 and "test" in element 5. This is worth mentioning, but we almost always store only values of the same type in an array. You can enforce that, for example, by declaring the array type Double in the `Dim` statement.

The index origin of zero is the default for arrays in VBA. We might call this a displacement index because its value indicates how many places from the origin the reference is. For many, perhaps most, engineering and science applications, we are used to subscripts related to mathematical descriptions and these typically have an origin of one. For this reason, it is possible to set the array origin to one with the declaration

```
Option Base 1
```

The difference is shown in Figure 11.5. The Sub has a For-Next loop that uses an index variable i as a subscript and assigns values of the square root of i to the elements of the array A. The array is declared as type Double in the `Dim` statement. The figure illustrates single-stepping through seven iterations ($i = 7$) and the values assigned to A up to that point. We will get into the details of the For-Next structure later in this chapter.

```
Option Explicit
Option Base 1
Sub ArrayExample()
Dim A(10) As Double, i As Integer
For i = 1 To 10
    A(i) = Sqr(i)
Next i
End Sub
```
⟹
```
Option Explicit
Option Base 1
Sub ArrayExample()
Dim A(10) As Double, i As Integer
For i = 1 To 10
    A(i) = Sqr(i)
⇨ Next i
End Sub
```
⟺

Locals

VBAProject.Module4.ArrayExample

Expression	Value	Type
⊞ Module4		Module4/Module4
⊟ A		Double(1 to 10)
— A(1)	1	Double
— A(2)	1.4142135623731	Double
— A(3)	1.73205080756888	Double
— A(4)	2	Double
— A(5)	2.23606797749979	Double
— A(6)	2.44948974278318	Double
— A(7)	2.64575131106459	Double
— A(8)	0	Double
— A(9)	0	Double
— A(10)	0	Double
i	7	Integer

FIGURE 11.5   Sub with `Option Base 1` and array referencing.

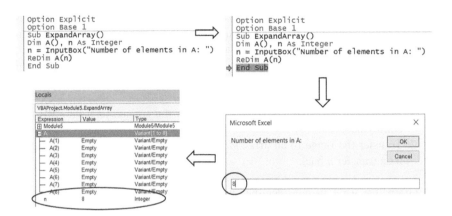

FIGURE 11.6   Expanding an empty array with user input.

Array sizes can be changed, increased, or decreased, during the execution of VBA code. This is accomplished with the ReDim statement. You have seen examples of this in Chapter 10 with user-defined functions which deal with variable-sized arguments. Often, we start by declaring the array as empty, as with

```
Dim A()
```

and then expand it with the ReDim statement. See Figure 11.6. An InputBox statement is used to get the array size, n, from the user. The Locals window shows that the array is expanded.

It should be noted that whenever using the ReDim statement, including expanding a current array of a given size and reducing it, the resulting array will not have any values stored. This issue is addressed in an example later with the ReDim Preserve statement.

As a last elementary example, a two-dimensional array is illustrated in Figure 11.7. We generally interpret the pair of indices in matrix terminology as row and column. So, the array A is declared as four rows and three columns. Then, the element in the second row and third column is assigned the value of 25. See how the branches of A are expanded in the Locals window to show the assignment.

FIGURE 11.7   Declaring a two-dimensional array and setting the value of one of its elements.

We encountered arrays as arguments to functions in Chapter 10. For most engineering and scientific applications, we use either one-dimensional or two-dimensional arrays.[2] Although these may relate to ranges of cells on the spreadsheet, they may be entirely within VBA. When dealing with one- or two-dimensional ranges on the spreadsheet, VBA presents two alternatives:

1. Manipulate the range contents directly on the spreadsheet with VBA commands, or
2. Transfer the range to an array variable in VBA for manipulations and then transfer it back.

Often, we want to consider the second alternative because manipulations are more efficient, and/or we want to modify the array without affecting its appearance on the spreadsheet. We will illustrate both alternatives with a simple example.

Figure 11.8 shows a brief column range of cells filled with entries and illustrates an operation we call *downward rotation*. We would like to move each value down one cell and return the last value to the top cell of the range. As the modified scheme on the right shows, to accomplish the rotation, it is necessary to store the value from the bottom of the range before the method moves the second-to-last value down; otherwise, we would lose the last value.

A way to accomplish the rotation is to store the last value in the Temp location or variable, then move up through the range moving the second-to-last value down one cell and so on up through moving the first value down to the second location. Finally, we can transfer the Temp value to the first location.

The VBA code in Figure 11.9 illustrates how to accomplish the rotation directly on the spreadsheet and shows the before and after appearance of the range of cells on the spreadsheet.

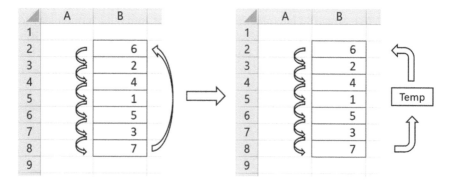

FIGURE 11.8 Downward rotation operation and the need for a temporary storage location.

```
Option Explicit
Sub Rotate1()
Dim n As Integer, i As Integer, TempVal
n = Range("datarange").Count   'find the number of cells in the range
Range("datarange").Select   'select the range, ActiveCell is the top cell
TempVal = ActiveCell.Offset(n - 1, 0).Value   'store the bottom value in TempVal
For i = n To 2 Step -1 'move values down starting at the bottom of the range
    ActiveCell.Offset(i - 1, 0).Value = ActiveCell.Offset(i - 2, 0).Value
Next i
ActiveCell.Value = TempVal 'assign the stored value to the top cell
End Sub
```

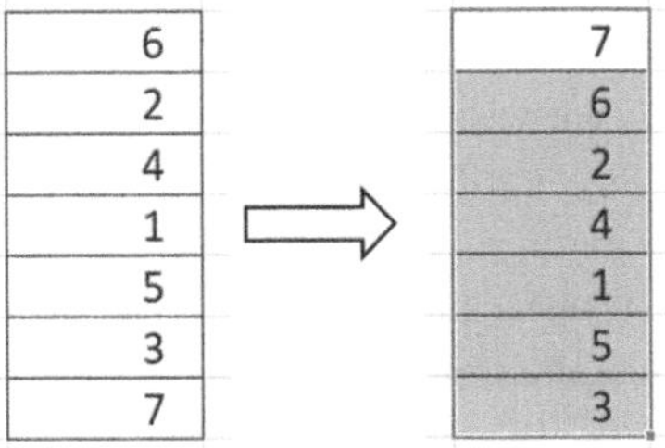

6
2
4
1
5
3
7

⇨

7
6
2
4
1
5
3

FIGURE 11.9   VBA Sub to implementing downward rotation on the spreadsheet.

You will note in the VBA code that we Dim the n and i variables as type Integer. This is because they are counting variables. We leave TempVal as a Variant type by default. The range of cells on the spreadsheet is named datarange. We could have used cell addresses B2:B8 instead. The Count property of the Range("datarange") object is used to determine the number of cells, n. The range of cells is then selected on the spreadsheet using the Select method.

When the range is selected, the top cell is the ActiveCell. We carry out our manipulations with respect to the ActiveCell using the Offset property. The syntax for the Offset property is as follows:

```
cell_reference.Offset(row offset, column offset).property
                                              (or .method)
```

The arguments of the Offset property specify how far to move from the cell reference. For example, Offset(n-1,0) references the last cell in datarange since that is n-1 rows from the ActiveCell and zero columns shifted. In this way, the VBA code stores the last value in datarange in the TempVal variable.

The following statements in Figure 11.9 represent a *count-controlled For-Next loop*. We will deal with this structure in detail later in this chapter. For now, understand that the loop cycles the index variable i from n To 2 in steps of -1. The i variable is used in the Offset property to reference the cells needed to assign values to cells from the cells located just above. Finally, the code assigns the TempVal value to the top cell in the range, the ActiveCell.

```
Option Explicit
Option Base 1
Sub Rotate2()
Dim n As Integer, i As Integer
Dim DataArray, TempVal
n = Range("datarange").count 'find the number of cells in the range
DataArray = Range("datarange").Value 'transfer the spreadsheet range to local array
TempVal = DataArray(n, 1) 'store the last value in TempVal
For i = n To 2 Step -1 'move values down starting at the last item in the array
    DataArray(i, 1) = DataArray(i - 1, 1)
Next i
DataArray(1, 1) = TempVal 'assign the stored value to the first location in the array
Range("datarange").Value = DataArray 'transfer the array back to the spreadsheet
End Sub
```

FIGURE 11.10   VBA Sub to carry out downward rotation by transferring cell range.

An alternate approach is to transfer the cell range to an array in VBA, carry out the rotation there, and return the result to the spreadsheet. This is shown in Figure 11.10. Apart from the `Option Explicit` declaration, there is `Option Base 1`. As mentioned previously, this provides for array indices/subscripts to have an origin of one rather than the default of zero.

Here, the entire `datarange` is transferred to the local variable `DataArray`, which is declared as a Variant (by default). `DataArray` is automatically sized to hold the n values from `datarange`. You will note that the subscripts for `DataArray` include a second 1, which represents the first column of the array. Of course, there is only one column, but doing it this way requires that we specify the second subscript. The code to store the final array value in `TempVal` and then move back up through the array is similar to the previous VBA Sub in Figure 11.9. Finally, the rearranged array is assigned back to the spreadsheet. One possibility is that it could be transferred to a different location on the spreadsheet. The results are the same as shown in Figure 11.9.

It is also possible to create arrays within VBA without referring to the spreadsheet. One possibility is to allow the user to enter values to build an array. This involves use of the `InputBox` function (more on that in Chapter 12) and is illustrated in Figure 11.11. There is some care taken in composing the `Dim` statement. The `NumVals` variable is declared as type Integer since it is a counting variable. `NewVal` is declared as Double because we expect the user to enter numeric quantities. An array variable, `MyArray`, is declared as empty with () and will expand as

```
Option Explicit
Option Base 1
Sub BuildArray()
  Dim NumVals As Integer, NewVal As Double, MyArray(), MyArrayMedian
  Do 'loop to allow user to enter values into array
    NewVal = InputBox("Enter number for array or -9999") 'get value from user
    If NewVal = -9999 Then Exit Do 'if user enters -9999, entry is complete
    NumVals = NumVals + 1 'increment number of values
    ReDim Preserve MyArray(NumVals) 'expand MyArray
    MyArray(NumVals) = NewVal 'assign the new value to the new, last element of MyArray
  Loop 'back to Do for more entry
  MyArrayMedian = Application.WorksheetFunction.Average(MyArray)
  MsgBox ("Array median = " & MyArrayMedian)
End Sub
```

FIGURE 11.11   VBA Sub to build an array from user input.

required with the user entries. The iterative structure here is the Do-Loop (more details on that later on in this chapter). The loop will repeat until the user enters –9999, called a sentinel value,[3] at which point the Exit Do command will be executed.

The InputBox statement includes a prompt in quotes and returns a String-type result. This is converted to type Double when it is assigned to NewVal. With a new value from the user, the NumVals variable is incremented, and then MyArray is expanded to NumVals entries. The ReDim Preserve statement keeps the earlier entries in the array. The NewVal value is then added as the new, final element of MyArray. There are two final statements of the Sub that calculate the median of the MyArray values and display the result in a message box as an illustration of use of the array.

We see many uses for arrays in VBA, and there will be examples later in this chapter and in Chapter 12.

11.3 OBJECTS, PROPERTIES, METHODS, AND EVENTS

Object-oriented programming is the basis of VBA. An object is a computer-based entity that has properties and can be manipulated. It is a very general definition. Excel has a hierarchy of objects and properties. This starts with Excel itself – the BIG object, which is called Application. There are other Application objects in the Microsoft Office and Windows environments, e.g., Word, PowerPoint, and Access. It is also possible to define new objects, classes of objects, and properties; however, we do not emphasize that in this book.

Figure 11.12 presents a schematic diagram of part of the Excel *object tree structure*. The overall object structure is complex and impossible to commit to memory. You will know objects that you use frequently, and occasionally you will need to expand your knowledge.

One way to explore the structure is with the *Object Browser* available in the Visual Basic Editor. You can view the Object Browser by selecting it from the View menu or pressing the F2 key. There is also an icon on the VBE toolbar.

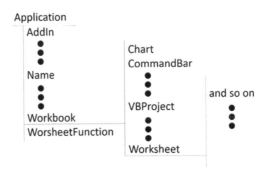

FIGURE 11.12   Schematic diagram of part of the Excel object structure.

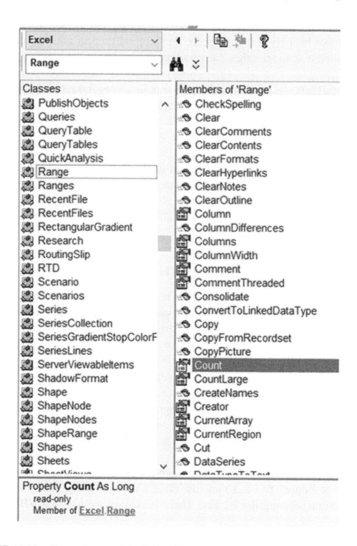

FIGURE 11.13   Example use of the Object Browser.

Figure 11.13 gives an example of the Object Browser. In this case, we have chosen the Excel library and searched for the Range object class. Under Range, we have located the Count property, and information on it is presented at the bottom of the Browser.

The Object Browser also finds use in communicating with other object-oriented software packages. In such cases, one selects the object library of the package from the Tools ⇨ References window. The library will then appear on the dropdown list in the upper left corner of the Browser. By selecting that library, you can identify objects and properties in the other package, acquire information, and send back changes from VBA.

In communicating with the spreadsheet environment from VBA, we often use object collections. The common ones are

- `Workbooks`   all currently open workbook objects
- `Sheets`   all sheets, including worksheets and chart sheets
- `Worksheets`   all worksheets
- `Charts`   all charts

Two examples of their use would be:

```
Worksheets("Data")
Workbooks("Project8")
```

A complete object reference might be

```
Application.Workbooks("Project8").Worksheets("Data").Range("B2:E10")
```

This reference can be simplified all the way to

```
Range("B2:E10")
```

as long as it is unambiguous. If the workbook has more than one worksheet, this last reference would be to the `ActiveSheet`, but to be sure, we might want to include a `Worksheets` specification.

There are three elements in VBA that relate to objects:

1. Properties   attributes of objects   `Range("A1").Width`
2. Methods   actions taken on objects   `Range("B2:E10").Clear`
3. Events   happenings that objects respond to `Open` (a workbook is opened)

There are thousands of these, but there is a manageable number that you will use frequently. When you need to reference a new one, you can usually find it via the Object Browser, Macro Recorder, or elsewhere.

Since we often work with cells on the spreadsheet, the Range type is commonly used. Here are several examples:

- `Range("XY22")`   a single cell, using its address
- `Range("B2:E10")`   a range of cells, using their address
- `Range("mydata")`   a named cell or block of cells
- `Range("D:D")`   all of column D
- `Range("3:5")`   all of rows 3, 4, and 5

Other references are

- `Columns(D:G)`   columns D through G
- `Rows(4:100)`   rows 4 through 100

We can also use the Cells and Offset properties:

- `Cells(4,5)`   row 4, col 5, same as E4
  [useful with variable arguments]
- `Range("xval").Offset(1,1)` down 1 row, over 1 column from the xval cell

Range object properties are

- `Value`   what's stored in the cell   [read/write][4]
  [This is the default property – it is what is referenced if you leave the specification out.]
- `Text`   a formatted string of what's stored in the cell   [read only]
- `Count`   the number of cells in a range   [read only]
- `Column`   index number of the column   [read only]
- `Row`   index number of the row   [read only]
- `Address`   absolute address of a cell or block of cells   [read only]

The common range object methods are

- `Select`   select a cell or block of cells
- `Copy`   `Range("B2").Copy Range("C2")` supply the destination cell
- `Copy/Paste`   `Range("mydata").Select`   copy a block of cells
                      `Selection.Copy`
                      `Range("E2").Select`
                      `ActiveSheet.Paste`
- `Clear`   erase the contents *and* formatting of a cell or block of cells
- `ClearContents`   erase only the contents of a cell or block of cells
- `Delete`   delete a cell or block of cells, shifting other cells into place

Of course, there are many other possibilities for objects, properties, methods, and events. We will see events used frequently in Chapter 12 in building user interfaces.

11.4   VARIABLE SCOPE

We encountered the concept of scope on the spreadsheet with named cells back in Chapter 3. By default, created names have global scope across all worksheets of the workbook, but it is possible to create a name with scope restricted to a single worksheet. The scope of variables in VBA is more involved but embodies the same concept.

Scope for a variable has to do with the reach of the variable, that is, from where it can a variable be seen, and its value referenced. There are three levels of scope in VBA:

Local   only within a single procedure (Sub or Function)
           declare with a Dim or Static statement within the procedure
Module   only within the current module
           declare with a Dim statement in the module outside of the Subs
           and Functions
Public   can be seen from everywhere in the project
           declare with a Public statement at the module level

Figure 11.14 provides a schematic diagram of the hierarchy of variable scope. As VBA tasks become more complicated, it is natural to subdivide them

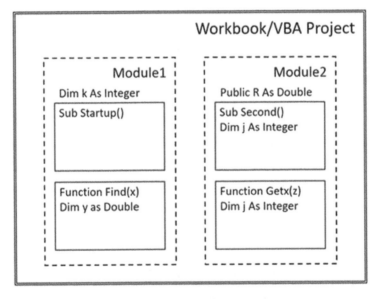

R can be referenced or set anywhere in the project
k can only be referenced or set in Module1
y is a local variable in Function Find and can't be seen
       anywhere else
j is a local variable in Sub Second and can't be seen
       anywhere else
x,z are function arguments and do not have storage
       locations in their respective functions

FIGURE 11.14   Scheme of variable scope.

into modules, and within the modules have multiple subprograms (Subs and Functions). Variables that are declared within subprograms are local to those and cannot be referenced elsewhere. The arguments to the Find and Getx functions are called formal parameters and have no storage local to the functions. Rather, their values or reference locations are provided when the function is invoked, whether from the spreadsheet or another subprogram.

As we deal with multiple subprograms and modules, an issue that can arise is whether a subprogram remembers the value of a local variable from the last time that the subprogram was invoked. That generally will not be the case as all Dim'd variables are zeroed out when the Sub or Function executes. There is an alternate declaration, Static, that provides that previous values stored will be retained for the next execution of the subprogram. An example would be

```
Static y_old as Double
```

This declaration is not often used, but it is worthwhile being aware of it. Other programming languages do not always provide this capability.

Another detail is the symbolic constant. This is an alternative to the Public declaration when the desired value doesn't change. This is declared at the module level, and an example is

```
Const Rgas As Double = 8314.46
```

After the declaration is put in place, the symbol Rgas becomes synonymous with the constant value. It is worth mentioning that you can also assign a symbolic constant on the spreadsheet by creating a name, for example, using Formulas ⇨ Defined Names ⇨ Define Name. Then, in the New Name window, type in the name and in the Refers to: field, type in the value instead of a cell reference.

11.5   VBA PROGRAMMING STRUCTURES

Algorithm structures in programming languages fall into three general categories:

1. **Sequence:** Stepwise execution of commands one at a time in sequential order
2. **Selection:** Branching of code sequence based on true/false decisions
3. **Repetition:** Iteration of code, also called looping

and VBA provides for these in a variety of ways. We will introduce these in this section. Sequence requires no introduction. Most of the VBA codes we have seen so far execute in a linear fashion step by step. That is, the program statements are executed line by line starting at the top of the procedure and moving down to the end. In contrast, as described next, selection and repetition empower programmers to develop algorithms that are much more complex and efficient.

11.5.1 SELECTION STRUCTURES

Selection in VBA makes use of various If commands that depend on the evaluation of a logical expression. *Logical expressions* can take various forms that evaluate True or False. These include

- logical/Boolean variable
- relational expression involving relational operators, $>$, $<$, $>=$, $<=$, $=$, $<>$
- logical expression involving a logical operator:
    - Not   logical negation
    - And   both True for the result to be True
    - Or   either True for the result to be True
    - XOr   one or the other is True (but not both) for the result to be True
    - Eqv   both True or both False for the result to be True
    - Imp   logical implication, compares two logical expressions (rarely used)

The first, and simplest structure is the *one-way If*, also called an *If-Then*. The standard syntax is

```
If condition Then
    .
    .
    .
    code
    .
    .
    .
End If
continue here if condition is False
```

If the logical condition is true, one or more lines of code are executed. If it is false, nothing happens, and the program moves directly to the statement following the End If. A simple example that sometimes occurs in calculations is to square a quantity and maintain the sign of the quantity in the result. This is shown in Figure 11.15.

FIGURE 11.15   The standard one-way If structure.

Observe that we indent the code between the If and the End If statements. This is not provided automatically by the VBE. Rather, we do it manually and intentionally to make the code visually easier to follow; that is, the If/Then and the End If statements clearly define the start and the end of the structure. Also note that we have included a comment. Although this is a non-executable statement, it is meant to indicate that the body of the If/Then can be several (and often many) statements. In those cases, the advantage of indentation comes to the fore.

If we omit the comment from the code, only one statement is executed. In those cases, it is often convenient to implement a subset of the structure, the *one-line If/Then*,

```
Function SgnSqr(x)
SgnSqr = x ^ 2
If x < 0 Then SgnSqr = -SgnSqr
End Function
```

The single-line version can have multiple statements, but they must be separated by : so that they can be written as a single line. For instance, a multi-line If/Then:

```
If a > 0 Then
    c = b / a
    MsgBox c
End If
```

can also be written as a single-line If/Then as

```
If a > 0 Then c = b / a: MsgBox c
```

The *two-way If structure*, also called *If-Then-Else*, allows for code to execute when the condition tests False. Its syntax is

```
If condition Then
    code executed if condition tests True
    then continue past the End If
Else
    code executed if condition tests False
End If
```

An example of an If/Then/Else is illustrated in Figure 11.16. The xLimit function constrains the value of the input x to its value between −1 and 1 but limits it to −1 or 1 outside those limits. As with Figure 11.15, the function takes advantage of the VBA Sgn function[5] to produce the limiting values. You will also notice that we continue the practice of indenting code within the clauses of the structure.

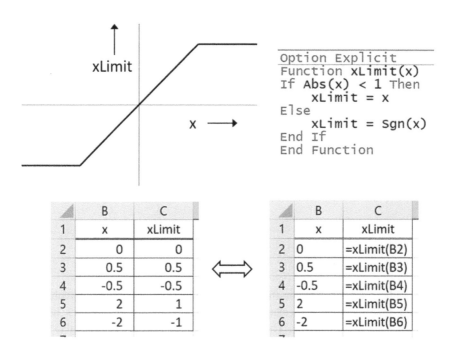

FIGURE 11.16 Example of two-way If with a limiting function.

VBA expands the two-way If to provide for testing a number of conditions in sequence. The structure is called the *multi-alternative If*, or *If-Then-ElseIf*, and its syntax is

```
If condition1 Then
      code block 1
ElseIf condition2 Then
      code block 2
         •
         •
         •
ElseIf condition n then
      code block n
Else
      else code block
End If
```

Once one of the If, ElseIf, or Else conditions is satisfied, the corresponding code is executed, and then execution jumps beyond the End If statement. Note that in order to reach a particular ElseIf statement, all the conditions before that must have tested False. Also, the Else clause of the structure, which executes if all previous conditions are False, is not required if there is no code needed there.

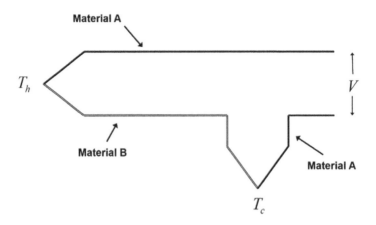

FIGURE 11.17 Thermocouple circuit.

A realistic example, where the multi-alternative If is appropriate, involves the use of a thermocouple to measure temperature. See Figure 11.17 for a schematic drawing of a thermocouple. This device generates a small voltage, also called an electromotive force or emf, that is proportional to the temperature difference, dT, between a "hot junction" where the temperature, T_h, is measured and a "cold junction" where a reference temperature, T_c, is known. However, the relationship between the emf and temperature is not quite linear. To provide an accurate scheme relating voltage to temperature, NIST[6] publishes a series of polynomials for different voltage ranges.

These are shown in Figure 11.18 for a Type J thermocouple (iron-constantan[7]). The voltages are in millivolt units, and the resulting temperature is in °C. The temperature computed is for a thermocouple with reference junction at 0°C, so a different reference junction temperature must be added to the result.

$$dT = 19.53V - 1.229V^2 - 1.075V^3 - 0.5909V^4$$
$$- 0.1726V^5 - 0.02813V^6 - 2.396 \times 10^{-3}V^7 \qquad -8.10 \leq V < 0.0 \; mV$$
$$- 8.382 \times 10^{-5}V^8$$

$$dT = 19.78V - 0.2001V^2 + 0.01037V^3 - 2.550 \times 10^{-4}V^4$$
$$+ 3.585 \times 10^{-6}V^5 - 5.344 \times 10^{-8}V^6 + 5.100 \times 10^{-10}V^7 \qquad 0.0 \leq V < 42.92 \; mV$$

$$dT = -3114.0 + 300.5V - 9.948V^2 + 0.1703V^3 - 1.430 \times 10^{-3}V^4$$
$$+ 4.739 \times 10^{-6}V^5 \qquad 42.92 \leq V < 69.55 \; mV$$

FIGURE 11.18 Polynomial relationships between voltage and temperature for a Type J thermocouple with 0°C reference junction.

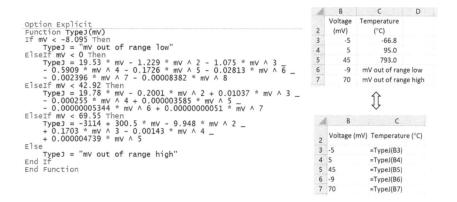

```
Option Explicit
Function TypeJ(mV)
If mV < -8.095 Then
    TypeJ = "mV out of range low"
ElseIf mV < 0 Then
    TypeJ = 19.53 * mV - 1.229 * mV ^ 2 - 1.075 * mV ^ 3 _
    - 0.5909 * mV ^ 4 - 0.1726 * mV ^ 5 - 0.02813 * mV ^ 6 _
    - 0.002396 * mV ^ 7 - 0.00008382 * mV ^ 8
ElseIf mV < 42.92 Then
    TypeJ = 19.78 * mV - 0.2001 * mV ^ 2 + 0.01037 * mV ^ 3 _
    - 0.000255 * mV ^ 4 + 0.000003585 * mV ^ 5 _
    - 0.00000005344 * mV ^ 6 + 0.00000000051 * mV ^ 7
ElseIf mV < 69.55 Then
    TypeJ = -3114 + 300.5 * mV - 9.948 * mV ^ 2 _
    + 0.1703 * mV ^ 3 - 0.00143 * mV ^ 4 _
    + 0.000004739 * mV ^ 5
Else
    TypeJ = "mV out of range high"
End If
End Function
```

	B	C	D
	Voltage	Temperature	
2	(mV)	(°C)	
3	-5	-66.8	
4	5	95.0	
5	45	793.0	
6	-9	mV out of range low	
7	70	mV out of range high	

	B	C
2	Voltage (mV)	Temperature (°C)
3	-5	=TypeJ(B3)
4	5	=TypeJ(B4)
5	45	=TypeJ(B5)
6	-9	=TypeJ(B6)
7	70	=TypeJ(B7)

FIGURE 11.19   Function for Type J thermocouple.

A VBA function, TypeJ, is shown in Figure 11.19 along with several examples of its application on the spreadsheet. The small coefficients from Figure 11.18 were entered in exponential format in the VBE, e.g., 2.396e-3, but they are reformatted automatically as shown in the figure. Since the type of the function is not declared, by default it is a Variant type. For this reason, the returns from the function can be either numerical or text, as shown.

The final selection structure in VBA is called *Select-Case*. It is similar to the multi-alternative If, but it doesn't require that the condition tests occur in any order. The general syntax is

```
Select Case test expression
    Case list1
        code block 1
    Case list2
        code block 2
            •
            •
            •
    Case Else
        else code block
End Select
```

A typical application for the Select-Case is that of a menu with a number of choices. In the example shown in Figure 11.20, a temperature in °C is provided by the user via an InputBox statement. Then, the user inputs a choice for another unit of temperature, and the Select-Case carries out the appropriate conversion to that unit. Finally, the result is displayed in a message box. Two examples of execution of the TempConvert Sub are shown in Figure 11.21.

```
Option Explicit
Sub TempConvert()
Dim Choice As String, TempIn As Double, TempOut
TempIn = InputBox("Enter temperature in degC: ")
Choice = InputBox("Enter your letter of choice" & vbCrLf & _
                  "A: degF" & vbCrLf & _
                  "B: K" & vbCrLf & _
                  "C: degR" & vbCrLf)
Choice = UCase(Choice)
Select Case Choice
    Case "A"
        TempOut = 1.8 * TempIn + 32
    Case "B"
        TempOut = TempIn + 273.15
    Case "C"
        TempOut = 1.8 * TempIn + 491.67
End Select
MsgBox TempOut
End Sub
```

FIGURE 11.20   Example of Select-Case for temperature conversion from °C.

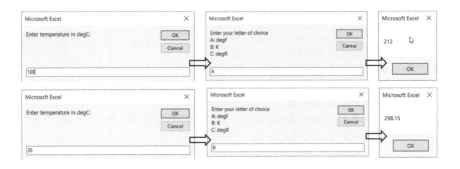

FIGURE 11.21   Two examples of the TempConvert Sub.

11.5.2 REPETITION STRUCTURES

Iteration or *looping* is an essential element in programming languages. VBA provides two structures to accomplish repetition: a general Do-Loop and a count-controlled For-Next. You have seen examples of these previously in the text, but now we will introduce them in detail. The distinction between the two, and what governs your choice of which to use, is whether you know the number of iterations that the loop will perform ahead of time or not.

11.5.2.1 The Do-Loop Structure

The form of the Do-Loop structure that we recommend has the following syntax:

```
Do
    pre-test code
    If condition Then Exit Do
    post-test code
Loop
```

```
Option Explicit
Sub TempConvert()
Dim Choice As String, TempIn As Double, TempOut
Do
    TempIn = InputBox("Enter temperature in degC: ")
    If TempIn > -273.15 Then Exit Do
    MsgBox "Temperature must be greater than -273.15 degC" _
                    & vbCrLf & "Try again"
Loop
Do
    Choice = InputBox("Enter your letter of choice" & vbCrLf & _
                        "A: degF" & vbCrLf & _
                        "B: K" & vbCrLf & _
                        "C: degR" & vbCrLf)
    Choice = UCase(Choice)
    If Choice = "A" Or Choice = "B" Or Choice = "C" Then Exit Do
    MsgBox "Choice letters must be A, B, or C. Try again."
Loop
Select Case Choice
    Case "A"
        TempOut = 1.8 * TempIn + 32
    Case "B"
        TempOut = TempIn + 273.15
    Case "C"
        TempOut = 1.8 * TempIn + 491.67
End Select
MsgBox TempOut
End Sub
```

FIGURE 11.22 `TempConvert` Sub with two input validation Do-Loop structures.

When the condition in the If statement tests True, the loop is exited, and execution continues below the Loop statement. The pre-test code or the post-test code may be missing, but there must be code that eventually causes the condition to test True; otherwise, there will never be an exit from the loop.[8] This is called an *infinite loop*.

Two common applications of the Do-Loop are input validation and convergence of numerical methods. As an example of *input validation*, we can modify the code from the TempConvert Sub of the previous section as shown in Figure 11.22. There are two Do-Loop structures. In each, an input value is obtained from the user and checked for validity. If the input value meets the test, the loop is exited. If not, a corrective message is generated using a message box, and the loop is repeated. The first checks that the temperature value entered is greater than absolute zero. The second checks that the user has entered one of the acceptable menu letter choices.

An example execution of the Sub in Figure 11.22 is illustrated in Figure 11.23. It includes making erroneous inputs for temperature and menu choice.

A second example of the Do-Loop structure involves *convergence*. We will consider transforming the spherical tank solution using the Newton-Raphson method from Chapter 7, Section 7.1.2, to a VBA Sub. In that example, we solved the equation for the tank volume for the liquid depth, h, given the tank radius, R, and the liquid volume, V,

$$V = \pi h^2 \left(\frac{3R - h}{3} \right) \quad \Rightarrow \quad f(h) = h^3 - 3Rh^2 + \frac{3V}{\pi} = 0 \quad (11.1)$$

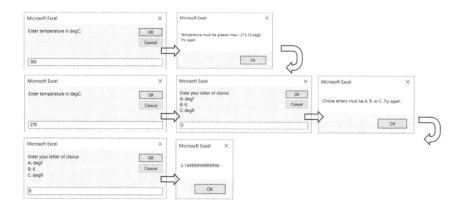

FIGURE 11.23   Execution of `TempConvert` Sub with input validation and input errors.

We also used the formula for the derivative

$$f'(h) = 3h^2 - 6Rh \qquad (11.2)$$

First, we will describe in words, in a pseudocode, the method.[9]

Set the error stopping criterion, e_s
Set or obtain values for R and V
Set the initial estimate for h at mid-tank level, R
Enter the loop
    For the current value of h, compute $f(x)$ and $f'(x)$
    Compute h_{new} using the Newton-Raphson formula,
        $h_{new} = h - f(h)/f'(h)$
    Compute the relative error, e, using
        $e = |(h_{new} - h)/h|$
    If $e < e_s$ exit the loop
    Otherwise, set h to the value of h_{new}
    Continue the loop
Loop exit here
Display the last h_{new} value as the solution

Next, we create the VBA code, as shown in Figure 11.24, which also shows the output of the execution of the Sub in a message box. An addition to the code that is not in the pseudocode is adding the variable `Pi` and assigning its value. Also, features have been added to the `MsgBox` command including text strings enclosed in quotation marks (") and the `FormatNumber` function to specify the number of decimal fraction digits to display.

A question that is not answered in this example is how many iterations were actually taken. We can explore that by single-stepping the code, and we could modify the code to add an iteration counter. By single-stepping, we observe rapid convergence in four iterations. There are other enhancements we can consider,

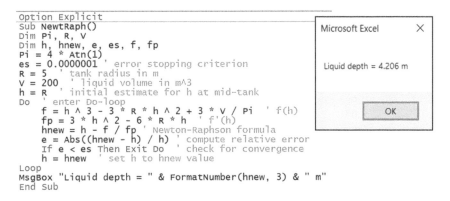

```
Option Explicit
Sub NewtRaph()
Dim Pi, R, V
Dim h, hnew, e, es, f, fp
Pi = 4 * Atn(1)
es = 0.0000001 ' error stopping criterion
R = 5   ' tank radius in m
V = 200  ' liquid volume in m^3
h = R   ' initial estimate for h at mid-tank
Do   ' enter Do-loop
    f = h ^ 3 - 3 * R * h ^ 2 + 3 * V / Pi   ' f(h)
    fp = 3 * h ^ 2 - 6 * R * h   ' f'(h)
    hnew = h - f / fp ' Newton-Raphson formula
    e = Abs((hnew - h) / h) ' compute relative error
    If e < es Then Exit Do  ' check for convergence
    h = hnew   ' set h to hnew value
Loop
MsgBox "Liquid depth = " & FormatNumber(hnew, 3) & " m"
End Sub
```

FIGURE 11.24 VBA Sub to implement Newton-Raphson method.

such as including a limit on the number of iterations and providing the method as a Function that can be invoked from the spreadsheet. These will be addressed later in this chapter.

11.5.2.2 The For-Next Structure

The *For-Next structure* is called a *count-controlled* loop because the number of repetitions of the loop is fixed and computed before the loop is entered. The general syntax is

```
For index = start To limit Step increment
        loop code
Next index
```

The *Step* part of the For statement is optional. If it is left out, the increment is one. If the limit is less than the start, the increment must be negative. The number of iterations is calculated prior to the For statement with the formula

$$n = \text{INT}\left(\frac{limit - start}{increment}\right) + 1 \qquad \text{INT indicates truncation to integer} \qquad (11.3)$$

and cannot be modified once in the loop. Examples are shown in Table 11.2. Note for the last entry in the table that the limit is not met; rather, the loop values are 1, 3, 6, and 9. Also, the third entry is interesting because one might expect the result to be 21. Actually, if you implement Equation 11.3 in VBA for these values, the result is 20. This is because of roundoff error in the division of the formula.

We often use an integer variable to represent the subscript of an array. Figure 11.25 shows a For-Next loop in a Sub that sets the initial values in an array to one.

An interesting idiosyncrasy is shown in the Locals window after the loop is exited via the Next statement. The index variable is one step beyond the final value that occurs in the loop. Note: This is not always just one beyond the limit. In the last example in the table, the exit value of the index would be 12.

TABLE 11.2

Example of Loop Counts

Start	Limit	Increment	i after Loop Exit
1	10	1	11
5	1	−1	0
0	1	0.05	20
0	10	3	12

```
Option Explicit
Option Base 1
Sub ForNextTest1()
Dim A(10), i As Integer
For i = 1 To 10
    A(i) = 1
Next
End Sub
```

```
Option Explicit
Option Base 1
Sub ForNextTest1()
Dim A(10), i As Integer
For i = 1 To 10
    A(i) = 1
Next
End Sub
```

Locals

Expression	Value	Type
Module1		Module1/Module1
A		Variant(1 to 10)
A(1)	1	Variant/Integer
A(2)	1	Variant/Integer
A(3)	1	Variant/Integer
A(4)	1	Variant/Integer
A(5)	1	Variant/Integer
A(6)	1	Variant/Integer
A(7)	1	Variant/Integer
A(8)	1	Variant/Integer
A(9)	1	Variant/Integer
A(10)	1	Variant/Integer
i	11	Integer

FIGURE 11.25 For-Next loop to set array values.

In general, we recommend against using the index variable and its value after the loop has terminated. There are notable exceptions, and we will illustrate one later in this chapter.

If you change the value of the limit or step (as variables) in the loop, it will have no effect on the number of times the loop is iterated because that is pre-calculated. There are also situations where we want to exit a For-Next loop before all iterations are complete. This is depicted in Figure 11.26. As the For-Next loop iterates, the sum variable takes on values of 1, 3, 6, and then 10 in iteration 4. At that point, the If statement executes the Exit For statement and execution continues beyond the Next statement. The message box confirms the values of i and sum when the exit takes place.

```
Option Explicit
Sub ForNextTest2()
Dim i As Integer, sum, limit As Integer
limit = 10
For i = 1 To limit
    sum = sum + i
    If sum > 6 Then Exit For
Next i
MsgBox i & " " & sum
End Sub
```

Microsoft Excel X

4 10

OK

FIGURE 11.26 Example For-Next loop with premature exit.

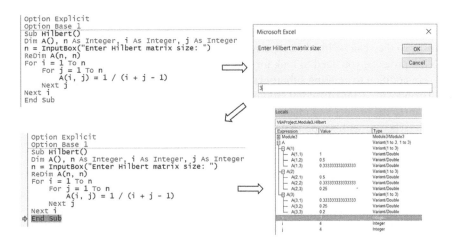

FIGURE 11.27   Nested For-Next loops to establish an $n\times$ Hilbert matrix.

For-Next loops commonly reference array subscripts and can be embedded one within another when referencing the subscripts of a two-dimensional array. An example of such *nesting* is based on establishing a Hilbert matrix, which has the following format:

$$\mathbf{H} = \begin{bmatrix} 1 & 1/2 & 1/3 & \cdot & 1/n \\ 1/2 & 1/3 & 1/4 & \cdot & 1/(n+1) \\ 1/3 & 1/4 & 1/5 & \cdot & 1/(n+2) \\ \cdot & \cdot & & \cdot & \cdot \\ 1/n & 1/(n+1) & 1/(n+2) & \cdot & 1/(n+n-1) \end{bmatrix} \qquad h_{ij} = \frac{1}{i+j-1} \qquad (11.4)$$

You can see that the individual elements of the matrix depend on the row (i) and column (j) indices. Figure 11.27 provides a VBA Sub that produces a Hilbert matrix, **H**, of dimension $n\times n$. The value of n is obtained from the user, and the H matrix is ReDim'd to that size. The outer For-Next loop iterates the i index, the row number, and the inner loop iterates j, the column number. The example execution inputs 3 as n, and ReDim's H to a 3×3 array. By single-stepping the code, the final version of the H array is shown in the Locals window.

11.6   VBA PROGRAMMING APPLICATIONS

The VBA examples you have seen so far have contained only a few statements. As you develop your own VBA applications, their length and complexity will increase, possibly to hundreds, even thousands of statements. Good programming practice is to break down the programming project into manageable sub-tasks. In this way, the sub-tasks can be tested independently before they are integrated into the overall project. Also, certain tasks may be repeated within the project, and, rather than repeating code segments, it makes sense to compartmentalize the tasks so they can be invoked repeatedly as Subs or Functions. In developing your VBA projects, you need to look for opportunities to *modularize* the code.

In addition to creating subprograms, Subs and Functions, it is advantageous to distribute these into different modules in the project according to themes that help organize the project and make it more manageable. One should also consider the data structure with scope in mind, considering which variables should be local, module level, and Public. We will see this more in Chapter 12 when we program event handlers associated with userform interfaces.

It is also worth mentioning that VBA code can, of course, be copied and pasted from one module to one in another project. Also, within the Project Explorer, a module can be dragged and dropped from one project to another as a copy.

To keep things manageable, we will use some small-scale examples to illustrate modular concepts. The primary goal is to illustrate how to invoke Functions and Subs from within other subprograms. A useful scenario for this is to consider a spreadsheet solution we have developed earlier in the text and translate (or elevate) that into VBA code, possibly to improve it and make its use on the spreadsheet more compact.

Example 11.1   Bisection Function

In Chapter 7, Example 7.1, we created the solution of a nonlinear equation, a model of liquid in a spherical tank, using the bisection method. Earlier in this chapter, we showed how the equation could be solved using the Newton-Raphson method within a VBA Sub. There were two disadvantages to our spreadsheet solution. First, it was dedicated to the spherical tank problem and only with some effort could be adapted to another equation. Second, it was bulky, taking up a block of 175 cells. There is an incentive to consider the spreadsheet solution a prototype and elevate this method into a VBA Function that could be used on the spreadsheet and, in fact, from multiple locations on the spreadsheet.

To start with, we will translate our spreadsheet solution from Chapter 7 into a pseudocode for a generic bisection solution:

Start with initial x_l and x_h values and the number of iterations, n
 If $f(x_l)$ and $f(x_h)$ have same sign
  Exit with error message
 Otherwise
  Start count-controlled loop from 1 to n
   Compute midpoint, $x_m = (x_l + x_h)/2$
   If $f(x_l)$ and $f(x_m)$ have same sign
    x_m becomes the new x_l
   Otherwise
    x_m becomes the new x_h
  Continue loop
  Return the last value of x_m as the solution

```
Option Explicit
Function Bisect(xl, xh, n)
Dim xm, i
If f(xl) * f(xh) > 0 Then
    Bisect = "initial estimates do not bracket solution"
Else
    For i = 1 To n
        xm = (xl + xh) / 2
        If f(xl) * f(xm) > 0 Then
            xl = xm
        Else
            xh = xm
        End If
    Next i
    Bisect = xm
End If
End Function
```

FIGURE 11.28   Function to implement the bisection method.

You note in the pseudocode that reference is made to a generic function, $f(x)$. This will have to be supplied separately. A VBA Function, based on the pseudocode, is shown in Figure 11.28. To check whether $f(x_l)$ and $f(x_h)$ have the same sign, we use their product, testing whether that is greater than zero. If this test is True, the function name is assigned an error message. Note: The function is a variant type by default, so Bisect can take on different type values. If the test is False, the For-Next loop is entered. It is a simple loop with the midpoint, x_m, computed. This is followed by testing whether $f(x_l)$ and $f(x_m)$ have the same sign. If so, x_l is assigned the value of x_m. If the test fails, it is assumed that $f(x_h)$ and $f(x_m)$ have the same sign, and x_h is assigned the value of x_m. The loop is then repeated. Finally, when the loop is exited, the last value of x_m is assigned to Bisect and the function is returned.

To test the Bisect function, we could do so from VBA, but it would be more common to invoke the function from the spreadsheet. Figure 11.29 illustrates the additional VBA code for the function f and the implementation on the spreadsheet.

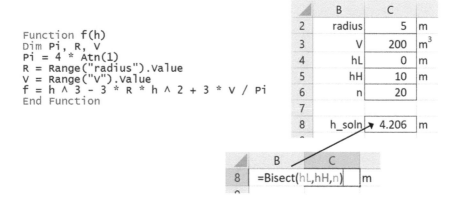

FIGURE 11.29   Function f for the spherical tank equation and implementation on the spreadsheet.

The function **f** here obtains values for the tank radius, R, and liquid volume, V, from the spreadsheet, and then computes the equation error, f, per Equation 7.2. You will note that we have included the number of iterations as the final argument to the `Bisect` function. If we would always be satisfied with 20, we could have set that internal to the function instead of allowing the user to specify it.

The generic function, f(x), is replaced by f(h) in function f to make it consistent with the problem being solved. Since the argument h, a formal parameter, refers back to where the function f was invoked, there is no problem with this change in variable name.

The result shown in Figure 11.29 in the h_soln cell on the spreadsheet agrees with the result from Example 7.1. Also, note that this is a live solution. If we change the V value, the h_soln value updates automatically. That also implies that we could use this scheme in a case study with the Data Table feature on the spreadsheet, as we did with the on-spreadsheet solution.

An alternative implementation would be to include the name of the f(x) function as an argument to the `Bisect` function. In other programming languages/software packages, e.g., MATLAB® and Python, this is straightforward. In Excel's VBA, it can be accomplished but is somewhat awkward. Figure 11.30 shows the modifications in the VBA code and on the spreadsheet.

The key changes include

- adding a `func` argument to `Function Bisect1` specified as String type
- within the `Bisect1` function, invoking the function named in `func` with the `Application.Run` method with arguments `func` and the input variable to `func`
- modifying the formula on the spreadsheet to specify, as a string, the function `fh` as a first argument

This formulation becomes useful in a workbook where we want to use the `Bisect1` function to solve more than one equation, where VBA functions are supplied for each equation.

```
Option Explicit
Function Bisect1(func As String, xl, xh, n)
Dim xm, i, fl, fh, fm
  fl = Application.Run(func, xl)
  fh = Application.Run(func, xh)
  If fl * fh > 0 Then
    Bisect1 = "initial estimates do not bracket solution"
  Else
    For i = 1 To n
      xm = (xl + xh) / 2
      fl = Application.Run(func, xl)
      fm = Application.Run(func, xm)
      If fl * fm > 0 Then
        xl = xm
      Else
        xh = xm
      End If
    Next i
    Bisect1 = xm
  End If
End Function
```

```
Function fh(h)
Dim Pi, R, V
  Pi = 4 * Atn(1)
  R = Range("radius").Value
  V = Range("v").Value
  fh = h ^ 3 - 3 * R * h ^ 2 + 3 * V / Pi
End Function
```

	B	C	
2	radius	5	m
3	V	200	m³
4	hL	0	m
5	hH	10	m
6	n	20	
7			
8	h_soln	4.206	m

	A	B
8	=Bisect1("fh",hL,hH,n)	

FIGURE 11.30 Modified bisection function with equation function name as an argument.

Example 11.2   Newton-Raphson Function

In implementing VBA code for the *Newton-Raphson* method here, we are going to illustrate the use of a Sub that is invoked or called from within the code. Although the method could be implemented in a similar fashion to the previous example, it is useful to see how Subs are used. An advantage we have here is that we already developed a VBA Sub in Section 11.5.2.1 for the method. We are going to transform that into a function that can be invoked on the spreadsheet. The function will have the syntax

```
Function NewtRaph(x0,es)
```

where
   x0:   the initial estimate for x
   es: the relative error threshold

Figure 11.31 shows the code for the NewtRaph function and an associated Getf_and_fp Sub. The function is made generic via a "call" to the Sub. The interesting feature of the Sub is that it takes one input argument, named h in the Sub, and provides two output arguments, named f and fp, for the function and its derivative, respectively. This illustrates how a Sub can be used with a mix of input and output arguments. Notice how the Sub is invoked with the Call command. The result displayed in the h_solv cell agrees with the previous solutions.

A further consideration of the use of the Do-Loop structure in convergence schemes is that there is no limit on the number of iterations taken. There are two ways to incorporate an iteration limit into, for example, the NewtRaph function. One is to increment a counter variable within the Do-Loop and

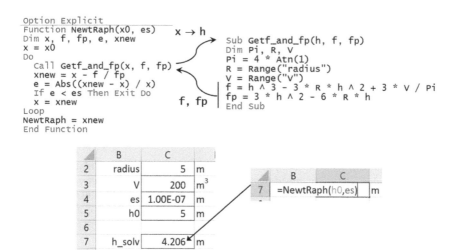

FIGURE 11.31   Function NewtRaph for solution of the spherical tank equation.

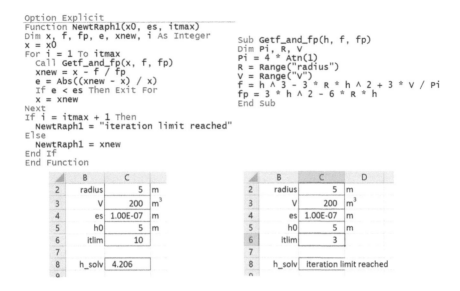

```
Option Explicit
Function NewtRaph1(x0, es, itmax)
Dim x, f, fp, e, xnew, i As Integer
x = x0
For i = 1 To itmax
   Call Getf_and_fp(x, f, fp)
   xnew = x - f / fp
   e = Abs((xnew - x) / x)
   If e < es Then Exit For
   x = xnew
Next
If i = itmax + 1 Then
   NewtRaph1 = "iteration limit reached"
Else
   NewtRaph1 = xnew
End If
End Function
```

```
Sub Getf_and_fp(h, f, fp)
Dim Pi, R, V
Pi = 4 * Atn(1)
R = Range("radius")
V = Range("v")
f = h ^ 3 - 3 * R * h ^ 2 + 3 * V / Pi
fp = 3 * h ^ 2 - 6 * R * h
End Sub
```

	B	C	
2	radius	5	m
3	V	200	m^3
4	es	1.00E-07	m
5	h0	5	m
6	itlim	10	
7			
8	h_solv	4.206	
9			

	B	C	D
2	radius	5	m
3	V	200	m^3
4	es	1.00E-07	m
5	h0	5	m
6	itlim	3	
7			
8	h_solv	iteration limit reached	
9			

FIGURE 11.32 Newton_Raphson function with iteration limit.

provide another test to exit the loop based on comparison of the counter to an iteration limit. Then, once out of the Do-Loop, the code must detect whether the iteration limit was reached or whether the error threshold was met.

Another approach is a bit simpler. It is to replace the Do-Loop with a For-Next loop counting to the iteration limit. If the error threshold is reached, there is an Exit For; otherwise, the loop exits at the limit. Again, the code should detect the nature of the exit from the For-Next loop. Figure 11.32 illustrates this latter strategy. We encourage you to try both. You will note in the figure that, with an iteration limit of 10, the method converges to the solution, but with a limit of 3, it reaches the limit first and returns an error message. We commented earlier in the text that the method requires four iterations to converge.

As a final example, we will emphasize the theme of elevating another spreadsheet solution as a prototype into a more flexible VBA function.

Example 11.3   A Function for Table Interpolation

In Chapter 4, Section 4.4.5, we developed a spreadsheet calculation that interpolated values of the density of water at temperatures between 0°C and 100°C from a table of values. The solution made use of Excel's lookup functions, INDEX and MATCH. This represents a good example of a spreadsheet prototype that can be elevated into a VBA Function and made more versatile.

The general syntax of our table interpolation function would be

```
Interp(xin,datatable)
```

and thus would be applicable to any table of properties. A pseudocode for this function is

```
Determine the length of the table, n
If xin is less than the first value in datatable column 1
    return an error message
Otherwise, if xin is greater than the last value in datatable column 1
    return an error message
Otherwise, if xin is equal to the last value in datatable column 1
    Interp = datatbase(n,2)
Otherwise
    proceed with the interpolation
    For i from 1 to n
        Is datatable(i,1) > xin
            Find the first datatable value > xin
            exit the loop
    Set lowrow = i - 1
    Set highrow = i
    xlo = datatable(lowrow,1)
    xhi = datatable(highrow,1)
    ylo = datatable(lowrow,2)
    yhi = datatable(highrow,2)
    Interp = (xin - xlow)/(xhi - xlo) * (yhi - ylo) + ylo
Return
```

This is translated into VBA code shown in Figure 11.33. The number of rows in the argument `datatable` is determined with the `Rows.Count` property.

```
Option Explicit
Option Base 1
Function Interp(xin, datatable)
Dim n As Integer, i As Integer
Dim lowrow As Integer, highrow As Integer
Dim xlo, xhi, ylo, yhi
n = datatable.Rows.Count
If xin < datatable(1, 1) Then
    Interp = "input value less than table range"
ElseIf xin > datatable(n, 1) Then
    Interp = "input value greater than table range"
ElseIf xin = datatable(n, 1) Then
    Interp = datatable(n, 2)
Else
    For i = 1 To n
        If datatable(i, 1) > xin Then Exit For
    Next i
    lowrow = i - 1
    highrow = i
    xlo = datatable(lowrow, 1)
    xhi = datatable(highrow, 1)
    ylo = datatable(lowrow, 2)
    yhi = datatable(highrow, 2)
    Interp = (xin - xlo) / (xhi - xlo) * (yhi - ylo) + ylo
End If
End Function
```

FIGURE 11.33 Function `Interp` for linear interpolation of a generic date table.

The datatable array is referenced with (row,column) subscripts where column 1 is the x variable and column 2 is the y variable. The If structure first tests for a low xin value less than the first entry in the table, then for a high xin value greater than the last value in the table. In either case, an error message is returned. The third test in the If structure is if the xin value is identical with the last value in the table. Then the y value in the last row is returned.

If none of these tests True, a For-Next loop searches down the first column of the table for the first value that is greater than the xin value. At that point, the loop is exited, and the i value is used to formulate the two rows for interpolation. The adjacent x and y values are assigned to variables, and those variables are used to implement the linear interpolation formula, identical to the spreadsheet solution.

The use of the Interp function on the spreadsheet is shown in Figure 11.34. An example interpolation for 33°C is shown with a result of 994.71 kg/m³. Also, low and high values of temperature outside the range of the table are shown with their respective error messages.

The resulting Interp function is much better than the original prototype on the spreadsheet. However, the latter established the scheme and

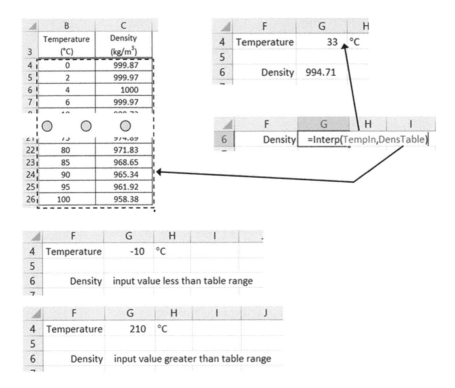

FIGURE 11.34 Implementation of the Interp function on the spreadsheet including error tests.

set the stage for the development of the VBA function, which has several advantages:

- It can be invoked several times on the spreadsheet with reference to the same table.
- It can be used easily with different tables.
- It is a more compact solution, not requiring as many cells on the spreadsheet.

This is a typical illustration of prototyping a calculation on the spreadsheet and then elevating it into a better VBA-based implementation. This opportunity occurs frequently in engineering and scientific problem solving.

We have illustrated several examples of VBA code development in this section. Developing similar efforts may not come easily to you at the start, but repetition will make the process easier. We can relate this to playing a sport or learning a foreign language, where initial participation is slow and frustrating. Persistence will overcome this, and eventually you may wonder why you had so much difficulty at the beginning. The problems at the end of this chapter will provide you some opportunities to develop your VBA programming skills.

In this text's concluding chapter, we will move on to building user interfaces that involve graphic design and event handlers. Much of what we have learned so far about VBA will be put to work in developing these projects.

PROBLEMS

11.1 Write a VBA Sub that declares a variable as type Double and then assigns either a large or a small numerical value to the variable. Using your Sub investigate by trial-and-error the limits of the Double type. Compare these to similar experiments in cells on the spreadsheet.

11.2 The Long integer type in VBA is based on 4-byte storage of the number. Write a VBA Sub that declares a variable as type Long and then assigns a large value, either positive or negative, to the variable. Then use a MsgBox command to display the variable's value. Use your Sub to investigate by trial-and-error the limits of the Long type. Show how this relates to the integer range of a 4-byte number.

11.3 A *Vandermonde matrix* in the variable x array has the general form

$$\begin{bmatrix} x_1^{n-1} & x_1^{n-2} & \cdot & x_1 & 1 \\ x_2^{n-1} & x_2^{n-2} & \cdot & x_2 & 1 \\ \cdot & \cdot & \cdot & \cdot & \cdot \\ x_n^{n-1} & x_n^{n-2} & \cdot & x & 1 \end{bmatrix}$$

Spreadsheet Problem Solving and Programming

Write a VBA Sub that declares a matrix V as an empty array and a vector x as an empty array. Obtain a value for an integer variable n from the user in the range of 2 to 5. Expand the declared arrays to V(n,n) and x(n). Obtain input from the user for the elements of the x array. Add code to set the values of the V matrix based on the x array values.

11.4 Examples of circular matrices are

$$\begin{bmatrix} 1 & 2 \\ 3 & 4 \end{bmatrix} \text{ and } \begin{bmatrix} 1 & 2 & 3 \\ 4 & 5 & 6 \\ 7 & 8 & 9 \end{bmatrix}$$

Write a VBA Sub that declares an empty array C and obtains an integer value n from the user in the range 2 to 5. Expand the C array to a size $n \times n$. Follow this by code that assigns values to the elements of C to establish it as a circular array.

11.5 Modify the downward Rotate Sub from Figure 11.9 to an upward Rotate Sub. Demonstrate this with a similar array on the spreadsheet to that illustrated in the figure.

11.6 An ascending bubble sort algorithm executes the following operations on a one-dimensional array of length n located on the spreadsheet:

Sets up a loop for a variable i from 1 to n-1
    Sets up a loop for a variable j from i+1 to n
        Finds the minimum value of the array at location j
    Compares the jth value with the ith value
    If the jth value is less than the ith value, swap the two values
    Otherwise don't swap

Demonstrate the before and after states of the array to confirm the sort.

Suggestion: You may want to establish the array twice on the spreadsheet so that, after an attempted sort, you can reestablish the original form of the array by copying.

11.7 Record a macro that changes the number format in a cell to Scientific with three significant figures. Based on the recorded macro, find the property modified using the Object Browser and document its location.

11.8 Write VBA Subs that modify a range of cells using the Clear, ClearContents, and Delete methods and contrast by evidence their differences.

11.9 Write a logical expression involving three variables, a, b, and c, of type Double that checks whether $a < b < c$ and returns True if this is the case and False otherwise.

11.10 *DeMorgan's Laws* can be summarized

$$not(A \text{ or } B) = not(A) \text{ and } not(B)$$
$$not(A \text{ and } B) = not(A) \text{ or } not(B)$$

for A and B as logical (Boolean) variables. Write VBA Subs that confirm these two laws.

11.11 The solutions of a quadratic polynomial,

$$ax^2 + bx + c = 0$$

are well known and given by the quadratic formula

$$x = \frac{-b \pm \sqrt{b^2 - 4ac}}{2a}$$

Different specific solutions arise depending on the coefficients a, b, and c and the resulting value of the discriminant, $b^2 - 4ac$. If the discriminant is negative, the solutions are complex numbers,

$$x = \frac{-b \pm i\sqrt{4ac - b^2}}{2a} \qquad \text{where} \qquad i = \sqrt{-1}$$

There are also special cases:
- If $a = 0$, this is not a quadratic equation, but has one solution $x = -c/b$.
- If $a = b = 0$, there is no equation and no valid solution.

Write a VBA Sub that obtains values of a, b, and c from the user, and presents the appropriate solutions in a message box.

11.12 Pressure is described in many different units. Six common ones are:

- pascals (Pa, also N/m²)
- kilopascals (kPa), 1000 Pa
- pounds per square inch (psi), 6894.76 Pa
- atmospheres (atm), 101325 Pa
- bars, 100000 Pa
- millimeters of mercury (mm Hg also torr), 133.322 Pa

Write a VBA Sub that performs the following:

- Obtains from the user their choice of units for the input pressure.
- Obtains from the user the value of pressure.
- Obtains from the user their choice of units for output pressure.
- Converts the input pressure to the output pressure.
- Displays in a message box the input pressure, input units, output pressure, and output units.

11.13 You have seen the bisection method implemented on the spreadsheet and in VBA to find the root of a nonlinear equation. A similar problem is to find an extremum, maximum, or minimum of a function, $f(x)$, over a domain $a \le x \le b$. One practical way to attempt this is a modification of the bisection method that goes as follows for finding a minimum:

compute $x_m = (a + b)/2$ and $f(x_m)$
compute $x_d = x_m * (1+delta)$ where *delta* is a small number, e.g., 1×10^{-7}
compute $f(x_d)$
if $f(x_d) < f(x_m)$ the minimum is $> x_m$, so replace set $a = x_m$
otherwise the minimum is $< x_m$, so set $b = x_m$
repeat this calculation until the relative absolute error

$$\left| \frac{x_{mnew} - x_{mold}}{x_{mold}} \right| < e_s$$

If you need to find a maximum instead of a minimum, you can use the same method but with $-f(x)$. Also, you could modify the logic of the code, but that is not necessary.

Develop a Function BiMin to implement this method with arguments

- the name of the function for $f(x)$ as a string
- the initial values of a and b
- the error threshold e_s

Test your function with the *humps function* from Example 5.1.

$$f(x) = \frac{1}{(x-0.3)^2 + 0.01} + \frac{1}{(x-0.9)^2 + 0..04} - 6$$

a. Find a minimum for $a = 0.4$ and $b = 0.8$.
b. Find a maximum for $a = 0.7$ and $b = 1.1$.
c. Modify your function to include an iteration limit and test it with part (a) to determine when the iteration limit is reached.

NOTES

1. There are also complex numbers that correspond to storing two real numbers, one for the real part and another for the imaginary part. Unfortunately, VBA doesn't have a data type or operators for complex numbers.
2. It is possible to use higher-dimensional arrays in VBA. We just don't see them that often.
3. A *sentinel value* (also referred to as a flag or signal value) is a special value used as a termination condition, typically in a loop. The value of the sentinel should be chosen so that it is guaranteed to be distinct from all legal data values. Hence, the value of -9999 used in Figure 11.11. Another example would be any negative integer for indicating the end of a sequence of non-negative integers.

4. "read/write" means that the property can be modified. "read only" means that the property can only be acquired.

5. The Sgn function returns +1 for a positive argument, −1 for a negative argument, and 0 for a zero argument.

6. U.S. National Institute for Standards and Technology

7. Iron-constantan is the most common thermocouple junction. Constantan is an alloy of 55% copper and 45% nickel.

8. We note here that there are additional versions of the Do-Loop that involve the keywords While and Until. We do not use these here because their function can be accomplished with the one Do-Loop syntax shown. We have observed that the additional versions cause more confusion than bring value.

9. Traditionally, these structures have been described using flowcharts. We chose to use pseudocode here.

12 User Interfaces

CHAPTER OBJECTIVES

- Learn how to place buttons that run macros on the Excel worksheet
- Develop more knowledge on the use of message boxes
- Expand your ability to use input boxes
- Understand what event handlers are and how they relate to userforms
- Learn how to design and implement userforms for an Excel/VBA application
- See how Excel can be linked to other software applications via VBA

There is frequent interest, especially in the professional setting, to create user interfaces for Excel spreadsheets and VBA code that encapsulate an application. This is especially pertinent when the applications are being prepared for other users within the organization that may not have significant skill and knowledge of Excel. It can also lead to the development of applications that are marketed. The latter often requires the design of new elements of the Ribbon, including tabs, groups, and commands. This is a detailed software development activity that we do not explore in this book.

In Chapter 9, we created useful macros and linked them to keystroke sequences involving the Ctrl key. These were primarily to streamline formatting of the spreadsheet. Another use for macros is to initiate VBA code that involves calculations and the spreadsheet. When preparing spreadsheets for others to use, it is convenient to place a labeled button on the spreadsheet that runs the macro. We will illustrate that here.

You have seen elementary use of message boxes and input boxes in previous chapters. In this chapter, we will add some detail that allows you to customize these boxes and use them more creatively in an application. The topic of userforms is a big one when it comes to user interfaces, and we will spend considerable time and effort to introduce you to these powerful tools.[1] Since userforms contain objects, such as buttons, which allow for user action, the VBA code that responds to these actions are called event handlers. We will introduce them.

The ability to link Excel to other software packages is varied. This often requires installing an add-in to Excel that modifies the Ribbon. It is also possible to communicate with certain packages via VBA. We will illustrate this with the MATLAB® package and Aspentech's Hysys simulator.

 DOI: 10.1201/9781003361053-12

12.1 ON-SHEET BUTTONS

Placing a button on the spreadsheet is a convenient way to initiate an application that involves user interfaces. In general, it allows a click of the button to run a macro. The following example illustrates how this is done.

First, we will create a "startup" macro that simply displays a message box with the word "startup" shown as depicted in Figure 12.1. You will note that it is a macro attached to the workbook project.

Next, we need to place a button on the spreadsheet. This is available from the Developer tab, the Controls group, and the Insert dropdown command. Figure 12.2 illustrates the procedure. After clicking on the button icon in the upper left of the Form Controls window, one uses the mouse to draw a button on the spreadsheet of reasonable size and shape. As soon as that drawing is complete, the button outline disappears, and the Assign Macro window appears. There, we select the Startup macro, as the figure shows, and click OK.

After assigning the macro to the button, it will appear on the spreadsheet as shown in Figure 12.3. If you click the button, the macro will run, as appears in the figure.

There is limited ability to format the button. Figure 12.4 illustrates how we can change the button caption to a relevant title. This is accomplished by right-clicking on the button to invoke the context-sensitive menu and selecting Edit Text.

The Format Control option provides ways to change the font of the button caption. Figure 12.5 illustrates an example where we increase the font size, change

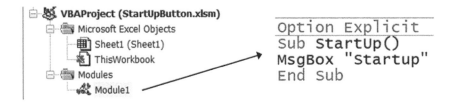

FIGURE 12.1 Simple startup macro.

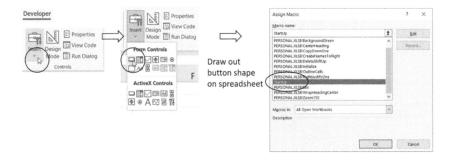

FIGURE 12.2 Procedure to install a button on the spreadsheet.

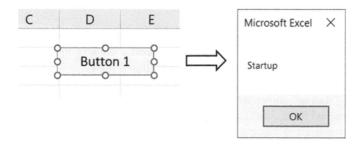

FIGURE 12.3   Testing the on-sheet button.

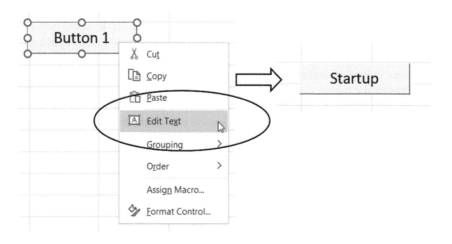

FIGURE 12.4   Changing the button text.

FIGURE 12.5   Changing the style of the button caption.

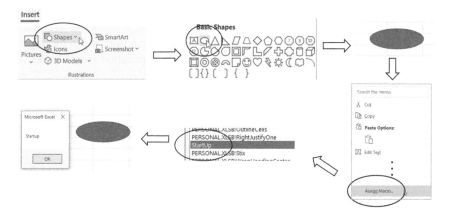

FIGURE 12.6   Adding a shape to the spreadsheet and assigning it to a macro.

the style to bold/italic, and change its color to blue. Notably, we cannot change the background color of the button.

Another option we have is to insert a shape object on the spreadsheet instead of a button control. This is accomplished in a similar fashion to installing a button and is shown in Figure 12.6. After selecting and placing the shape, an oval is used here, right-clicking the shape allows for the Assign Macro command to be selected. This links the oval to the `Startup` macro.

There is more control over the appearance of the shape. First, there are many shapes from which to choose. Then, text can be added, the background color and outline can be changed, and the color and style of the font modified. A possible result is shown in Figure 12.7. Here, a gradient fill is selected, text is added, and then formatted in style, size, and color.

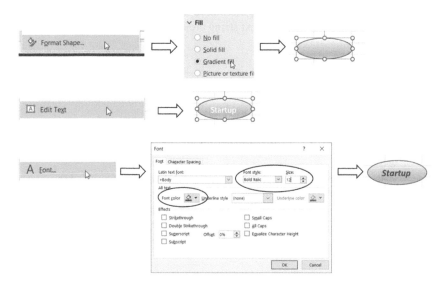

FIGURE 12.7   Customizing the appearance of the shape.

An issue to consider with the use of on-sheet buttons is that, if the view of the worksheet is changed to obscure the location of a button, it cannot be seen and then used to run a macro. For this reason, buttons are often installed in the upper left of the initial worksheet and used to start up an application based on VBA code and other user interface features, such as userforms.

We suggest moderation in using on-sheet buttons. It is easy to clutter the worksheet with them and lose their aesthetic appeal. Also, to keep things functional, we prefer to use the standard button form control, as illustrated in Figures 12.2–12.4.

12.2   MESSAGE BOXES

Message boxes provide a convenient device for a simple display of information in a dialog box format on the spreadsheet and obtain a response from the user, such as OK, Cancel, Yes, No, and Retry. Up to this point, we have illustrated the MsgBox command in VBA to provide a simple display of variable values or perhaps an error message. The MsgBox statement takes on two forms in VBA:

MsgBox *text string, additional arguments*   as a command

                                                                       or

variable = MsgBox(*text string, additional arguments*) as a function

In the latter case, the function is assigned to a variable using the = sign.

The *text string* can be composed of several elements joined with the *concatenation*[2] operator, the ampersand (&). The MsgBox function returns a result code corresponding to the button clicked by the user. Here is the specific syntax for the MsgBox function:

MsgBox(*text string, button codes, title, helpfile, context*)

where *text string* is what is displayed in the message box.

The four arguments following text string are optional. Table 12.1 lists the various button codes.

TABLE 12.1
Button Codes and Numerical Equivalents

Code	Number	Code	Number
vbOKOnly	0	vbExclamation	48
vbOKCancel	1	vbInformation	64
vbAbortRetryIgnore	2	vbDefaultButton1	128
vbYesNoCancel	3	vbDefaultButton2	256
vbYesNo	4	vbDefaultButton3	512
vbRetryCancel	5	vbDefaultButton4	768
vbCritical	16	vbSystemModal	4096
vbQuestion	32		

TABLE 12.2
Button Result Codes and
Numerical Equivalents

Code	Number
vbOK	1
vbCancel	2
vbAbort	3
vbRetry	4
vbIgnore	5
vbYes	6
vbNo	7

Certain button codes can be combined by addition, and the combination is unambiguous. For example,

vbYesNo + vbExclamation = 4 + 48 = 52

To check how this is possible, we use binary numbers for eight-bit bytes:

$00000100_2 + 00110000_2 = 00110100_2$

Since the bits superimpose without *interference*,[3] there is no ambiguity in the meaning.

The MsgBox function returns a code result that corresponds to the button clicked by the user. Table 12.2 defines the response code and related numerical value.

Figure 12.8 shows a VBA macro that illustrates how button codes and result codes can be used. A simple InputBox function is used to obtain a button code from the user, and then that code is used in a MsgBox function to display the box with the corresponding buttons. The MsgBox function returns a code that is

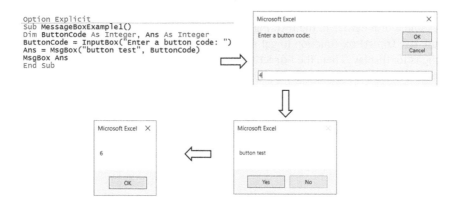

FIGURE 12.8   VBA macro to illustrate button codes and result codes.

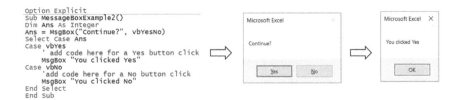

```
Option Explicit
Sub MessageBoxExample2()
Dim Ans As Integer
Ans = MsgBox("Continue?", vbYesNo)
Select Case Ans
Case vbYes
    ' add code here for a Yes button click
    MsgBox "You clicked Yes"
Case vbNo
    'add code here for a No button click
    MsgBox "You clicked No"
End Select
End Sub
```

FIGURE 12.9   Use of MsgBox buttons to direct a Select Case structure.

assigned to the Ans variable. The Ans value is then displayed in a second message box. The example illustrates the use of the Yes and No buttons and shows the result value when the user clicks the Yes button.

A more practical structure for use of MsgBox buttons is shown in Figure 12.9. Here, the selection of the Yes or No button is used to direct the code in a Select Case structure. You will notice that it is not necessary to use numerical codes for buttons and button results. In fact, the use of code names like vbYes makes the VBA Sub easier to understand.

There are three formatting features of message boxes that we find useful. The first is to change the box title to something more relevant than Microsoft Excel. This can be done by adding the title argument after the button code argument. If only the default button code, OK, is used, that specification can be left blank in the MsgBox command. This is shown in Figure 12.10. Notice the pair of adjacent commas in the MsgBox command. This corresponds to the blank for the button code, which causes the default OK button to be displayed.

We have seen before the use of the vbCrLf keyword to cause a new line for MsgBox text. Figure 12.11 illustrates this using the & symbol to concatenate the elements of the text. Note here that the continuation of the MsgBox command with the underscore character (_) to a second line has nothing to do with the two-line format in the message box. This is created by interposing the vbCrLf code between the two text strings.

Another useful feature of the MsgBox is controlling the display of real numbers. This is accomplished with the *FormatNumber function*. We have seen an example of it earlier in the text. Figure 12.12 illustrates it again. In this case, we use the InputBox function to get a number and specify the number of decimal digits for display. Then, the FormatNumber function is used in the MsgBox command. Note by this example that the display is rounded from the actual number.

```
Option Explicit
Sub MessageBoxWithTitle()
MsgBox "see custom title above", , "Custom Title"
End Sub
```

FIGURE 12.10   Establishing a custom title for the message box.

FIGURE 12.11   Use of the vbCrLf code to create new lines in MsgBox test.

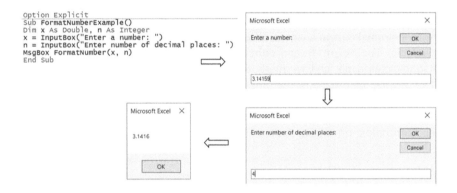

FIGURE 12.12   Use of the FormatNumber function to control the display of decimal digits.

For many user interfaces developed with VBA we choose userforms over message boxes; however, message boxes can be used to advantage in various situations. One of these is corrective messages for erroneous user input as will be illustrated later in this chapter when we develop userforms.

12.3   INPUT BOXES

As we have already seen, input boxes are a very convenient interface element for user input. They can be programmed quickly without much design or complexity. The simplest implementation is the InputBox function, and there is a more sophisticated alternative, the InputBox method.

Input boxes are designed to allow a single entry. The result returned by the input box is a text string. If that string represents a number to be used later in a calculation, it will have to be converted into a numerical type. That typically occurs via assignment of the input box result to a variable of a specific type, like Double, or a Variant variable that can adapt to subsequent calculations.

We use the *InputBox function* commonly. Its simplified syntax is

```
InputBox(prompt,title,default value)
```

We have not seen the latter two arguments before. They are optional. The *title* is, as was the case for the message box, used to customize the window title to something relevant rather than the standard: "Microsoft Excel." The *default value* argument can be useful. It is the value that appears in the InputBox field when the box appears. In case the user just clicks OK or presses Enter, this is the value that will be returned.

The syntax for the complete InputBox function is

InputBox(*prompt, title, default, xpos, ypos, helpfile, context*)

Again, all but the first argument is optional. The *xpos, ypos* arguments allow the positioning of the InputBox on the monitor screen. These arguments are dependent on the resolution of the computer monitor, and therefore it is not possible to make them adapt to different screen sizes without more information available. The *helpfile* and *context* are similar to those arguments in the MsgBox command, and we rarely use them. We note that there is no control over the size and shape of both message and input boxes. Simplicity brings its limitations.

Figure 12.13 shows an example Sub with InputBox and MsgBox commands. The InputBox, titled Item Count, appears in the upper left of the display screen

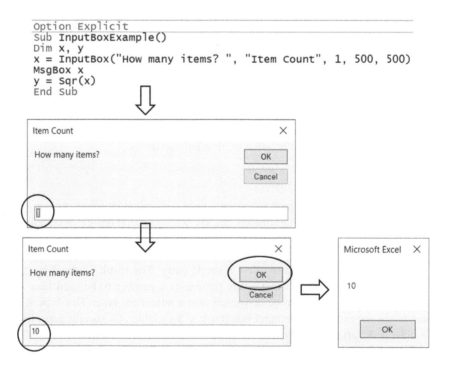

FIGURE 12.13   InputBox example with title, default value, and positioning.

TABLE 12.3

Type Codes for `InputBox` **Method**

Type Code	Description
0	formula
1	number
2	string
4	logical
8	cell reference
16	error value
64	array of values

(in this case, for 1920 × 1080 screen resolution) and appears with the default value of 1 displayed. Here, the user changes the default value to 10 and clicks OK. The message box simply displays the value entered in the InputBox. Although the x value has adapted type Variant/String, if it were to be used in a numerical calculation, the Variant type would adapt it to the required numeric type.

The *InputBox method* is different and represents an alternative to the corresponding function. The syntax is

```
objectname.InputBox(prompt,title,default,xpos,ypos,helpfile,
context,type)
```

The additional argument is *type*. The type of input required is specified, and Table 12.3 shows the type codes and their description.

The codes in the table can be combined to allow for inputs of multiple types, e.g., code 3 allows a number or a string.

Figure 12.14 shows a Sub using the `InputBox` method. This example is a little more involved than previous ones. First, the result of the `InputBox` method is defined as a `Range` type. Then, the `DefaultRange` string variable is set equal to the `Address` property of the `Selection` on the spreadsheet. An addition to this Sub is the `On Error` event handler that jumps to the empty `Canceled:` statement if any error occurs and exits the Sub.

The main statement of the Sub is the `Set UserRange` with the `InputBox` method. You note that, since the `Range` type is an object, the `Set` keyword must be used in an assignment. An additional feature is the use of keywords for the arguments with the := assignment operator. Using this style, arguments may be entered in any order or left out. The `type` argument is 8, specifying a cell reference, per Table 12.3.

Once the `UserRange` variable is set, it is used with the `ClearContents` and `Select` methods to take action on the spreadsheet. The figure shows the results of executing the Sub on a sub-range of a block of random numbers on the spreadsheet.

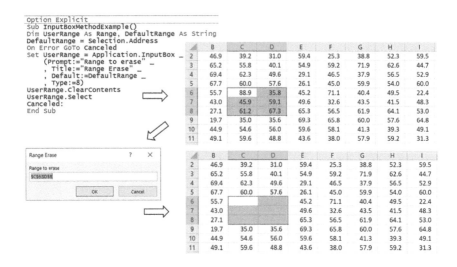

FIGURE 12.14 Example of the `InputBox` method with cell reference input.

The implementation of the `InputBox` method, as shown in Figure 12.14, is perhaps more involved than we would usually see. And, in fact, our use of the `InputBox` is more common as the function and not the method. Finally, as we move to applications, we will provide for user input with userforms rather than the `InputBox`.

12.4 EVENT HANDLERS

VBA code can be written that executes in response to an event that occurs. For example, as we have just seen, if a button is place on the spreadsheet, a macro can be executed in response to the user clicking the button. The button click is an event.

As we will see in Section 12.5, userforms present many scenarios where events occur, and code executes as a result. Events can be classified by the level or environment in which they occur.

- workbook, such as `Open`
- worksheet, such as `Calculate`
- chart, such as `SeriesChange`
- application, such as `NewWorkbook`
- userform
- other, such as `OnKey` and `On Error`

Event handler code is most often located with the object with which it is associated. For example, in the VBE Project Explorer, for a project, there is a

Microsoft Excel Objects branch that will contain objects such as `ThisWorkbook` and `Sheet1`. That is where event handler code will be located. An exception we have already seen is the on-sheet button where the response code is a Sub located in a module.

It is possible to disable all events, except those that are in response to actions on userforms, by issuing the command

```
Application.EnableEvents = False
```

and then reverse that with

```
Application.EnableEvents = True
```

Disabling events will apply to all open workbooks.

Event handlers are Subs with specified, corresponding names; for example, `Workbook_Open`. The VBE provides the initial framework for the event handler, and the programmer then adds the code. Figure 12.15 illustrates the initial step in creating this Sub. By double-clicking the ThisWorkbook icon under the Microsoft Excel Objects branch of the VBAProject, the code window opens with two dropdown categories at the top, (General) and (Declarations). The dropdown menu under (General) shows the `Workbook` object, and selecting that automatically generates the Sub, `Private Sub Workbook_Open()` where code can be added. For example, this code could be VBA statements that launch an initial userform instead of using an on-sheet button to do this. You will note the keyword `Private` is generated automatically. This indicates that this Sub can only be activated within the context of the `Workbook` object. *Private Subs* are commonly used as event handlers.

The (Declarations) dropdown list to the right of the General category contains the `Open` option, but this is generated by default after we select `Workbook` from the General category dropdown, as depicted in the lower bottom of Figure 12.15.

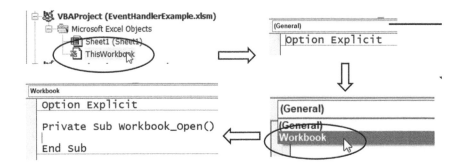

FIGURE 12.15 Creating the framework for a `Workbook_Open` event handler.

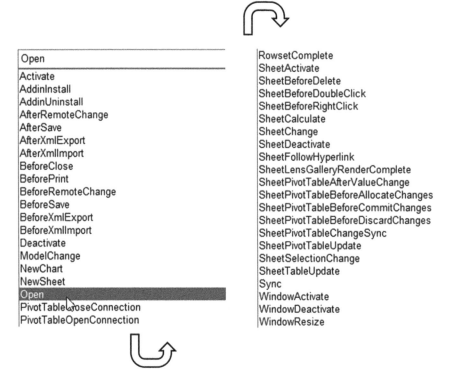

Open	RowsetComplete
Activate	SheetActivate
AddinInstall	SheetBeforeDelete
AddinUninstall	SheetBeforeDoubleClick
AfterRemoteChange	SheetBeforeRightClick
AfterSave	SheetCalculate
AfterXmlExport	SheetChange
AfterXmlImport	SheetDeactivate
BeforeClose	SheetFollowHyperlink
BeforePrint	SheetLensGalleryRenderComplete
BeforeRemoteChange	SheetPivotTableAfterValueChange
BeforeSave	SheetPivotTableBeforeAllocateChanges
BeforeXmlExport	SheetPivotTableBeforeCommitChanges
BeforeXmlImport	SheetPivotTableBeforeDiscardChanges
Deactivate	SheetPivotTableChangeSync
ModelChange	SheetPivotTableUpdate
NewChart	SheetSelectionChange
NewSheet	SheetTableUpdate
Open	Sync
PivotTableCloseConnection	WindowActivate
PivotTableOpenConnection	WindowDeactivate
	WindowResize

FIGURE 12.16   List of events associated with the Workbook object.

In case we do not want a Workbook_Open Sub, it can simply be deleted. The entire (Declarations) list is shown in Figure 12.16. It is lengthy, and we certainly will not illustrate all or many of the items given the scope of this chapter.

Another example of a Workbook event handler that can be selected from the list from Figure 12.16 is illustrated in Figure 12.17. This one, SheetActivate, requires an argument. The argument name by default is Sh, which could be changed. A ByVal keyword is used to indicate that the Sh value produced by the event will be supplied as a literal value and not a reference to a variable elsewhere in the Project. The Sh value is also declared to be an object type, as it represents a sheet in the workbook.

The MsgBox command is not particularly useful but illustrates the nature of the event handler. As Sheet2 is added in the workbook and activated, the first message box shown in the figure appears. The command shows the use of the TypeName function, which generates the "Worksheet" text on the message box. Then, on the next line, the Name property of the Sh object, "Sheet2" in this case, is displayed. When we click OK, the message box disappears, and when we then click the Sheet1 tab, the second message box in the figure appears.

```
Private Sub Workbook_SheetActivate(ByVal Sh As Object)

End Sub
```

```
Private Sub Workbook_SheetActivate(ByVal Sh As Object)
MsgBox TypeName(Sh) & vbCrLf & Sh.Name
End Sub
```

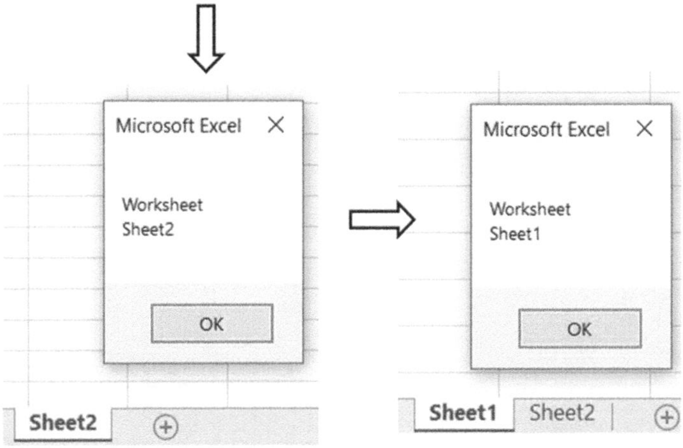

FIGURE 12.17   Example `Workbook` event handler: `SheetActivate`.

Another event handler example has a two-way argument instead of the `ByVal` type. This is shown in Figure 12.18. In this case, the `Workbook_BeforePrint` handler will execute automatically whenever, in the spreadsheet environment, we attempt a print action. The Cancel argument is True/False type, and its value can be transmitted by the Sub. This is called a *by-reference*, or `ByRef`, argument and is the default when `ByVal` is not specified. If the user clicks the No button on the message box, the print action is canceled; otherwise, it proceeds.

To complete this section, we will now review several event handlers that are most frequently encountered in application development. The first is the `Workbook_Open` handler that we showed above. This is commonly used to start up a user interface for spreadsheet applications that are developed for others to use. Other purposes include

- displaying a welcome message using the `MsgBox` command
- opening other supplementary workbooks
- activating a particular worksheet
- checking that initial conditions are satisfied

```
Private Sub Workbook_BeforePrint(ByRef Cancel As Boolean)
Dim Ans As String
Ans = MsgBox("Are you ready to print? ", vbYesNo)
If Ans = vbNo Then Cancel = True
End Sub
```

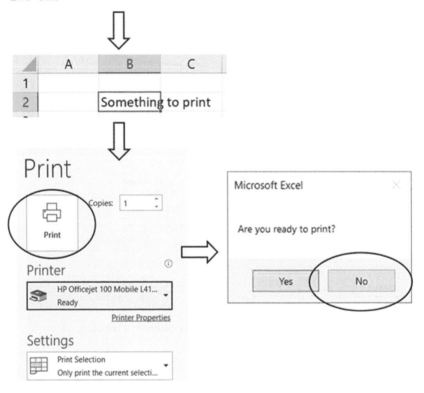

FIGURE 12.18 `Workbook_BeforePrint` event handler with a two-way Cancel argument.

Note: If you hold down the Shift key while a workbook is opened, the `Workbook_Open` Sub will not execute. Two simple examples of this event handler are

```
Private Sub Workbook_Open()
Worksheets("Sheet2").Activate
End Sub
```

and

```
Private Sub Workbook_Open()
MyUserForm.Show
```

The first ensures that Sheet2 is activated and displayed, and the second causes a userform to appear.

The BeforeClose event handler causes its Sub to execute just before the workbook is closed. A more sophisticated, combined application of the Open and BeforeClose handlers is to activate a new menu tab on the Ribbon with groups and commands relevant to the application upon the Open event and then remove that when the BeforeClose event occurs.

It might be tempting to use the BeforeClose handler to remind the user to save the workbook before closing, but Excel already has this reminder built in, so that is not necessary. There is even a way to bypass the Excel reminder if absolutely necessary, but we do not advise this.

The NewSheet workbook event handler will execute whenever a sheet is added to the workbook. The handler code can make initial settings on the worksheet, such as placing a header, and adjusting column widths. Here is an example code:

```
Private Sub Workbook_NewSheet(ByVal Sh As Object)
If TypeName(Sh) = "Worksheet" Then
    Sh.Range("A1") = Now()
    Sh.Range("A1").NumberFormat = "dd-mm-yy"
End If
End Sub
```

This event handler places the current date in cell A1 and formats its display. The time is also included, but given the format, it is not displayed.

After the Workbook category, a next level of event relates to worksheets. By selecting a particular worksheet object in the Project Explorer, you can find the list of available events in the (Declarations) dropdown on the right of the code window. An example handler is Worksheet_Change. The associated code is triggered whenever the contents or other aspects of the worksheet are changed by the user. A specific example is illustrated in Figure 12.19. The Target argument is provided in ByVal form, and the MsgBox command displays the Address property, indicating which cell was changed.

A more detailed worksheet event handler can be programmed to react only if certain cells or ranges of cells are changed. This is shown in Figure 12.20. In this case, the Target represents the cell on the spreadsheet where a change was made, and the CheckRange is the Range object that is set to a particular block of cells. The If statement is different from what we have seen so far. It uses the Intersect function to compare CheckRange with Target. This will evaluate to empty if there are no cells in common. When compared to Nothing that would then be True. If there are cells in common, it will evaluate False. The logical Not operator reverses this, so the result will be True if the Target cell is part of the CheckRange cells. That causes the MsgBox command to execute.

```
Private Sub Worksheet_Change(ByVal Target As Range)
MsgBox "Range " & Target.Address & " was changed"
End Sub
```

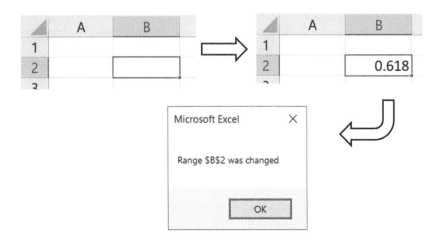

FIGURE 12.19 An example of the Worksheet_Change event handler.

```
Private Sub Worksheet_Change(ByVal Target As Range)
Dim CheckRange As Range
Set CheckRange = Range("InputCells")
If Not Intersect(CheckRange, Target) Is Nothing Then
    MsgBox "an input cell was changed"
End If
End Sub
```

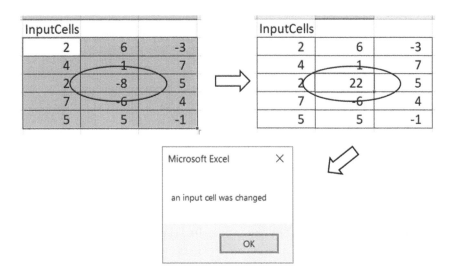

FIGURE 12.20 Worksheet_Change event handler that is restricted to certain cells.

With these event handlers, you can observe that there are command syntaxes that are new to us. This raises the question of how to find these if we have a situation different from the examples shown here. Since our coverage is not comprehensive, this is a dilemma. A good way to approach this is to look for examples on the Internet or make use of a good reference. An example of the latter is the book by Alexander and Kusleika (2019) or previous editions by these authors or Walkenbach, choosing the edition that matches your version of Excel.

To finish this section, we want to introduce three specialized event handlers. The first is OnTime and it is coded in a module, neither within a Sub or Function nor in the code associated with a Microsoft Excel object. It specifies an event that occurs at a given time of day. The event causes a procedure to execute. Here is the syntax for an example time of 8 a.m.

```
Application.OnTime TimeValue("08:00:00"), "myProcedure"
```

This statement would be coupled with a Sub such as

```
Sub MyProcedure()
MsgBox "Morning coffee over. Time to get to work!"
End Sub
```

We should mention in passing another command in VBA that is inherited from early versions of the BASIC language. It is the *Timer function*, which returns the number of seconds since midnight of the current day. This can be used to execute a task at a given time interval.

A simple example of Timer use is illustrated in Figure 12.21. Here an initial time, InitTime, is set from the Timer and an interval, DelTime, is set to five seconds. A future time, NewTime, is set to the initial time plus the interval. In the Do Loop, Timer is checked against NewTime until it equals or exceeds it. When that occurs, the loop is exited and the message box displayed. An example result is shown, and you can see that the interval is five seconds to the nearest 1/100 seconds.

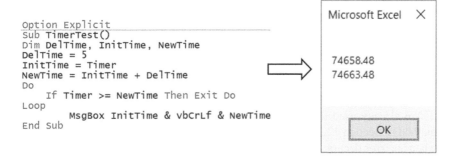

FIGURE 12.21   Use of the Timer command.

While the Sub in Figure 12.21 is executing, you cannot do other work on the spreadsheet or in VBE. A modification of the code provides a structure for timed execution of a Sub called TaskSub at a five-second interval.

```
Sub TimerTest()
Dim DelTime, InitTime, NewTime
DelTime = 5
InitTime = Timer
NewTime = InitTime + DelTime
Do
    If Timer >= NewTime Then
        NewTime = NewTime + DelTime
        Call TaskSub()
    EndIf
Loop
End Sub
```

This code forms an infinite loop that will never be exited until the user aborts execution. A possible use for this structure would be timed acquisition of information external to Excel, such as live data acquisition from a process or experiment.

The *OnKey command* is also programmed in a module. It can be used to set up the execution of a procedure in response to a keystroke or combination keystrokes. The syntax for the command is

```
Application.OnKey "keycode", "key procedure"
```

and to cancel the event response

```
Application.OnKey "(keycode)"
```

with no procedure specified. The key codes are many. Several examples are

- "A"   The A key
- "+A"   Shift-A key combination
- "^A"   Ctrl-A key combination
- "%A"   Alt-A key combination
- "{ESC}" Escape key
- "{F12}" F12 key

The key procedure must be enclosed in quotation marks. More information is available from VBA Help.

We already used the *On Error command* in the example of the InputBox method in Section 12.3. It can be placed in any VBA procedure and will respond when an error occurs. There are two common implementations.

```
On Error Resume Next
```

This statement causes VBA to ignore the error condition and continue execution on the next line of code. It is used in special circumstances where errors are anticipated and need to be ignored. There is some danger in its use, and so it should not be used indiscriminately.

The second implementation is the one we have seen.

```
On Error GoTo line_label
```

This is one of the few times we consider using the old BASIC *GoTo statement*.[4] It identifies a segment of code that will execute when an error condition occurs at any point within the procedure.

Perhaps the most common use of event handlers is in the development of userforms for applications. This will be considered in the next section.

12.5 USERFORMS

Userforms are the primary feature in VBA for the creation of user interfaces. They are more flexible and capable than message and input boxes. Along with this power comes complexity. We will introduce userforms via examples typical of engineering and scientific calculations. They have broad applications that are more general in nature, and one can find other examples in references, such as Alexander and Kusleika (2019).

The typical functionality of one or more userforms in an engineering application is to establish choices and set parameter values for a VBA program, cause the code to execute, and display results in message boxes, on the spreadsheet as tables and charts, or on the userform itself. A schematic depiction of a typical userform is shown in Figure 12.22. The userform itself is an object and all the

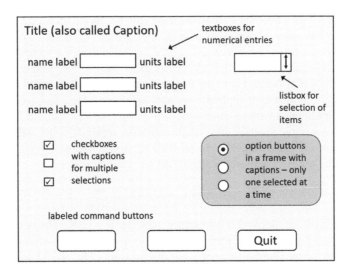

FIGURE 12.22 Depiction of typical userform.

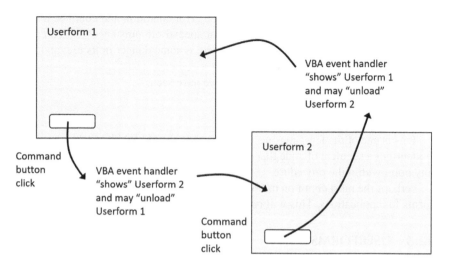

FIGURE 12.23   Sequential arrangement of two userforms.

elements placed on it are objects. There are a myriad of properties associated with each object that are displayed and adjusted in the Properties Window of the VBE.

Event handler VBA code is commonly associated with the command buttons to respond to their selection/click, but it can be associated with a change in any object. The size of the userform can be adjusted to accommodate more or fewer objects, but some judgment must be applied to make the layout attractive and manageable by the user. This is especially true if too many objects are placed.

As applications become extensive, they call for more than one userform. Figure 12.23 shows a sequential arrangement of two userforms. A command button click on one form initiates an event handler that activates a second userform and typically deactivates the first. A challenge we have to face and overcome is that information associated with the first userform is typically needed in the second. Again, with judgment applied, most applications involve a few userforms but are not overwhelmed with too many.

As a first, simple example of a single userform application, we will consider the conversion of temperature from one scale to another, including °C, °F, K, and °R. The relationships we want to implement are summarized as

$$°F = 1.8°C + 32 \qquad K = °C + 273.15 \qquad °R = °F + 458.67$$

plus, corresponding derived relationships, such as °R = 1.8K. The design of the userform layout is shown in Figure 12.24.

There are many alternative designs for this userform. One would be to remove the Convert button and have an output textbox that always shows the converted

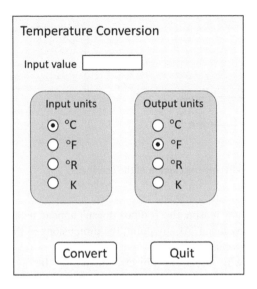

FIGURE 12.24   Design of temperature conversion userform.

value. You will notice that we have set up the option buttons to show the conversion of °C to °F as the default arrangement. Another approach would be possible. One of the reasons we have used this design is that it introduces elements we will use in a more complicated application later in this chapter.

Next, we move on to implementing our design. Pausing, we comment that it is advisable to sketch out userform designs prior to implementing them. Changes in layout and function can be made, but it is a good idea to have a starting point.

Starting with a new workbook, it is useful to save it right away as a macro-enabled workbook file, such as TemperatureConversion.xlsm. Then, in the VBE, with our project selected in the Project Explorer window, select Insert ⇨ Userform from the menu. Figure 12.25 shows this. The userform appears with a default title caption, UserForm1, and adjacent to it is a Toolbox palette of controls that can be

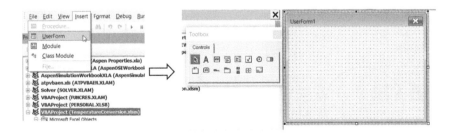

FIGURE 12.25   Inserting a blank userform into the project.

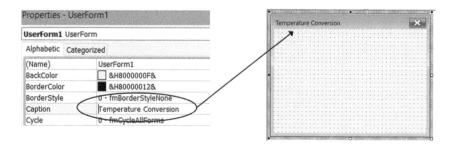

FIGURE 12.26   Modifying the caption of the userform.

placed on it. If, for any reason, the Toolbox doesn't appear, it can be made to show again from the View menu. At any point, the dimensions of the userform can be adjusted, stretched, or shrunk.

As we work with userform objects, including the userform itself, the Properties window becomes important. For example, we note the initial caption is UserForm1. This can be changed to Temperature Conversion, as shown in Figure 12.26. It should be noted that we do not change the name of the userform, but its caption. There are cases where we will change the name of a control or object on the userform, but that is unnecessary here.

As we go forward with the development of the userform, there will be many changes in the Properties window, but we will not show all these in figures – just note them in description. There are many cosmetic adjustments we could make to the userform including caption font, the background color, adding pictures, etc. In fact, one can spend an inordinate amount of time on the "cosmetic" appearance of the userform. Again, some judgment should be used of the return on investment of that time.

Following our design from Figure 12.24, we will add a textbox to the userform for the input temperature value. Figure 12.27 illustrates this. After clicking the TextBox icon on the Toolbox, we draw out the textbox on the userform. Its shape can be stretched/shrunk to suit. You will note that when the textbox is selected, it is highlighted with a border. Also, the Properties window now displays its properties and not those of the userform object.

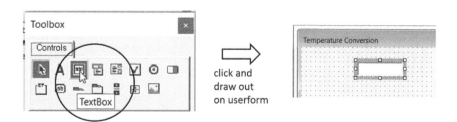

FIGURE 12.27   Placing a textbox control on the userform.

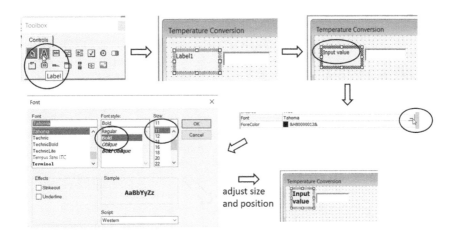

FIGURE 12.28 Placing a label object and formatting it.

Since we will use the value placed in the textbox in the conversion calculation, we change the name of the textbox to TempIn. This is the first field of the Properties window entitled (Name). We can make further changes later to this object.

Next, we add a label control to the left of the textbox and format it. This is shown in Figure 12.28. After the label is drawn out on the userform, the caption, Label1, is modified to Input value, as per our design. This can be done by clicking in and typing directly or by changing the Caption field in the Properties window. The format of the text is then changed via the Font field and the ••• button on its right side. Finally, the sizes and positions of the label and textbox are adjusted to suit.

Next, we will consider installing the frame and option buttons for the input units. The reason we use a frame is so the buttons within it are isolated from other buttons on the userform. Figure 12.29 illustrates the frame and the first option button.

After these are installed, the following customizations are applied:

- remove the frame caption by deleting it in the Properties window
- change the name of the option button object to degCin
- change the caption of the option button to degC or °C (° copied from elsewhere)
- adjust the font of the option button caption to bold 10 pt

The remaining three buttons in the frame are added similarly and appear as shown in Figure 12.30. Here the three option buttons have been named degFin, Kin, and degRin, respectively. Also, the Value property of the degCin button is set to True. Consequently, since they are in the same frame, the Value property of the remaining buttons are all False. You can see that

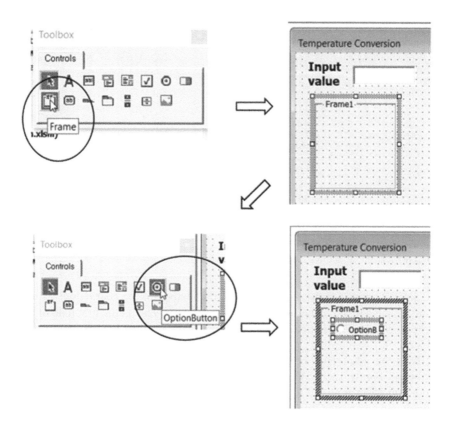

FIGURE 12.29   Installation of the frame and first option button.

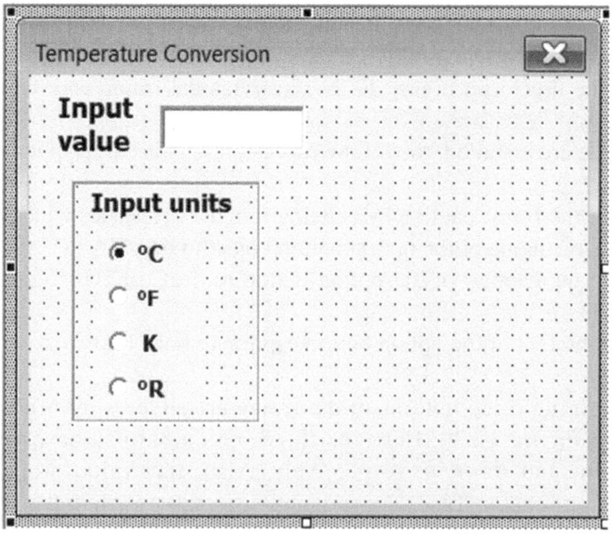

FIGURE 12.30   Userform with input units frame complete.

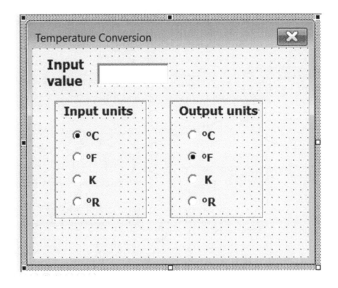

FIGURE 12.31   Userform with both frames complete.

a label, Input units, has been added to the top of the frame. To make everything appear well, the frame has been stretched vertically and the option buttons shifted downward. And then the entire userform has been stretched downward a bit.

Without going into as much detail, and following the pattern of the Input units frame, the Output units frame is added to the right. This is shown in Figure 12.31. A convenient way to do this is to copy and paste the Input units frame and then make the necessary adjustments. The names of the four buttons become degCout, degFout, Kout, and degRout, respectively, and the Value property of the degFout button is set to True. This establishes a default conversion from °C to °F. The label is obviously edited to Output units.

Again, referring to our design in Figure 12.24, we install two command buttons in the lower part of the userform and change the names and captions of these to Convert and Quit respectively. This is depicted in Figure 12.32.

The controls (objects) on the userform are now complete, as per our design. There are two important tasks left. First, we must add the event handler code that runs when either of the two command buttons is clicked. Second, we need to add a Sub in a module to cause the userform to appear, in essence a startup Sub.

If we double-click the Convert button, a code window opens with the framework statements

```
Private Sub Convert_Click()
End Sub
```

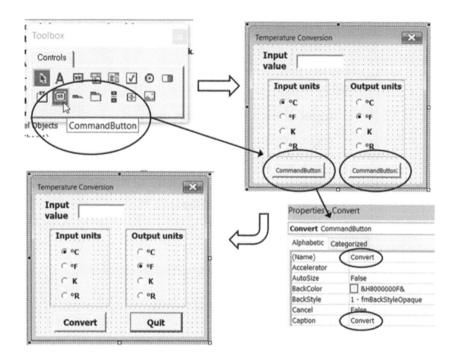

FIGURE 12.32   Adding command buttons to the userform.

Once code is in place, you can switch back and forth from the userform to the code window with F7 (userform ⇨ code window) and Shift-F7 (code window ⇨ userform). At this point, we want to add VBA statements to carry out the necessary calculations and display, depending on what is entered in the Input value textbox and which choices are indicated in the two frames. The general scheme is described:

- obtain the value from the TempIn textbox and convert it to a number
- check which Input units button is selected
- depending on the Input units button, check which Output units button is selected
- make the appropriate conversion based on the selected Input and Output units buttons
- display the resulting conversion in a message box

Figure 12.33 shows the code that is created. Since the TempIn.Value is a string, it is converted to a number using the built-in Val function and assigned to a local variable TempVal. Next there is an overall multi-alternative If statement that checks which Input unit button is selected. The option buttons have a Boolean value, so the If statement just checks the object name (default property is Value)

```
Option Explicit
Private Sub Convert_Click()                          ElseIf Kin Then 'here if input units are K
Dim TempVal As Double, TempOut As Double                If degCout Then
TempVal = Val(TempIn) 'convert TempIn to number            TempOut = TempVal - 273.15
If degCin Then 'here if Input units are degC            ElseIf degFout Then
    If degFout Then                                        TempOut = (TempVal - 273.15) * 1.8 + 32
        TempOut = TempVal * 1.8 + 32                    ElseIf degRout Then
    ElseIf Kout Then                                       TempOut = TempVal * 1.8
        TempOut = TempVal + 273.15                      End If
    ElseIf degRout Then                              ElseIf degRin Then 'here if input units are degR
        TempOut = (TempVal + 273.15) * 1.8              If degCout Then
    End If                                                 TempOut = (TempVal - 458.67) / 1.8
ElseIf degFin Then 'here if input units are degF        ElseIf degFout Then
    If degCout Then                                        TempOut = TempVal - 458.67
        TempOut = (TempVal - 32) / 1.8                  ElseIf Kout Then
    ElseIf Kout Then                                       TempOut = TempVal / 1.8
        TempOut = (TempVal + 458.67) / 1.8              End If
    ElseIf degRout Then                              End If
        TempOut = TempVal + 458.67                   MsgBox "Converted temperature =" & FormatNumber(TempOut, 2)
    End If                                           TempIn.Value = ""
                                                     End Sub
```

FIGURE 12.33 Initial event handler code for `Convert` button click.

which will be True or False. Within each If/ElseIf clause there is another multi-alternative If statement that determines which Output unit button is selected and computes the appropriate conversion, assigning it to the `TempOut` variable. Finally, the message box with the `TempOut` value is produced and, once that is clicked OK by the user, the `TempIn` value is set to an empty string, leaving the Input value textbox empty for the next entry.

The code for the `Quit` button event handler is very simple and is shown in Figure 12.34. Also, the `Startup` Sub is located in a module in the project. Conveniently, we add a Start Convert button to the worksheet that launches the `Startup` macro. The `Quit` event handler uses the `Unload` command to remove the userform and then the `TempConvert` object is set to `Nothing`. The latter is done to free memory associated with the object and prevent possible "Out of memory" error warnings. The `Startup` Sub applies the `Show` method to the `TermConvert` object to make the userform appear. And then we add the on-sheet button and link it to the `Startup` macro.

When we click the Start Convert button, the userform appears on the worksheet, and we can try it out. Figure 12.35 illustrates this. A value of 100 is entered. Note: the font type and size have been adjusted for the `TempIn` textbox to make it more visible and consistent with the rest of the userform. The default conversion of °C to °F is left in place, and, upon clicking the `Convert` button, the converted temperature value is displayed.

```
Private Sub Quit_Click()
Unload TempConvert
Set TempConvert = Nothing
End Sub
```
in Event Handlers

```
Option Explicit
Sub Startup()
TempConvert.Show
End Sub
```
in Module1

Start Convert

on Sheet1

FIGURE 12.34 `Quit` event handler, `Startup` Sub, and Start Convert on-sheet button.

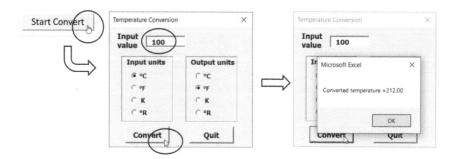

FIGURE 12.35   First demonstration of the userform.

There are many ways we can improve this userform including design improvements. For now, we will focus on making it less susceptible to errors on the part of the user. We imagine four possible error conditions:

1. The Convert button is clicked with nothing in the Input value textbox
2. The Convert button is clicked with something other than a number in the Input value textbox
3. The value entered is improper – it is less than absolute zero (K or °R)
4. The user requests a conversion to the same Output unit as the chosen Input unit

We can start by addressing the error condition if the user fails to enter valid information in the Input value textbox. The code that can handle this is a multi-alternative If structure that imbeds our current conversion calculation code, as in

```
If TempIn.Value = "" Then
    MsgBox "You must enter an input value"
ElseIf Not IsNumeric(TempIn.Value) Then
    MsgBox "You must enter a numeric input value"
Else
    'conversion calculations here
End If
```

If this is done, the first two error conditions above will be detected, and the user will be given the opportunity to correct them. The third error condition related to absolute zero can be tested in each main branch of the multi-alternative If structure shown in Figure 12.33. For the first branch, this would look like

```
    If TempVal < -273.15 Then
        MsgBox "Must enter a temperature > absolute zero"
        GoTo ErrorExit
    End If
•
•
```

```
ErrorExit:
TempIn.Value = ""
End Sub
```

There would be another, more elegant way to handle this with an imbedded multi-alternative If structure, but, for the sake of simplicity, we have used the GoTo directed to the end of the Sub.

The last of the error conditions can be handled by including an Else clause in the inner multi-alternative If structures where the only way the Else clause executes is if the Input units and Output units choices are the same. Again, for the first branch, the code added would be

```
Else
    MsgBox "Cannot select same units for input and output"
    GoTo ErrorExit
```

The final code for the Convert event handler is shown in Figure 12.36.

To validate the application, we should test it for all conversions and error modes. Conversions can be checked by using well-known temperatures, e.g., $100°C \Rightarrow 212°F$. Three error mode tests are illustrated in Figure 12.37. These do not include the test for invalid temperature below absolute zero. That should be checked too. It is good practice to "poke and prod" the application in all ways.

One last note before we move on to a more in-depth case study. As in this example, when a userform appears via the Show method, other activities on the spreadsheet are suspended. This is formally described as the userform appearing "modally." If you want to have spreadsheet operations available while the userform is present, you can modify the Show method command as follows:

```
TempConvert.Show vbModeless
```

```
Option Explicit
Private Sub Convert_Click()
Dim TempVal As Double, TempOut As Double
If TempIn.Value = "" Then
    MsgBox "You must enter an input value"
ElseIf Not IsNumeric(TempIn.Value) Then
    MsgBox "You must enter a numeric input value"
Else
    TempVal = Val(TempIn)
    If degCin Then
        If TempVal < -273.15 Then
            MsgBox "Must enter a temperature > absolute zero"
            GoTo ErrorExit
        End If
        If degFout Then
            TempOut = TempVal * 1.8 + 32
        ElseIf Kout Then
            TempOut = TempVal + 273.15
        ElseIf degRout Then
            TempOut = (TempVal + 273.15) * 1.8
        Else
            MsgBox "Cannot select same units for input and output"
            GoTo ErrorExit
        End If
    ElseIf degFin Then
        If TempVal < -458.67 Then
            MsgBox "Must enter a temperature > absolute zero"
            GoTo ErrorExit
        End If
        If degCout Then
            TempOut = (TempVal - 32) / 1.8
        ElseIf Kout Then
            TempOut = (TempVal + 458.67) / 1.8
        ElseIf degRout Then
            TempOut = TempVal + 458.67
        Else
            MsgBox "Cannot select same units for input and output"
            GoTo ErrorExit
        End If
    ElseIf Kin Then
        If TempVal < 0 Then
            MsgBox "Must enter a temperature > absolute zero"
            GoTo ErrorExit
        End If
        If degCout Then
            TempOut = TempVal - 273.15
        ElseIf degFout Then
            TempOut = (TempVal - 273.15) * 1.8 + 32
        ElseIf degRout Then
            TempOut = TempVal * 1.8
        Else
            MsgBox "Cannot select same units for input and output"
            GoTo ErrorExit
        End If
    ElseIf degRin Then
        If TempVal < 0 Then
            MsgBox "Must enter a temperature > absolute zero"
            GoTo ErrorExit
        End If
        If degCout Then
            TempOut = (TempVal - 458.67) / 1.8
        ElseIf degFout Then
            TempOut = TempVal - 458.67
        ElseIf Kout Then
            TempOut = TempVal / 1.8
        Else
            MsgBox "Cannot select same units for input and output"
            GoTo ErrorExit
        End If
    End If
    MsgBox "Converted temperature =" & FormatNumber(TempOut, 2)
End If
ErrorExit:
TempIn.Value = ""
End Sub
```

FIGURE 12.36 Convert command button event handler with error corrections added.

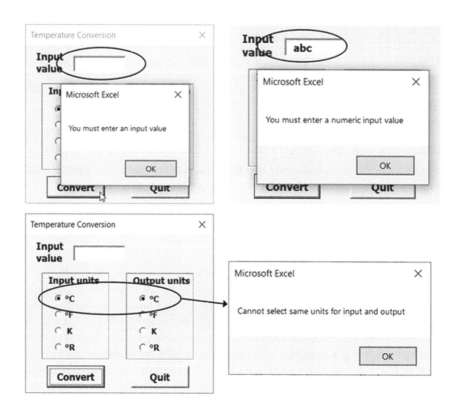

FIGURE 12.37   Userform error conditions.

12.6   USER INTERFACE CASE STUDY

For a case study that includes more involved userforms, we choose an example based on an extension of our projectile trajectory model in Chapter 3, Problem 3.4. That model is

$$y = \tan(\theta)x - \frac{g}{2 v_0^2 \cos^2(\theta)} x^2 + y_0 \qquad (12.1)$$

where x = horizontal distance or range, m, y = vertical distance or elevation, m, v_0 = initial velocity, m/s, y_0 = initial elevation, m, θ = launch angle, radians, and g = gravitational acceleration, ≅ 9.81 m/s².

As illustrated in Figure 12.38 (which is the same as Figure P3.4), this model produces a parabolic trajectory that neglects the frictional effects on the projectile, and so its predictions are suspect.

To include the effect of frictional *drag* on the projectile, the model description becomes a set of differential equations. As more features are included, for example, projectile spin, the model would become even more complicated.

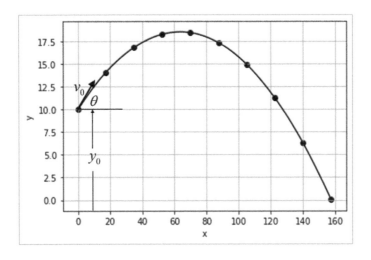

FIGURE 12.38 Projectile trajectory without friction.

Also, if we include crosswind effects, the number of equations increases. In our example here, we will take the first step by including drag. The differential equations then are

$$\frac{dx}{dt} = v_x \qquad x(0) = x_0 \qquad (12.2)$$

$$\frac{dy}{dt} = v_y \qquad y(0) = y_0 \qquad (12.3)$$

$$m\frac{dv_x}{dt} = -c_d \frac{v_x}{2}\sqrt{v_x^2 + v_y^2} \qquad v_x(0) = v_{x0} = v_0 \cos(\theta) \qquad (12.4)$$

$$m\frac{dv_y}{dt} = -mg - c_d \frac{v_y}{2}\sqrt{v_x^2 + v_y^2} \qquad v_y(0) = v_{y0} = v_0 \sin(\theta) \qquad (12.5)$$

where additional variables are, with typical SI units, v_x = component of the velocity in the x direction, m/s; v_y = component of the velocity in the y direction, m/s; x_0 = initial value of x, often zero, m; t = time of flight, s; m = mass of the projectile, kg; and c_d = dimensional drag coefficient, kg/m.

The dimensional drag coefficient, c_d, is related to a dimensionless drag coefficient, C_d, by

$$c_d = \rho A C_d \qquad (12.6)$$

where ρ = fluid density, kg/m³; A = cross-sectional area of the projectile as viewed from the direction of travel, m²; and C_d = dimensionless drag coefficient, typically between 0.3 and 0.7 for spherical projectiles.

The general algorithm we imagine for the application is

1. specify the gas (typically air) parameters so the density, ρ, can be computed,
2. establish the parameters of the projectile, m, A, and C_d,
3. set the launch parameters, x_0, y_0, v_0, and θ,
4. compute the trajectory by solving the differential equations until $y = 0$ is reached, and
5. display the trajectory in a table on the spreadsheet and in a plot.

This leads to consideration of the design involving one or more userforms. Here, we choose a hybrid design to illustrate various userform features. Figure 12.39 provides a schematic diagram of the design. It shows two userforms and a Start button on the worksheet. The first userform computes the air density given the conditions of pressure, temperature, and humidity. Then, via a command button, it links to a userform with three tabs. The first tab, Projectile, establishes the projectile parameters, including looking up the dimensionless drag coefficient based on a list of projectile types. The second, Conditions, specifies the initial conditions for the launch. The third tab, Results, provides the command to solve the model and displays two key solution parameters, the projectile travel time and its range (when and where it returns to ground). It is anticipated that the solution will include updating a table of values on the spreadsheet and updating a plot of the trajectory.

The Results tab has a command button that returns to the Air Density userform to allow for solving another case. Both userforms contain a Quit button that terminates the application.

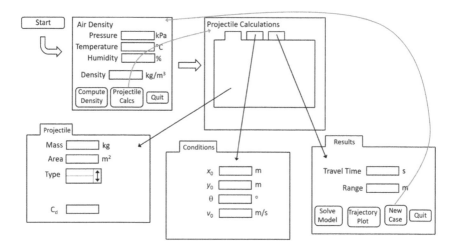

FIGURE 12.39   Schematic design of userforms for the projectile trajectory application.

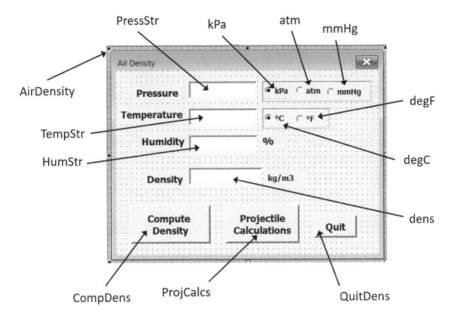

FIGURE 12.40 Air Density userform with names of objects noted.

Next comes the creation of the userforms and the corresponding event handlers. We will start with the Air Density userform. This is shown in Figure 12.40. Where specific names have been created for objects, these are noted in the periphery of the figure.

To compute the density of humid air, we make use of the ideal gas law in the following way:

$$\frac{n}{V} = \frac{m}{MW \cdot V} = \frac{\rho}{MW} = \frac{P}{RT} \Rightarrow \rho = \frac{P \cdot MW}{RT} \tag{12.7}$$

The formula is valid for a single component (or dry air with its constant composition and average MW). When we account for water vapor, the formula is modified to

$$\rho = \frac{P_a \cdot MW_a}{RT} + \frac{P_w \cdot MW_w}{RT} \tag{12.8}$$

where
P_a = partial pressure of the dry air
P_w = partial pressure of the water vapor

The partial pressure of the water vapor is

$$P_w = Rh \cdot P_v \qquad\qquad (12.9)$$

where
Rh: relative humidity, $0 \rightarrow 1$
P_v: vapor pressure of water at T

The vapor pressure of water at a given T can be computed from the *Antoine equation* from Equation 8.35 in Chapter 8:

$$\log_{10}(P_V) = A - \frac{B}{C+T} \qquad or \qquad P_V = 10^{A - \frac{B}{C+T}} \qquad (12.10)$$

The coefficients, for T in °C and P in kPa, were determined to be

$$A = 7.568, \ B = 1950.4, \ C = 250.98 \qquad\qquad (12.11)$$

The remaining quantity to determine is P_a, which is the difference between P and P_w.

We can use these relationships for the Compute Density event handler shown below. All local variables are declared in Dim statements. There is only one result variable, the density, rho, which will be required in the subsequent event handler that computes the solution. For this reason, rho will be declared Public, and that is done in the module, not in an event handler.

```
Option Explicit
Private Sub CompDens_Click()
Dim Press As Double, Tmp As Double
Dim P, T, Tabs, Rh, Pv, Pw, Pa, Mwa, Mww
Dim A, B, C, Rgas
A = 7.568: B = 1950.4: C = 250.98
Rgas = 8.3145 ' kPa*m3/[K*kmol]
Mwa = 28.96: Mww = 18.02   ' kg/kmol
Press = Val(PressStr)
If kPa Then
    P = Press
ElseIf atm Then
    P = Press * 101.325
ElseIf mmHg Then
    P = Press / 760 * 101.325
End If
Tmp = Val(TempStr)
If degC Then
    T = Tmp
ElseIf degF Then
    T = (Tmp - 32) / 1.8
End If
```

```
Tabs = T + 273.15
Rh = Val(HumStr) / 100
Pv = 10 ^ (A - B / (T + C))
Pw = Rh * Pv
Pa = P - Pw
rho = (Pa * Mwa + Pw * Mww) / Rgas / Tabs
dens = FormatNumber(rho, 2)
End Sub
```

After assigning values to the various constant variables, the values of pressure, temperature, and humidity are acquired from the textboxes on the userform. Since these are string quantities, they are converted to numbers with the built-in Val function. There are multi-alternative If statements for the pressure and temperature entries to deal with the selection of units in the adjacent frames. The If conditions use the names of the option buttons because those return a Boolean result. The remaining statements implement Equations 12.8–12.10 to compute the density of the humid air and assign it to the rho variable (declared Public). Also, the result is returned to the dens textbox on the userform with a display to two decimal places.

The next event handler on the Air Density userform is associated with the Projectile Calculations command button, named ProjectileCalcs. Its purpose is to shut down the current userform and cause the next one to appear. The code for this is

```
Private Sub ProjectileCalcs_Click()
ProjCalcs.Show vbModeless
ProjCalcs.MultiPage1.Value = 0
Unload AirDensity
Set AirDensity = Nothing
End Sub
```

Apart from causing the Projectile Calculations userform to appear with the Show method, the multipage object is set to show the first tab, the one on the left. After that, the AirDensity userform is removed, unloaded, and its memory released.

The final event handler for the AirDensity userform is related to the Quit button. The code for this is short and simple. The AirDensity userform is unloaded and its memory released by setting the object name to Nothing. Also, to ensure the spreadsheet display returns to the worksheet named Main, where the Startup button will be located, that worksheet is selected.

```
Private Sub QuitDens_Click()
Unload AirDensity
Set AirDensity = Nothing
Sheets("Main").Select
End Sub
```

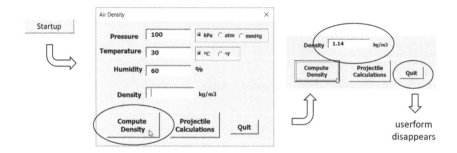

FIGURE 12.41 Test operations with the Air Density userform.

Although we have not created the `ProjCalcs` userform yet, we can test the `AirDensity` userform by creating the `Startup` macro in a module in the project. This code will be

```
Sub Startup()
AirDensity.Show vbModeless
End Sub
```

An on-sheet button with the caption Startup is then added and linked to this macro.

Figure 12.41 illustrates this, where we exercise the calculation of the humid air density and then the Quit button. For now, we leave the Projectile Calculations button alone. As the figure indicates, the conditions specified are 100 kPa, 30°C, and 60% humidity, and the density is computed to be 1.14 kg/m³.

Before we proceed with the design and coding of the Project Calculations userform, we comment that the entries on the Air Density userform are not protected against erroneous values. For example, we cannot accept a negative pressure value (absolute pressure here, not gauge). In fact, we need to protect against the user leaving a textbox entry blank. The code segments below handle this.

```
Press = Val(PressStr)
If Press <= 0 Then
    MsgBox "Pressure must be > 0"
    Exit Sub
End If
•
•
Tmp = Val(TempStr)
If Tmp < -40 Then
    MsgBox "Temperature must be >= -40"
    Exit Sub
End If
•
•
```

```
Rh = Val(HumStr) / 100
If Rh < 0 Or Rh > 1 Then
    MsgBox "Humidity must be >= 0 and <= 100 %"
End If
```

Each segment tests for an invalid entry and displays a corrective message box. Execution of the event handler Sub is then ended via the Exit Sub command. This is an alternative to the previous GoTo technique. The user can then click OK on the message box and make the necessary correction. Note that the value for an empty textbox is zero, so the third test allows this and uses the zero humidity ⇨ dry air.

Next, we move on to the implementation of the Project Calculations userform. Here, we take advantage of the MultiPage control. Installing this is illustrated in Figure 12.42, including adding a third page to match our design.

The userform's name is named ProjCalcs and its caption is changed to Projectile Calculations. The MultiPage format is useful when there are many parameters to enter as part of a calculation, and they can be neatly organized on separate tabs of the control. Each of these tabs can then be selected and its name and caption customized to the application. Figure 12.43 shows the first tab with its controls added and indicates the names associated with the objects.

Notice that the tab is named ProjSpecs and its caption changed to Projectile. The only other new element here is the ListBox control that has been installed and named ShapeType.

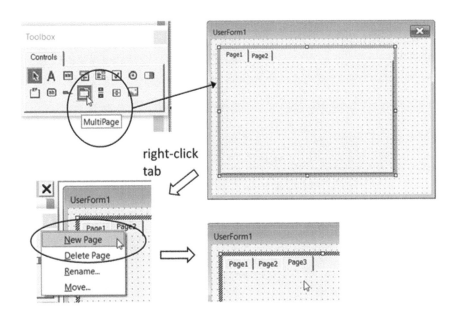

FIGURE 12.42 Installing the MultiPage control on the userform and adding a tab.

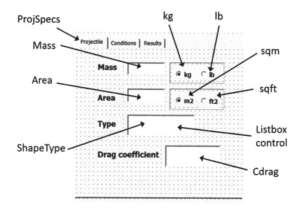

FIGURE 12.43 Projectile tab with controls and associated names.

The second tab is shown in Figure 12.44 with its controls and their associated names. The tab is named CondSpecs and its caption Conditions. Of course, these names are arbitrary, but they are chosen to relate the objects to the problem being solved rather than using the default names assigned by the VBE. Here, we have appended Str, meaning string, to several names to distinguish these from variable names we will use later in the code.

Figure 12.45 shows the third tab of the MultiPage control both named Results and captioned Results. Much of the "work" of the application takes place in the SolveModel event handler.

With the MultiPage control and its tabs formulated, we return to the Projectile tab and deal with the ListBox control. To function, this control requires a reference

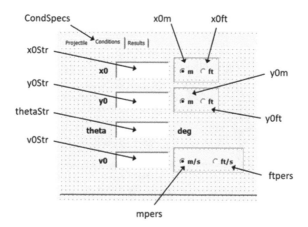

FIGURE 12.44 Conditions tab with controls and associated names.

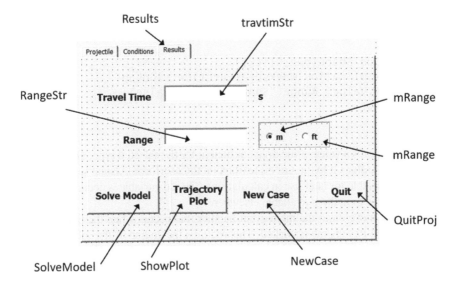

FIGURE 12.45   Results tab with controls and associated names.

to a list of selectable options. Because these are related to their corresponding dimensionless drag coefficients, Cd, we create a table on a separate worksheet of the workbook. It is not necessary to use a separate worksheet, but it is convenient from an organizational perspective. Figure 12.46 shows this worksheet. The worksheet tab (and worksheet itself) is named Proj Type, and the table has two named ranges. The shape range contains just the projectile types, and the shapeCd range includes the types and the drag coefficient values. We will use the latter in a table lookup.

With the shape table in place on the Proj Type worksheet, we can adjust the Properties window of the ListBox by entering "shape" for the RowSource property. When the userform is displayed, it will then present the list with the up-down

	A	B	C	D
1	Projectile Type and Drag Coefficient			
2		sphere	0.5	
3		airfoil	0.05	
4		bullet	0.3	
5		golf ball	0.4	
6				

Proj Type   shape   shapeCd

FIGURE 12.46   New worksheet named Proj Type with two named ranges, shape and shapeCd.

arrows for the user to select. We can enter an event handler that responds to a
selection in the ListBox as

```
Private Sub ShapeType_Click()
Cdrag = Application.WorksheetFunction.Vlookup( _
ShapeType.Value, Range("ShapeCd"), 2, False)
Cd = Cdrag
End Sub
```

This acquires a value of the drag coefficient for the selected type, ShapeType.
Value, and assigns it to the Cdrag textbox on the userform. Also, it sets the value
of the Cd variable for later use in calculations. The Cd variable is declared Public
in the module to make it available elsewhere.

We now move on to the main event handler that occurs when the Solve Model
button on the Results tab is clicked. This requires the coding of the solution of the
four differential equations modeling the trajectory. The first part of handler obtains
the key data from the userforms and assigns these to variables for later use. There
are checks on the validity of the entries. This is the leading code and is similar to
previous applications. If there is an invalid entry, an error is displayed in a message
box and the Sub is exited. The user can then click OK on the message box and return
to the userform to enter a correct value. In each case, a positive entry is required so,
if an entry is not made, which is equivalent to entering a zero, it is flagged as an error.

```
Private Sub SolveModel_Click()
Dim m, a, x0, y0, thetad, theta, v0, Pi
Dim TravTime, TravRange
Pi = 4 * Atn(1)
m = Val(Mass)
If m <= 0 Then
    MsgBox "Mass must be > 0"
    Exit Sub
End If
If lb Then
    m = m * 0.45359
End If
a = Val(Area)
If a <= 0 Then
    MsgBox "Area must be > 0"
    Exit Sub
End If
If sqft Then
    a = a * 0.0929
End If
x0 = Val(x0Str)
If x0ft Then
    x0 = x0 * 0.3048
End If
```

```
y0 = Val(y0Str)
If y0 < 0 Then
    MsgBox "y0 must be >= 0"
    Exit Sub
End If
If y0ft Then
    y0 = y0 * 0.3048
End If
thetad = Val(thetaStr)
If thetad <= 0 Or thetad > 75 Then
    MsgBox "Theta must be > 0 and <= 75 degrees"
    Exit Sub
End If
theta = thetad / 180 * Pi
v0 = Val(v0Str)
If v0 < 0 Then
    MsgBox "v0 must be > 0"
    Exit Sub
End If
If ftpers Then
    v0 = v0 * 0.3048
End If
```

The calculations that follow consist primarily of solving the differential equations. Here, we make the choice of placing that in the module and using a Sub Call from the event handler. This, including the final statements of the event handler, is

```
Call SolvMod1(m, a, x0, y0, theta, v0, TravTime, TravRange)
travtimStr = FormatNumber(TravTime, 2)
If ftRange Then
    TravRange = TravRange / 0.3048
End If
RangeStr = FormatNumber(TravRange, 2)
End Sub
```

The arguments to the SolvMod1 Sub include all the input variables required to solve the equations and two additional arguments, TravTime and TravRange, which are the results returned for display on the userform. The final statements use these two arguments, checking for units for the display of the range. Note that the variables Cd and rho are declared Public in the module and therefore will be available there for the calculations.

We consider the SolvMod1 Sub in the module next. This involves developing a strategy for solving the differential equations. The independent variable is time. We would like to solve the equations to a final time that allows the trajectory elevation, y, to reach zero where the projectile has encountered the ground. A dilemma is that, for each set of parameters, this will be a different time. If we use the Euler method to solve the equations (as shown in Chapter 7), we need to use a step size sufficiently small to provide an acceptable solution.

The strategy we adopt is to use Equation 12.1, describing a projectile with no air resistance, to solve for the range distance, x_{max}, at which $y = 0$. Given that value and the component of the launch velocity in the x direction, we can calculate a t_{max} value. Once we include air resistance and solve the differential equations, we know that the actual time and distance at which the projection returns to ground will be less than that predicted based on Equation 12.1, but those predictions will provide a practical limit to which we can solve the equations. Equation 12.1 is a quadratic polynomial, so, for $y = 0$, we can write it as

$$y = 0 = \tan(\theta)x_{max} - \frac{g}{2v_0^2\cos^2(\theta)}x_{max}^2 + y_0 \tag{12.12}$$

or, in traditional format,

$$-\frac{g}{2v_0^2\cos^2(\theta)}x_{max}^2 + \tan(\theta)x_{max} + y_0 = 0 \quad \Leftrightarrow \quad ax_{max}^2 + bx_{max} + c = 0 \tag{12.13}$$

where the solution is well known as

$$x_{max} = \frac{-b \pm \sqrt{b^2 - 4ac}}{2a} \quad \text{where} \quad a = -\frac{g}{2v_0^2\cos^2(\theta)} \quad b = \tan(\theta) \quad c = y_0 \tag{12.14}$$

where the correct result for the $\pm$ is the minus sign, $-$.

The initial statements of the event handler are:

```
Sub SolvMod1(m, Ar, x0, y0, theta, v0, TravTime, TravRange)
Dim cdd, a, b, c, xmax, tmax, dt, vx0, vy0
Dim i As Integer, imax As Integer, j As Integer
Dim tm(1000), x(1000), y(1000), vx(1000), vy(1000)
Dim dx(1000), dy(1000), dvx(1000), dvy(1000), Soln()
g = 9.81
cdd = Cd * rho * Ar
' based on no-drag, parabolic model,
' solve for x and t when y ==> 0
a = -g / 2 / v0 ^ 2 / Cos(theta) ^ 2
b = Tan(theta)
c = y0
xmax = (-b - Sqr(b ^ 2 - 4 * a * c)) / 2 / a
vx0 = v0 * Cos(theta)
tmax = xmax / vx0
dt = tmax / 1000
```

The Dim statements include the local variables used to determine the x_{max} and t_{max} values. There are other Dim statements for variables later in the event handler code. You can see how x_{max} is determined with Equation 12.14 and then t_{max} is determined by dividing x_{max} by the x component of the launch velocity, $v_x(0)$.

Then, the step size for the solution is computed to allow for 1,000 steps. Thus, we have now defined the extent of the Euler method solution of the differential equations. Based on Equations 12.2 through 12.5, the following code generates that solution

```
' solve 4 ode's using the Euler method
' establish initial conditions, i = 0
vy0 = v0 * Sin(theta)
tm(0) = 0
x(0) = x0
y(0) = y0
vx(0) = vx0
vy(0) = vy0
For i = 0 To 999
    'compute derivatives
    dx(i) = vx(i)
    dy(i) = vy(i)
    dvx(i) = -cdd * vx(i) / 2 / m * Sqr(vx(i) ^ 2 + vy(i) ^ 2)
    dvy(i) = -g -cdd * vy(i) / 2 / m * Sqr(vx(i) ^ 2 +
    vy(i) ^ 2)
    'integrate
    tm(i + 1) = tm(i) + dt
    x(i + 1) = x(i) + dx(i) * dt
    y(i + 1) = y(i) + dy(i) * dt
    vx(i + 1) = vx(i) + dvx(i) * dt
    vy(i + 1) = vy(i) + dvy(i) * dt
Next i
```

The solution populates the arrays declared in the Dim statements. You will note that we have not used the Option Base 1 declaration here. We find it convenient to use the 0 index for the initial condition and then 1,000 steps forward from there. The For loop only iterates to $i = 999$ because the last iteration computes the solution for the index $i + 1 = 1,000$.

When the solution is complete, it includes time values out to t_{max}, which will include y values that go below zero because of the manner that t_{max} was determined based on the no-friction model. We need to determine where the y values go negative. The following code steps back from the end of the y array to determine the i index where that happens. This will provide the last positive y value before it goes negative, and, given the small step size, it will be close enough to the ground for a well-described plot of the trajectory.

```
'find travel time and range
'step back from end of solution
'until y >= 0
For i = 1000 To 0 Step -1
    If y(i) >= 0 Then Exit For
Next i
imax = i
```

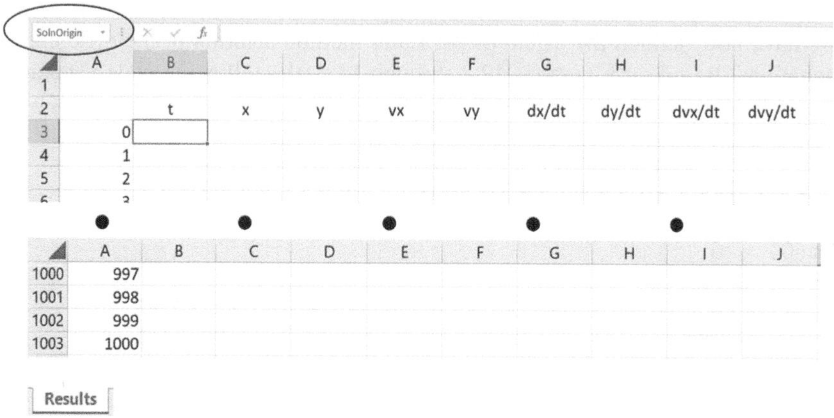

FIGURE 12.47   Results worksheet with headings, index column to 1,000, and SolnOrigin name.

Having determined the imax value, we can set the travel time and range values that will be returned to the userform event handler when Sub SolvMod1 is complete. You will recall these are the last two arguments of the Sub.

```
'return these to event handler
'for display on userform
TravTime = tm(imax)
TravRange = x(imax)
```

Finally, we complete the Sub by transferring the solution to a table on a Results worksheet in Excel. We create this worksheet as shown in Figure 12.47. A set of table headings are added along with an index column from 0 to 1,000. The solution will not take up the 1,000 rows and will vary from case to case. The B3 cell is named SolnOrigin for use in transferring the solution from the SolvMod1 Sub. There is another name created, Solution, which includes the entire interior of the table, B3:J1003.

The code used to transfer the numerical solution to the Results worksheet and complete the SolvModel Sub is

```
'transfer results to spreadsheet table
'will produce chart of trajectory
Worksheets("Results").Range("Solution").ClearContents
ReDim Soln(imax + 1, 8)
For i = 0 To imax
    Soln(i, 0) = tm(i)
    Soln(i, 1) = x(i)
    Soln(i, 2) = y(i)
    Soln(i, 3) = vx(i)
    Soln(i, 4) = vy(i)
    Soln(i, 5) = dx(i)
```

```
    Soln(i, 6) = dy(i)
    Soln(i, 7) = dvx(i)
    Soln(i, 8) = dvy(i)
Next i
For i = 0 To imax
    For j = 0 To 8
        Range("SolnOrigin").Offset(i, j).Value = Soln(i, j)
    Next j
Next i
End Sub
```

First, any previous solution is cleared out. Then, a two-dimensional Soln array is ReDim'd to the number of rows in this particular solution, and the values from the individual arrays are transferred to the Soln array. This step is not mandatory, but it creates compact code for the actual transfer. This is accomplished with nested For loops, the outer loop iterating the row index from 0 to imax, and the inner loop iterating the column index from 0 to 8.

Based on data in the Results table in Excel, there is a chart of the solution trajectory, y versus x. This chart is based on the entire Solution range of 1,000 steps, but it will only display the part filled by the transfer from VBA. The empty chart, named Trajectory, is shown in Figure 12.48. The horizontal and vertical scales will adjust automatically to the solution presented.

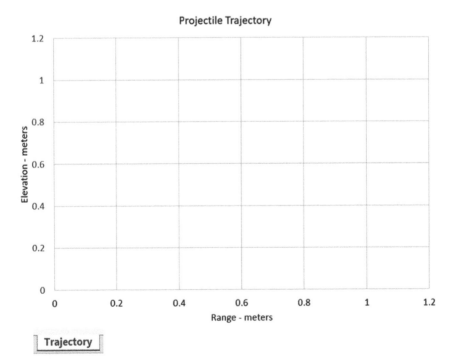

FIGURE 12.48   Trajectory chart based on empty data.

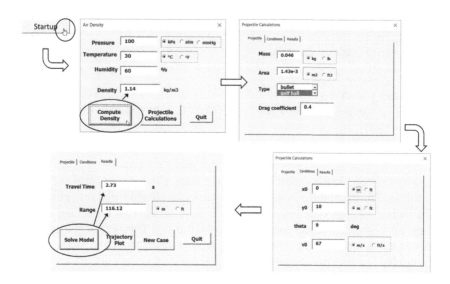

FIGURE 12.49   Example case for the trajectory application: golf ball with typical parameters.

What remains is to test the application. Figure 12.49 illustrates this for a golf ball object. The Results tab of the userform shows a time of travel of about 2.7 s and a range of about 116 m. When the Trajectory Plot button is clicked, the chart in the workbook updates to that shown in Figure 12.50.

At this point, we can try another case or quit the application. The solution obtained will be on the Result worksheet, and additional plots or calculations could be made manually. The Trajectory plot remains in place.

The Trajectory Plot command button causes a brief event handler to execute that activates the Trajectory chart in the workbook so that it is in view behind the userform. The event handler code is

```
Private Sub ShowPlot_Click()
Sheets("Trajectory").Select
End Sub
```

There are a few summary comments regarding the development of this application that are worthwhile. First, models of phenomena, such as the projectile trajectory equations used here, are always subject to improvement to provide more realistic results. Drag coefficients can change with the velocity of the projectile. For spheres, the spin, called the *Magnus effect*, influences the behavior of projectile flight. The rotation of linear projectiles, like bullets, has an effect. And then there are wind effects, including crosswinds. In other words, our model can quickly become more complex. But, for our purposes here, the model we have used has allowed us to illustrate important features of application development with userforms. We are certain that your projects will differ from ours, but this example should provide you with a good start.

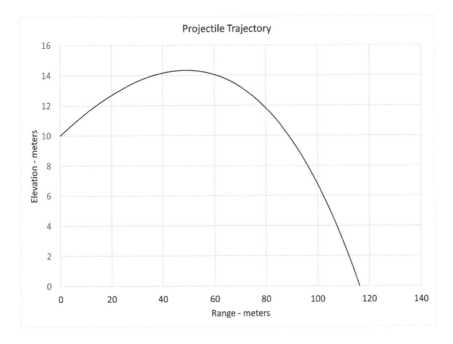

FIGURE 12.50   Trajectory plot of test case.

When using VBA to create userforms and event handlers, it is common that "gremlin" empty handler Subs will appear. For example,

```
Private Sub UserForm_Click()
•
End Sub
```

These can occur when you press the F7 key to switch to the handler code from the userform display. These can be ignored and best deleted as they accumulate.

If you see "Out of memory" warnings, these may not prevent your application from running, but they are bothersome. We have found that releasing userform memory with commands like those we have used,

```
Set ProjCalcs = Nothing
```

counters these warnings. You may wish to follow our lead here.

A final comment in this section is the old adage, "Don't bite off more than you can chew." It is best to develop your skills in creating user interfaces and applications incrementally. Try simpler projects first. When the projects become more complicated, build them step-by-step, testing along the way. When you are at a loss as to how to accomplish a task, look to others and other resources for help. We have mentioned more than once the well-tested reference by Alexander and

Kusleika (2019) and its prior editions including those by Walkenbach that go all the way back to Excel 5.0. Of course, the Internet provides a great resource, as do contacts, colleagues, and for older readers, grandchildren, who are also developing applications.

12.7 LINKING EXCEL TO OTHER SOFTWARE APPLICATIONS

There are many external software packages that are attractive to link to Excel to take advantage of the features of those packages and spreadsheet-based calculations. The manner in which the link is established varies by the software product. It would not serve our purpose here to present extensive coverage including many software packages because the reader may not have access to them. We would like to share two examples that illustrate different modes of interaction so that you will become oriented to what you may find as you consider linking Excel to a software package of particular interest to you. You may well not have access to (or interest in) these packages. If so, just review these as examples. A common thread to many of these package interfaces is the use of VBA.

12.7.1 MATLAB®

MATLAB® is a widely used software package for numerical calculations.[5] It is available in many academic settings and also in certain areas of professional practice. It is a robust software package with superb capabilities in numerical methods and scores of optional "toolboxes" that focus on particular numerical topics. One feature of interest to Excel users is that it has excellent plotting capabilities.

Mathworks provides in its toolboxes an interface to Excel in the form of an add-in called *Spreadsheet Link*. One obtains this product from Mathworks and installs it into Excel. The steps for doing this are illustrated in Figure 12.51.

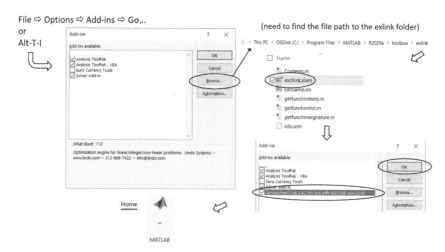

FIGURE 12.51   Installation of the MATLAB Spreadsheet Link add-in.

First, one must find the location of a folder called exlink. Then, launching the Add-ins window in Excel, one browses to that folder and selects the add-in file exclink.xlam and clicks OK. That installs the add-in and a MATLAB group on the Home tab of the Ribbon.[6]

The MATLAB command icon on the Home tab has several options on its dropdown list that afford manual communication back and forth between Excel and MATLAB. The first item on the list is Start MATLAB, which is necessary for any other features. Once MATLAB is launched, we can illustrate examples of the Spreadsheet Link add-in.

In the first example, shown in Figure 12.52, we have a two-way data table on the spreadsheet that evaluates the mathematical function

$$f(x,y) = \sin(2\pi x) \cdot \sin(3\pi y) \qquad 0 \le x \le 1 \qquad 0 \le y \le 1 \qquad (12.15)$$

We would like to have MATLAB create a three-dimensional plot of the data. We could use individual commands from the drop-down list to do this, but instead we place formulas using ML functions from the add-in in cells with the same effect. The resulting plot is overlaid on the spreadsheet. It has an attractive color scheme that doesn't show here, and it combines a surface plot with an underlying contour plot. In this case, there are ranges on the spreadsheet named xval (B3:B23), yval (C2:W2), and wval (C3:W23). The first three commands in cells B25:B27 transfer these ranges to similarly named variables in MATLAB, and the last formula in cell B28 executes the MATLAB *surfc* command. The plot is far superior to what we might create with Excel's capabilities.

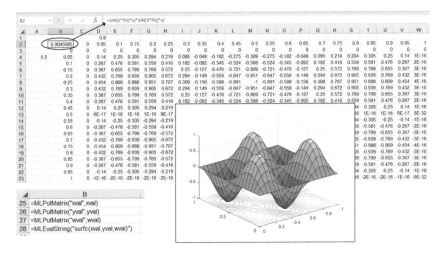

FIGURE 12.52 Two-dimensional case study in Excel, on-sheet MATLAB commands, and plot.

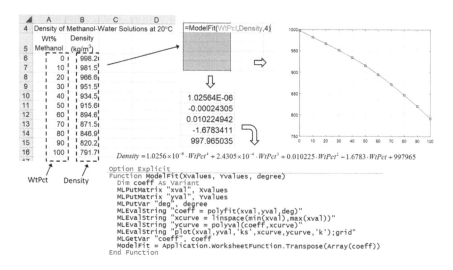

FIGURE 12.53 ModelFit array function to fit a polynomial to methanol-water density data.

A second MATLAB example employs a user-defined function to carry out a live calculation on the spreadsheet using MATLAB features. Figure 12.53 illustrates this. The VBA code creates a ModelFit array function that transfers the data to MATLAB, uses the *polyfit function* to fit a polynomial of order according to the degree argument. Then, the MATLAB commands create a plot of the data and the polynomial curve. Finally, the polynomial coefficients are returned to the spreadsheet in the range of the array formula. The number of cells covered by the array formula has to agree with the degree argument (4) value plus one (here, 5).

An advantage to programming a user-defined function is that the results are live on the spreadsheet. In this case, the data in the property table wouldn't change, but in other cases where the data or other involved parameters might change, the live results would be important.

Here we have illustrated two different examples of interfacing Excel with MATLAB. There are certainly many other possibilities. One would be taking advantage of MATLAB's advanced functions for solving differential equations, e.g., *ode45* and *ode15s*, for situations where the simple approach with the Euler method is insufficient.

12.7.2 ASPEN TECHNOLOGY'S HYSYS PROCESS SIMULATION SOFTWARE

Chemical and mechanical engineers in the fluid process industries use simulation software to model process units and overall processes. This software is particular and may not be of interest to other engineers and scientists. The Hysys software from Aspen Technology (Cambridge, MA) is commonly used in the

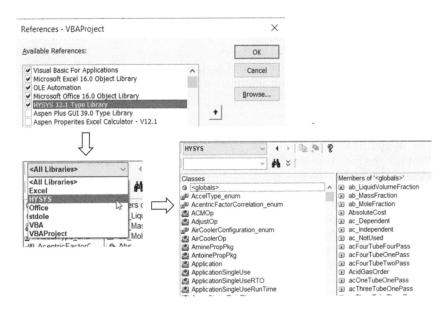

FIGURE 12.54 Providing access to the Hysys object library via the VBE.

oil and gas industry. Our point in using it here as an example is to illustrate the interface via VBA to software that is built upon an object library, similar to Excel and VBA.

In order to facilitate the interaction between VBA and the Hysys software, we need to add the Hysys Type Library to Tools ⇨ References in the VBE. This is shown in Figure 12.54. Once this is complete, the figure indicates that this library can be accessed via the Object Browser and objects of interest, their properties, and methods can be identified.

To illustrate how the intercommunication works, we use a simple simulation set up in Hysys for a flash drum.[7] This is shown in Figure 12.55. A multicomponent

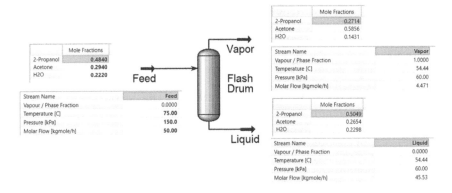

FIGURE 12.55 Hysys simulation of flash drum with stream conditions.

feed at 75°C and 150 kPa is reduced in pressure in the flash drum to 60 kPa and achieves an equilibrium temperature of 54.4°C. The compositions of the liquid and vapor streams are provided by Hysys.

To interact with the simulation in Excel, we create a VBA Sub to acquire information from and transmit information to the Hysys simulation. The code for this is

```
Option Explicit
Sub HysysLink()
'Notice below the declaration of variables of types
associated with Hysys
Dim hyApp As Hysys.Application
Dim HysysCase As SimulationCase
Dim Feed As ProcessStream
Dim Liquid As ProcessStream
Dim Vapor As ProcessStream
Dim FlashDrum As Separator
'Establish the open Hysys case
Set hyApp = CreateObject("Hysys.Application")
Set HysysCase = hyApp.ActiveDocument
HysysCase.Visible = True
'Assign the Hysys stream objects to local object variables
Set Feed = HysysCase.Flowsheet.MaterialStreams("Feed")
Set Liquid = HysysCase.Flowsheet.MaterialStreams("Liquid")
Set Vapor = HysysCase.Flowsheet.MaterialStreams("Vapor")
'Transfer the Feed Molar Flow to Hysys
Feed.MolarFlow.Value = Range("FeedFlow").Value
'Transfer the Feed mole fractions to Hysys
Feed.ComponentMolarFraction.Values = Application.
WorksheetFunction.Transpose(Range("FeedComposition").Value)
'Acquire the Liquid flow rate from Hysys and
'transfer to spreadsheet
Range("LiquidFlow").Value = Liquid.MolarFlow.Value
'Acquire the Liquid mole fractions from Hysys and
'transfer  to spreadsheet
Range("LiquidComposition").Value = _ Application.
WorksheetFunction.Transpose(Liquid.ComponentMolarFraction.
Values)
'Acquire the Vapor flow rate from Hysys
'and transfer to spreadsheet
Range("VaporFlow").Value = Vapor.MolarFlow.Value
'Acquire the Vapor mole fractions from Hysys
'and transfer to spreadsheet
Range("VaporComposition").Value = _ Application.
WorksheetFunction.Transpose(_
Vapor.ComponentMolarFraction.Values)
End Sub
```

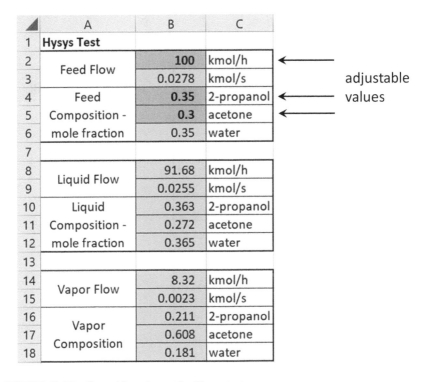

FIGURE 12.56 Spreadsheet layout for Hysys test.

The layout of the spreadsheet is shown in Figure 12.56.

You will see that we Dim several local variables as object types from the Hysys library. An example is the Feed object as a ProcessStream in Hysys. After setting the Hysys case, the Set commands assign the process streams to local object variables. Then, the Feed flow rate is transferred from the FeedFlow cell on the spreadsheet (cell B3, kmol/s units) to the simulation. The Feed compositions (cells B4:B6) are transferred. Because Hysys updates changes automatically, any transmissions back to the simulation are dealt with immediately. Similar software packages, e.g., the *Aspen Plus* product, require that we "run" the simulation each time we make a change. At this point, the code obtains the values for the Liquid and Vapor streams and returns these to the corresponding cells on the spreadsheet.

If we change the values in the adjustable cells shown in Figure 12.56, we can then use a Hysys Link on-sheet button to run the HysysLink Sub and, as shown in Figure 12.57, the results on the spreadsheet readjust with very little delay.

Of course, there is more detail involved in setting up this Hysys simulation. We do not take the time to go into that here because it isn't our point of emphasis. With this second example, we have illustrated how another object-oriented software package can be interfaced with Excel using VBA and the object library of that package.

	A	B	C	D	E	F
1	**Hysys Test**					
2	Feed Flow	75	kmol/h			
3		0.0208	kmol/s		**Run Hysys Link**	
4	Feed	0.4	2-propanol			
5	Composition -	0.2	acetone			
6	mole fraction	0.4	water			
7						
8	Liquid Flow	69.94	kmol/h			
9		0.0194	kmol/s			
10	Liquid	0.409	2-propanol			
11	Composition -	0.179	acetone			
12	mole fraction	0.413	water			
13						
14	Vapor Flow	5.06	kmol/h			
15		0.0014	kmol/s			
16	Vapor	0.282	2-propanol			
17	Composition	0.491	acetone			
18		0.227	water			

FIGURE 12.57   Spreadsheet with modified values after running the HysysLink Sub.

Using a vendor-supplied add-in, as with MATLAB, or object referencing through libraries represent two common ways of interfacing separate software packages with Excel. Even with the MATLAB interface, we found it convenient to program a user-defined array function to provide a connection. As you consider other software packages, often there are information and software options available from the vendor to facilitate the spreadsheet interface. For example, Aspen Technologies offers an Aspen Simulation Workbook add-in. This can be useful but is more dedicated than our freely coded interface.

Additionally, it is valuable to work from examples, and these may be found from the software vendor or out on the Internet. Like many endeavors we have described, there may be some struggles getting started, but resourcefulness and persistence will win out, and, in the end, you may wonder what the difficulty was.

PROBLEMS

Comment: The problems here are intended to be capstone in nature and not just particular to this chapter. Their purpose is to integrate material from various chapters in the text into single-problem solutions. If you have not studied Chapter 8, for example, you may not find the problems here that refer to that chapter to be pertinent.

12.1 In Chapter 8, we used the Data Analysis Regression tool to fit a polynomial model to water viscosity versus temperature. These data are repeated
here in Table P12.1. Using the Input X Range shown in Figure 8.27, record
a macro that performs the regression calculation and produces results
similar to those in Figure 8.28, on a separate sheet but unformatted.

Note: For the Regression tool to function in VBA, you must activate Analysis Toolpak – VBA on the Add-ins window in Excel (Alt-T-I
shortcut).

Challenge: Augment the macro to include code with neat formatting
of the regression output. The extra code should take into account that the
extent of the output will change depending on the number of coefficients
in the model and the length of the data series.

TABLE P12.1

Viscosity of Water (mPa·s) versus Temperature (°C)

Temperature (°C)	Viscosity (mPa·s)
0.0	1.794
4.4	1.546
10.0	1.310
15.6	1.129
21.1	0.982
26.7	0.862
32.2	0.764
37.8	0.682
48.9	0.559
60.0	0.470
71.1	0.401
82.2	0.347
93.3	0.305

12.2 In Chapter 7, Section 7.4, we used the Solver to find the solutions of the
equations

$$f_1(x_1, x_2) = x_1^2 - x_2 + 1 = 0$$
$$f_2(x_1, x_2) = 2\cos(x_1) - x_2 = 0$$

Following the method there, set up the problem, and record a macro that
follows the Solver procedure.

Note: For the Solver to execute in VBA, it needs to be included in the
References list of the VBE under the Tools menu item.

Hint: When the macro executes, it will show the Solver Results window, and you have to click OK to accept the solution. If you want the code to skip that display and just accept the solution, modify this line of the macro to

```
SolverSolve UserFinish:=True
```

12.3 In Section 8.2.2, we illustrated how to use the Data Analysis Histogram tool to produce a frequency table and bar chart based on a data set and a set of bin values. Based on a similar set of random numbers, record a macro that automates the histogram process.

Note: For the Histogram tool to function in VBA, you must activate Analysis Toolpak – VBA on the Add-ins window in Excel (Alt-T-I shortcut).

Challenge 1: Round the data to one decimal point. Modify the macro so that it generates its own number of bins and bin values based on the strategy put forth in Section 8.2.2. Include the calculation of bin center values and the modification of the bar chart to reflect these.

Challenge 2: Modify the macro produced in Challenge 1 to include determination of the resolution of the data, and use that information to set the bin boundaries to the half-resolution point so that no data will fall on a boundary.

12.4 Create a VBA macro similar to that in Figure 11.20 that uses an Input Box function to accept from the user a value of pressure in kPa and another Input Box function to obtain a desired unit for conversion, e.g., atm, mm Hg, psi. Using a Select Case structure, develop code to make the appropriate conversion and display the result in a message box. Include the units of the converted quantity in the message box. Add an on-sheet button that causes the macro to run.

12.5 Record a macro that uses the Data Analysis Descriptive Statistics tool to provide sample statistics on a new worksheet for a range of data. The first version of the macro should work with the data selected when the macro is run. Then, modify the macro to use the `InputBox` method to obtain from the user the cell range that should be used for the statistics. Refer to Figure 12.14 and the surrounding description.

Challenge: Add code to the macro to format the descriptive statistics worksheet.

12.6 Modify the Temperature Conversion workbook described in Section 12.5 so that the userform appears automatically when the workbook is opened and unloads the userform when the workbook is closed.

12.7 For the cash flow table produced in Chapter 6 (Figure 6.14), create a userform and associated event handlers that allow entry of the rate of inflation and cost of capital and displays the NPV, IRR, and BCR in textboxes. Other data, such as capital cost and first-year sales should be left constant.

12.8 Consider the table lookup example presented in Figure 4.38. Create a userform and associated event handlers that allow entry of nominal pipe diameter and pipe schedule in listboxes and present the inside diameter of the pipe in a textbox on the userform. The inside diameter display should respond to any changes in the listbox choices with no command button required.

12.9 A Sub was created in Chapter 11 (see Figure 11.24) to find the liquid depth in a spherical tank given the inside radius of the tank and the liquid volume. This was based on Chapter 7, Section 7.1.2. Create a userform interface that allows entry of radius and volume and displays the liquid depth. The event handler should "call" the Sub, which is located in a module.

12.10 In Example 7.3, we illustrated how to solve for the equilibrium of water and steam based on the ideal gas law and the *Antoine equation* for vapor pressure using an iterative technique. There, we used Excel's iterative solver. For this problem, set up a userform interface where there are entries for the vapor volume, V, and mass, mV, and a command button to compute the equilibrium. The event handler for the command button will check the entries and call a Sub in the module that uses the iterative technique. Stop iteration when the temperature is changing less than a relative error tolerance. This should be based on the formulas from Equations 7.30, which provide a convergent scheme. The resulting equilibrium temperature and pressure should be displayed in textboxes on the userform.

12.11 As a project, pick an area or topic of interest that involves calculations. Design a user interface that implements these calculations and includes at least two userforms. Implement your design and provide checks on error conditions.

12.12 In Figure 12.36, we described a convert event handler employing GoTo statements to prevent users from entering a temperature below absolute zero. Because GoTo statements should be avoided, develop a more elegant version of Figure 12.36 by employing an imbedded multi-alternative If structure.

NOTES

1. It is worth commenting here that userforms are not part of the Excel version for Office on the Mac. With apology, this section of this chapter will not be of much utility to Mac users. We consider this to be a major deficiency of the Mac version of Excel.
2. *Concatenation* means linking things together in a chain or series.
3. In this context, interference means that the ones in the numbers being added do not occur in the same position.
4. GoTo statements are frowned upon as their frequent use can lead to so-called "spaghetti code." This perjorative term refers to code that is extremely convoluted and, as a consequence, difficult to follow, maintain, and debug. Nevertheless, as described here, there are a few limited applications where they can perform a useful action without hurting code clarity.

5. MATLAB is a product of Mathworks, Inc., Cambridge, MA. It was pioneered by Cleve Moler in the 1970s and is an acronym for "Matrix Laboratory."
6. If you do not have administrator privileges on your computer, you may need to obtain assistance to do this from your information technology (IT) support group.
7. A *flash drum* is a vessel used for rapid separation of a feed stream into a vapor stream rich in more volatile component(s) and a liquid stream with lower concentrations of more volatile components.

Appendix A: Matrix Algebra Review

Knowledge of matrices is essential for understanding the solution of linear algebraic equations with Excel. The following sections outline how matrices provide a concise way to represent and manipulate linear algebraic equations.

A.1 MATRIX NOTATION

A *matrix* consists of a rectangular array of elements represented by a single symbol. As depicted in Figure A1, The symbol **A** is the shorthand notation for the matrix, and a_{ij} designates an individual *element* of the matrix.

A horizontal set of elements is called a *row*, and a vertical set is called a *column*. The first subscript i designates the number of the row in which the element lies. The second subscript j designates the column. For example, element a_{23} is in row 2 and column 3.

The matrix in Figure A1 has m rows and n columns and is said to have a dimension (also called *order*) of m by n (or $m \times n$). It is referred to as an m by n matrix.

Matrices with row dimension $m = 1$, such as

$$\mathbf{b} = \begin{bmatrix} b_1 & b_2 & \cdots & b_n \end{bmatrix} \tag{A1}$$

are called *row vectors*. Note that for simplicity, the first subscript of each element is dropped.

Matrices with column dimension $n = 1$, such as

$$\mathbf{c} = \begin{bmatrix} c_1 \\ c_2 \\ \cdot \\ \cdot \\ \cdot \\ c_m \end{bmatrix} \tag{A2}$$

are referred to as *column vectors*. As was the case with the row vector, the second subscript is dropped.

Column 3
$$\downarrow$$

$$\mathbf{A} = \begin{bmatrix} a_{11} & a_{12} & a_{13} & \cdots & a_{1n} \\ a_{21} & a_{22} & \boxed{a_{23}} & \cdots & a_{2n} \\ \vdots & \vdots & \vdots & \ddots & \vdots \\ a_{m1} & a_{m2} & a_{m3} & \cdots & a_{mn} \end{bmatrix} \longleftarrow \text{Row 2}$$

FIGURE A1 A general matrix A.

Matrices where $m = n$ are called *square matrices*. For example, a 3×3 square matrix is

$$\mathbf{A} = \begin{bmatrix} a_{11} & a_{12} & a_{13} \\ a_{21} & a_{22} & a_{23} \\ a_{31} & a_{32} & a_{33} \end{bmatrix} \tag{A3}$$

The diagonal consisting of the elements a_{11}, a_{22}, and a_{33} is termed the *principal* or *main diagonal* of the matrix.

Square matrices are particularly important when solving sets of simultaneous linear equations. For such systems, the number of equations (corresponding to rows) and the number of unknowns (corresponding to columns) must be equal for a unique solution to be possible. Consequently, square matrices of coefficients are often encountered when dealing with such systems.

There are several special forms of square matrices that are important and should be noted. A *symmetric matrix* is one where $a_{ij} = a_{ji}$ for all i's and j's. For example,

$$\mathbf{A} = \begin{bmatrix} 5 & 1 & 2 \\ 1 & 3 & 7 \\ 2 & 7 & 8 \end{bmatrix} \tag{A4}$$

is a 3×3 symmetric matrix.

A *diagonal matrix* is a square matrix where all elements off the main diagonal are equal to zero,

$$\mathbf{A} = \begin{bmatrix} a_{11} & 0 & 0 & \cdots & 0 \\ 0 & a_{22} & 0 & \cdots & 0 \\ 0 & 0 & a_{33} & \cdots & 0 \\ \vdots & \vdots & \vdots & \ddots & \vdots \\ 0 & 0 & 0 & \cdots & a_{nn} \end{bmatrix} \tag{A5}$$

An *identity matrix* is a diagonal matrix where all elements on the main diagonal are equal to 1, as in

$$\mathbf{I} = \begin{bmatrix} 1 & 0 & 0 & \cdots & 0 \\ 0 & 1 & 0 & \cdots & 0 \\ 0 & 0 & 1 & \cdots & 0 \\ \vdots & \vdots & \vdots & \ddots & \vdots \\ 0 & 0 & 0 & \cdots & 1 \end{bmatrix} \qquad (A6)$$

The symbol $\mathbf{I}$ is used to denote the identity matrix. The identity matrix has properties similar to unity. That is, just as $a \times 1 = 1 \times a = a$, so also does

$$\mathbf{A} \cdot \mathbf{I} = \mathbf{I} \cdot \mathbf{A} = \mathbf{A} \qquad (A7)$$

An *upper triangular matrix* is one where all the elements below the main diagonal are zero, as in

$$\begin{bmatrix} a_{11} & a_{12} & a_{13} & \cdots & a_{1n} \\ 0 & a_{22} & a_{23} & \cdots & a_{2n} \\ 0 & 0 & a_{33} & \cdots & a_{3n} \\ \vdots & \vdots & \vdots & \ddots & \vdots \\ 0 & 0 & 0 & \cdots & a_{nn} \end{bmatrix} \qquad (A8)$$

A *lower triangular matrix* is one where all elements above the main diagonal are zero, as in

$$\begin{bmatrix} a_{11} & 0 & 0 & \cdots & 0 \\ a_{21} & a_{22} & 0 & \cdots & 0 \\ a_{31} & a_{32} & a_{33} & \cdots & 0 \\ \vdots & \vdots & \vdots & \ddots & \vdots \\ a_{m1} & a_{m2} & a_{m3} & \cdots & a_{mm} \end{bmatrix} \qquad (A9)$$

A *banded matrix*, $\mathbf{B}$, has all elements equal to zero, with the exception of a band centered on the main diagonal:

$$\begin{bmatrix} a_{11} & a_{12} & 0 & 0 & 0 & \cdots & 0 \\ a_{21} & a_{22} & a_{23} & 0 & 0 & \cdots & 0 \\ 0 & a_{32} & a_{33} & a_{34} & 0 & \cdots & 0 \\ 0 & 0 & a_{43} & a_{44} & a_{45} & \cdots & 0 \\ 0 & 0 & 0 & \ddots & \ddots & \ddots & 0 \\ \vdots & \vdots & \vdots & a_{n-1,n-3} & a_{n-1,n-2} & a_{n-1,n-1} & 0 \\ 0 & 0 & 0 & 0 & a_{n,n-2} & a_{n,n-1} & a_{nn} \end{bmatrix} \qquad (A10)$$

The above matrix has a bandwidth of 3 and is given a special name – the *tridiagonal matrix*.

A.2   MATRIX OPERATING RULES

Now that we have specified what we mean by a matrix, we can define some operating rules that govern its use. Two $m \times n$ matrices are equal if, and only if, every element in the first is equal to every element in the second, that is, $\mathbf{A} = \mathbf{B}$ if $a_{ij} = b_{ij}$ for all i and j.

Addition of two matrices, say, $\mathbf{A}$ and $\mathbf{B}$, is accomplished by adding corresponding elements in each matrix. The elements of the resulting matrix $\mathbf{C}$ are computed,

$$c_{ij} = a_{ij} + b_{ij} \tag{A11}$$

for $i = 1, 2, \ldots, m$ and $j = 1, 2, \ldots, n$. Similarly, the subtraction of two matrices, say, $\mathbf{E}$ minus $\mathbf{F}$, is obtained by subtracting corresponding terms, as in

$$d_{ij} = e_{ij} - f_{ij} \tag{A12}$$

for $i = 1, 2, \ldots, m$ and $j = 1, 2, \ldots, n$. It follows directly from the above definitions that addition and subtraction can be performed only between matrices having the same dimensions.

Both addition and subtraction are *commutative*,

$$\mathbf{A} + \mathbf{B} = \mathbf{B} + \mathbf{A} \tag{A13}$$

and *associative*,

$$(\mathbf{A} + \mathbf{B}) + \mathbf{C} = \mathbf{A} + (\mathbf{B} + \mathbf{C}) \tag{A14}$$

The multiplication of a matrix $\mathbf{A}$ by a scalar g, or *scalar multiplication*, is obtained by multiplying every element of $\mathbf{A}$ by g. For example, for a 3×3 matrix,

$$g \cdot \mathbf{A} = \begin{bmatrix} ga_{11} & ga_{12} & ga_{13} \\ ga_{21} & ga_{22} & ga_{23} \\ ga_{31} & ga_{32} & ga_{33} \end{bmatrix} \tag{A15}$$

Matrix multiplication is less intuitive than the preceding operations. The product of two matrices is represented as $\mathbf{C} = \mathbf{A} \cdot \mathbf{B}$, where the elements of $\mathbf{C}$ are defined as

$$c_{ij} = \sum_{k=1}^{n} a_{ik} b_{kj} \tag{A16}$$

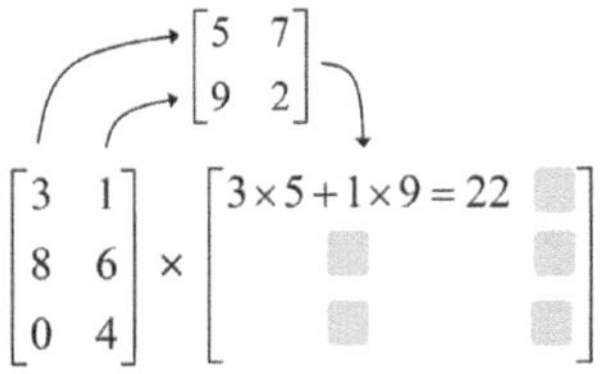

$$\begin{bmatrix} 5 & 7 \\ 9 & 2 \end{bmatrix}$$

$$\begin{bmatrix} 3 & 1 \\ 8 & 6 \\ 0 & 4 \end{bmatrix} \times \begin{bmatrix} 3\times5+1\times9=22 & \square \\ \square & \square \\ \square & \square \end{bmatrix}$$

FIGURE A2 Depiction of matrix multiplication for a row 1-column 1 combination.

where n = the column dimension of **A** and the row dimension of **B**. That is, the c_{ij} element is obtained by adding the product of individual elements from the ith row of the first matrix, in this case **A**, by the jth column of the second matrix **B**. Figure A2 depicts how the rows and columns line up in matrix multiplication for a specific row and column combination. This is illustrated in more detail in Chapter 7, Figure 7.18.

According to this definition, matrix multiplication can be performed only if the first matrix has as many columns as the number of rows in the second matrix. Thus, if **A** is an $m \times n$ matrix, **B** could be an $n \times l$ matrix. For this case, the resulting **C** matrix would have a dimension of m by l. However, if **B** were an $l \times n$ matrix, the multiplication could not be performed. Figure A3 provides an easy way to check whether two matrices can be multiplied.

If the dimensions of the matrices are suitable, matrix multiplication is *associative*,

$$(\mathbf{A} \cdot \mathbf{B})\mathbf{C} = \mathbf{A}(\mathbf{B} \cdot \mathbf{C}) \tag{A17}$$

and *distributive*,

$$\mathbf{A}(\mathbf{B} + \mathbf{C}) = \mathbf{A} \cdot \mathbf{B} + \mathbf{A} \cdot \mathbf{C} \tag{A18}$$

or

$$(\mathbf{A} + \mathbf{B})\mathbf{C} = \mathbf{A} \cdot \mathbf{C} + \mathbf{B} \cdot \mathbf{C} \tag{A19}$$

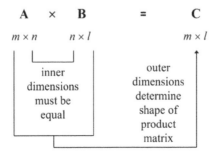

FIGURE A3 Matrix multiplication restrictions: can only be performed if the inner dimensions are equal to the result having the outer dimensions.

However, multiplication is not generally *commutative*:

$$\mathbf{A} \cdot \mathbf{B} \neq \mathbf{B} \cdot \mathbf{A} \tag{A20}$$

That is, the order of multiplication is important.

Although multiplication is possible, matrix division is not a defined operation. However, if a matrix $\mathbf{A}$ is square and nonsingular, there is another matrix $\mathbf{A}^{-1}$, called the inverse of $\mathbf{A}$, for which

$$\mathbf{A} \cdot \mathbf{A}^{-1} = \mathbf{A}^{-1} \cdot \mathbf{A} = \mathbf{I} \tag{A21}$$

Thus, the multiplication of a matrix by the inverse is analogous to division, in the sense that a number divided by itself is equal to one. That is, multiplication of a matrix by its inverse leads to the identity matrix.

The inverse of a two-dimensional square matrix can be represented simply by

$$\mathbf{A}^{-1} = \frac{1}{a_{11}a_{22} - a_{12}a_{21}} \begin{bmatrix} a_{22} & -a_{12} \\ -a_{21} & a_{11} \end{bmatrix} \tag{A22}$$

Similar formulas for higher-dimensional matrices are much more involved.

One other matrix manipulations that will have utility in our discussion is the *transpose* of a matrix. The transpose of a matrix involves transforming its rows into columns and its columns into rows. For example, for the 3×3 matrix,

$$\mathbf{A} = \begin{bmatrix} a_{11} & a_{12} & a_{13} \\ a_{21} & a_{22} & a_{23} \\ a_{31} & a_{32} & a_{33} \end{bmatrix} \tag{A23}$$

the *transpose*, designated $\mathbf{A}^{T}$, is defined as

$$\mathbf{A}^{T} = \begin{bmatrix} a_{11} & a_{21} & a_{31} \\ a_{12} & a_{22} & a_{32} \\ a_{13} & a_{23} & a_{33} \end{bmatrix} \tag{A24}$$

In other words, the element a_{ij} of the transpose is equal to the a_{ji} element of the original matrix. If you prefer, rows become columns and vice versa.

The transpose has a variety of functions in matrix algebra. One simple advantage is that it allows a column vector to be expressed as a row vector. For example, if

$$\mathbf{c} = \begin{bmatrix} c_1 \\ c_1 \\ c_1 \end{bmatrix} \tag{A25}$$

then

$$\mathbf{c}^T = \begin{bmatrix} c_1 & c_2 & c_3 \end{bmatrix} \tag{A26}$$

In addition, the transpose has numerous mathematical applications.

The final matrix manipulation that will have utility in our discussion is *augmentation*. A matrix is augmented by the addition of a column (or columns) to the original matrix. For example, suppose we have a 3×3 matrix of coefficients. We might wish to augment this matrix $\mathbf{A}$ with a 3×3 identity matrix to yield a 3-by-6-dimensional matrix:

$$\left[\begin{array}{ccc|ccc} a_{11} & a_{12} & a_{13} & 1 & 0 & 0 \\ a_{21} & a_{22} & a_{23} & 0 & 1 & 0 \\ a_{31} & a_{32} & a_{33} & 0 & 0 & 1 \end{array} \right] \tag{A27}$$

Such an expression has utility when we must perform a set of identical operations on two matrices. Thus, we can perform the operations on the single augmented matrix rather than on the two individual matrices.

A.3   REPRESENTING LINEAR ALGEBRAIC EQUATIONS IN MATRIX FORM

It should be clear that matrices provide a concise notation for representing simultaneous linear equations. For example, a 3×3 set of linear equations,

$$a_{11}x_1 + a_{12}x_2 + a_{13}x_3 = b_1$$

$$a_{21}x_1 + a_{22}x_2 + a_{23}x_3 = b_2 \tag{A28}$$

$$a_{31}x_1 + a_{32}x_2 + a_{33}x_3 = b_3$$

can be expressed as

$$\mathbf{A} \cdot \mathbf{x} = \mathbf{b} \tag{A29}$$

where $\mathbf{A}$ is the matrix of coefficients,

$$\mathbf{A} = \begin{bmatrix} a_{11} & a_{12} & a_{13} \\ a_{21} & a_{22} & a_{23} \\ a_{31} & a_{32} & a_{33} \end{bmatrix} \tag{A30}$$

b is the column vector of constants,

$$\mathbf{b} = \begin{bmatrix} b_1 \\ b_2 \\ b_3 \end{bmatrix} \qquad (A31)$$

and **x** is the column vector of unknowns,

$$\mathbf{x} = \begin{bmatrix} x_1 \\ x_2 \\ x_3 \end{bmatrix} \qquad (A32)$$

Recall the definition of matrix multiplication (Equation A16) to convince yourself that Equations A28 and A29 are equivalent. Also, realize that Equation A29 contains a valid matrix multiplication because the number of columns, n, of the first matrix **A** is equal to the number of rows, n, of the column vector, **x**.

Part of Chapter 7 is devoted to solving Equation A29 for **x**. A formal way to obtain a solution using matrix algebra is to multiply each side of the equation by the inverse of **A** to yield

$$\mathbf{A}^{-1} \cdot \mathbf{A}\mathbf{x} = \mathbf{A}^{-1}\,\mathbf{b} \qquad (A33)$$

Because $\mathbf{A}^{-1} \cdot \mathbf{A}$ equals the identity matrix, Equation A33 becomes

$$\mathbf{x} = \mathbf{A}^{-1}\,\mathbf{b} \qquad (A34)$$

Therefore, the equation has been solved for **x**. This is another example of how the inverse plays a role in matrix algebra that is similar to division. It should be noted that this is not a very efficient way to solve a system of equations. Thus, other approaches are employed in numerical algorithms. However, as discussed elsewhere (Chapra and Canale 2022), the matrix inverse itself has great value in the engineering and scientific analyses of such systems.

Finally, we will sometimes find it useful to augment **A** with **b**. For example, if $n = 3$, this results in a 3×4-dimensional matrix:

$$\left[\begin{array}{ccc|c} a_{11} & a_{12} & a_{13} & b_1 \\ a_{21} & a_{22} & a_{23} & b_2 \\ a_{31} & a_{32} & a_{33} & b_3 \end{array} \right] \qquad (A35)$$

Expressing the equations in this form is useful because several of the techniques for solving linear systems perform identical operations on a row of coefficients and the corresponding right-hand-side constant. When the equations are expressed as in Equation A35, these operations can be performed once on an individual row of the augmented matrix rather than separately on the coefficient matrix and the right-hand-side vector.

Appendix B: Shortcut Keys and Key Combinations

| Alt-F11 | switch back and forth between Excel and the Visual Basic Editor |

EXCEL

F1	help
F2	edit cell
F4	edit mode: change address to absolute and mixed modes (rotates through options) otherwise: transfers formatting from one cell to another
F5	go to
F9	calculate the spreadsheet (when calculations stalled with iterative solver)
F12	save as
Alt-F8	display the Macro dialog window
Alt-T-I	display Add-Ins window
Ctrl-C	copy
Ctrl-V	paste
Ctrl-X	cut
Ctrl-arrow	jump in the arrow direction
Ctrl-A	select all
Ctrl-*	select all
Ctrl-I	format italics or reverse
Ctrl-B	format bold or reverse
Ctrl-Z	undo

VBA

F1	VBA help
Ctrl-R	display Project Explorer
F4	display the Properties Window
F5	run Sub
F7	view userform object code

Shift-F7	view userform object
F8	single-step code
Ctrl-F8	run to cursor
Shift-F8	step over subprogram
F9	set/clear breakpoint
Ctrl-Shift-F9	clear all breakpoints

References

Alexander, M. and Kusleika D., previously by Walkenbach, J., 2019. *Excel 2019 Power Programming with VBA*, Wiley, Indianapolis, IN.

Chapra, S.C. and Canale, R.P., 2020. *Numerical Methods for Engineers with Software Applications and Programming*, 8th Ed., WCB/McGraw-Hill, New York, NY.

Chapra, S.C. and Clough, D.E., 2022. *Applied Numerical Methods with Python for Engineers and Scientists*, WCB/McGraw-Hill, New York, NY.

Clough, D.E. and Chapra, S.C., 2023. *Introduction to Engineering and Scientific Computing with Python*, CRC Press Taylor & Francis Group, Boca Raton, FL.

Higgins, R.C., 2007. *Analysis for Financial Management*, McGraw-Hill/Irwin, New York, NY.

Montgomery, D.C. and Runger, G.C., 2018. *Applied Statistics and Probability for Engineers*, 7th Ed., Wiley, New York, NY.

Olsson, J. et al., 1997, *Thermophysical properties of aqueous NaOH-H2O solutions at high concentrations*, International Journal of Thermophysics, Vol. 18, Issue 3, pp. 779–793.

Zwillinger, D., Ed., 2018. *CRC Standard Mathematical Tables and Formulas*, 33rd Ed., Chapman and Hall, London, UK.

Index

Index of Excel-VBA Terminology